Virtual
Manufacturing

Virtual Manufacturing

PRASHANT BANERJEE
University of Illinois at Chicago

DAN ZETU
General Motors Technical Center

JOHN WILEY & SONS, INC.

New York • Chichester • Weinheim • Brisbane • Singapore • Toronto

ALIAS/WAVEFRONT®, GRAPHICS LIBRARY/GL™ IRIX™, OPENGL®, OPEN INVENTOR™, and PERFORMER™ are all trademarks of SGI Inc. CAVE™ and IMMERSADESK R2™ are trademarks of Electronic Visualization Laboratory, University of Illinois at Chicago. CONAC™ is a trademark of MTI Research. MOTIONSTAR® is a registered trademark of Ascension Technology Corporation. SOFTIMAGE® is a registered trademark of SoftImage/Avid Technology. TELEGRIP is a trademark of Deneb Robotics Inc. 3D STUDIO MAX® and 3D STUDIO VIZ® are registered trademarks of AutoDesk Inc. WORLDTOOLKIT® and the Sense8® Product Line are registered trademarks of Engineering Animation, Inc.

This book is printed on acid-free paper. ∞

This publication is designed to provide accurate and authoritative information in regard to the subject matter covered. It is sold with the understanding that the publisher is not engaged in rendering professional services. If professional advice or other expert assistance is required, the services of a competent professional person should be sought.

Library of Congress Cataloging-in-Publication Data:

Banerjee, Prashant
    Virtual manufacturing / Prashant Banerjee, Dan Zetu
      p. cm.
    Includes index.
    ISBN: 0-471-35443-0 (cloth : alk. paper)
      1. Manufacturing processes—Computer simulation. 2. Virtual reality—Industrial applications. 3. Computer vision—Industrial applications. I. Zetu, Dan. II. Title.

TS183.B42 2001
670'.1'13—dc21

                  00-068504

Printed in the United States of America.
10 9 8 7 6 5 4 3 2 1

CONTENTS

v

3

Principles of Virtual Reality   55

4

Telemetry-Based Depth Recovery   101

5

Viewpoint-Based Shape Recovery from Multiple Views　　　**155**

6

Hybrid Tracking for Manufacturing Systems Automation　　　**179**

7

Exact Collision Detection　　　**197**

8

Motion Modeling　　　**219**

9

Telecollaborative Virtual Manufacturing Architecture   **231**

10

**Specialized Room Airflow Design Using Computational
Fluid Dynamics and Virtual Reality**   **253**

PREFACE

With the advent of high-resolution graphics, high-speed computing, and user interaction devices, virtual reality (VR) has emerged as a major new technology in recent years. An important new concept introduced by many VR systems is *immersion,* which refers to the feeling of complete immersion in a three-dimensional computer-generated environment by means of user-centered perspective achieved through tracking the user. This is a huge step forward compared to classical modeling and CAD/CAM packages, which inherently impose major limitations on intuitive user interaction. VR technology is currently used in a broad range of applications, the best known being flight simulators, walkthroughs, video games, and medicine (virtual surgery). From a manufacturing standpoint, some of the attractive applications include training, collaborative product and process design, facility monitoring, and management. Moreover, recent advances in broadband networks are also opening up new applications for telecollaborative virtual environments in these areas.

In this book we have chosen to cover certain areas which we believe to have a potentially significant impact on virtual manufacturing. This is still an evolving list and includes areas such as facility layout and work-cell management, real-time exact collision detection, motion modeling, avatar modeling, and virtual-real environment interaction in training, based on our years of experience in this field. In addition to these areas, the book explores some connections between VR and computer vision (especially camera self-calibration and stereo vision) in the context of depth recovery in virtual manufacturing. Some of the automation techniques resulting from these concepts can potentially reduce a lot of time and boredom for users involved in manually creating CAD-based virtual environments. Lately, with the emergence of complementary areas of VR such as augmented reality (AR), one can address crucial problems of registration between virtual and real worlds.

Chapter 1 provides an introduction to virtual manufacturing in the context of the book, provides illustrative applications and highlights research issues. In Chapters 2 to 5 we present some basic principles of computer graphics, virtual reality, depth extraction, and shape reconstruction, respectively. Numerous applications in virtual manufacturing are presented in Chapters 6 to 10. These applications have been developed based on work carried out at the Industrial Virtual Reality Institute at the University of Illinois at Chicago. They include the thesis work of A. Banerjee, R. Tesic, D. Zetu, and V. Giallorenzo. The contributions of the following individuals are also gratefully acknowledged: A. Akgunduz, C. Luciano, D. Thompson, P. Schneider, S. Anantharaman, and K. Gierach. The book is meant as an introduction to the basic concepts of virtual manufacturing from a VR perspective, not only for engineering students, but also for professionals interested in developing applications in this field. Last, but not the least, we would also like to thank T. DeFanti, D. Sandin, M. Brown, J. Leigh, D. Pape. D. Plepys, G. Dawe, A. Verlo, and numerous other members of the Electronic Visualization Laboratory at the University of Illinois at Chicago for providing constant encouragement and support over a number of years to make some of the work possible.

P. BANERJEE
D. ZETU

Chicago
January 2001

Virtual
Manufacturing

1

INTRODUCTION TO VIRTUAL MANUFACTURING AND AUTOMATION

1.1 VIRTUAL MANUFACTURING AND AUTOMATION

The term *virtual manufacturing* first came into prominence in the early 1990s, in part as a result of the U.S. Department of Defense Virtual Manufacturing Initiative. Both the concept and the term have now gained wide international acceptance and have somewhat broadened in scope. For the first half of the 1990s, pioneering work in this field has been done by a handful of major organizations, mainly in the aerospace, earthmoving equipment, and automobile industries, plus a few specialized academic research groups. Recently, accelerating worldwide market interest has become evident, fueled by price and performance improvements in the hardware and software technologies required and by increased awareness of the huge potential of virtual manufacturing. Virtual manufacturing can be considered one of the enabling technologies for the rapidly developing information technology infrastructure.

Virtual manufacturing is used loosely in a number of contexts. It refers broadly to the modeling of manufacturing systems and components with

1

effective use of audiovisual and/or other sensory features to simulate or design alternatives for an actual manufacturing environment, mainly through effective use of computers. The motivation is to enhance our ability to predict potential problems and inefficiencies in product functionality and manufacturability before real manufacturing occurs. Another term that is sometimes mentioned in the context of virtual manufacturing is *agile manufacturing,* sometimes defined as a structure within which agility is achieved through the integration of three primary resources: organization, people, and technologies. A way to achieve this is through innovative management structures and organization, a skill base of knowledgeable and empowered people, and flexible and intelligent technologies. Whereas agility focuses on the ability to make rapid changes in products and processes based on the voice of the customer, virtual manufacturing provides a means for doing so. One area in which virtual manufacturing has made an impact is that of rapid prototyping machines, building prototypes by precise deposition of layer upon layer of powdered metal, a process known as stereolithography. *Virtual reality* (VR) has been used by companies such as General Motors and Caterpillar to build electronic prototypes of vehicles, instead of physical prototypes. This process reduces product development time significantly.

Virtual manufacturing and automation often concentrate on an interface between VR technology and manufacturing and automation theory and practice. In this book we concentrate on the role of VR technology in developing this interface. It is our belief that the direction of evolution of manufacturing theory and practice will become clearer in the future once the role of VR technology is understood better in developing this interface.

Some areas that can benefit from development of virtual manufacturing include:

- Product design
- Hazardous operations modeling
- Production modeling
- Process modeling
- Training
- Education
- Information visualization
- Telecommunications and teletravel

To develop virtual manufacturing technology effectively it is essential that an interface layer be created between VR software and manufacturing software. Current examples of VR software are based on hardware specifications such as 3D Studio® by AutoDesk, Open Inventor™ by SGI, WorldToolKit® by Sense8 (now acquired by EAI), Performer™ by SGI, CAVE™ (Computer Assisted Virtual Environment) software library by VRCo, and Division library (now acquired by PTC), which is geared toward a room-sized VR implementation driven by multiple SGI processors. More recently, implementations of Virtual Reality Modeling Language (VRML) have been used, and VRML has also become an ISO standard. Cosmo Player is a useful VRML browser conceived by SGI and acquired subsequently by Computer Associates. Noncommercial highly parallel graphics multicomputer known as pixel-planes are also available, mainly for research purposes. Examples of manufacturing and automation software include simulation software, control software, and layout design software, among others.

1.2 BRIEF TOUR OF VIRTUAL REALITY

Definitions
Virtual reality. VR can be broadly defined as the ability to create and interact in cyberspace, i.e. a space that represents an environment which has a lot of similarity to the environment around us. VR is closely associated with an environment commonly known as *virtual environment* (VE). VE systems differ from previously developed computer-centered systems in the extent to which real-time interaction is facilitated and in terms of several characteristics: that the perceived visual space is three-dimensional rather than two-dimensional, the human–machine interface is multimodal, and the operator is immersed in the computer-generated environment.

Virtual environment. A commonly used definition for VE is an interactive, virtual image display enhanced by special processing and by non-visual display modalities, such as auditory and haptic, to convince users that they are immersed in a synthetic space. The term *immersion* refers to the fact that the user gets the feeling that he or she is immersed in the computer environment — the screen separating the user and the computer appears non-existent to the user.

Means of Simulation
The means of simulating a VE today is through immersion in computer graphics coupled with acoustic interface and domain-independent interacting devices such as wands and domain-specific devices such as the steering and brakes for cars or earthmovers or instrument clusters for airplanes. Immersion gives the feeling of depth, which is essential for a three-dimensional effect. Techniques for creating immersion in current VR systems are:

1. *Head-mounted displays (HMDs).* A spatial tracking device incorporating liquid-crystal displays (LCDs) or miniature cathode-ray tubes (CRTs) mounted on the head of the user provides information on head movements to update the visual images. These devices are bulky and mobility is restricted. The BOOM (binocular omni-orientation monitor) per-

Figure 1.1 BOOM personal immersive display. *Source:* picture courtesy of Fakespace Inc.

sonal immersive display (Figure 1.1) offers stereoscopic visualization on a counterbalanced, highly accurate, motion-tracking support structure for practically weightless viewing, with resolution unsurpassed by any alternative technology. High-resolution immersion CRT technology generates up to 1280 × 1024 pixels per eye for richly detailed full-color imagery. Image sharpness, realistic shading, and incredibly fast image updates add to the realism. A choice of several optics sets, with fields of view up to a panoramic 140°, lets you optimize the display for your application and desired degree of immersion.

2. *Stereoscopic projectors.* A three-dimensional sensation of depth is obtained from screen projections of two distinct views of an object separated by a small angle based on viewing the object with the left eye only and with the right eye only. The users normally wear LCD shutter glasses and a resolution of 1280 × 1024 pixels has been achieved. Examples of stereoscopic displays are provided in Figures 1.2 to 1.5. Unlike an HMD, multiple users can share the same experience using stero projectors. Usually one user gets the best perspective projection. Using active CRT technology it is currently possible to approach a resolution of 1600 × 1200 pixels, although the cost seems to ecalate as well. Using passive digital stereo technology it is potentially possible to achieve an even higher resolution of 1600 × 1600. The Electronic Visualization Laboratory at University of Illinois at Chicago is currently conducting research in this direction. Recently, Fakespace has introduced RAVE, a fully reconfigurable display that can quickly be configured to a flat wall, a cave-like immersive theater, or other formats, depending upon the available space and the audience.

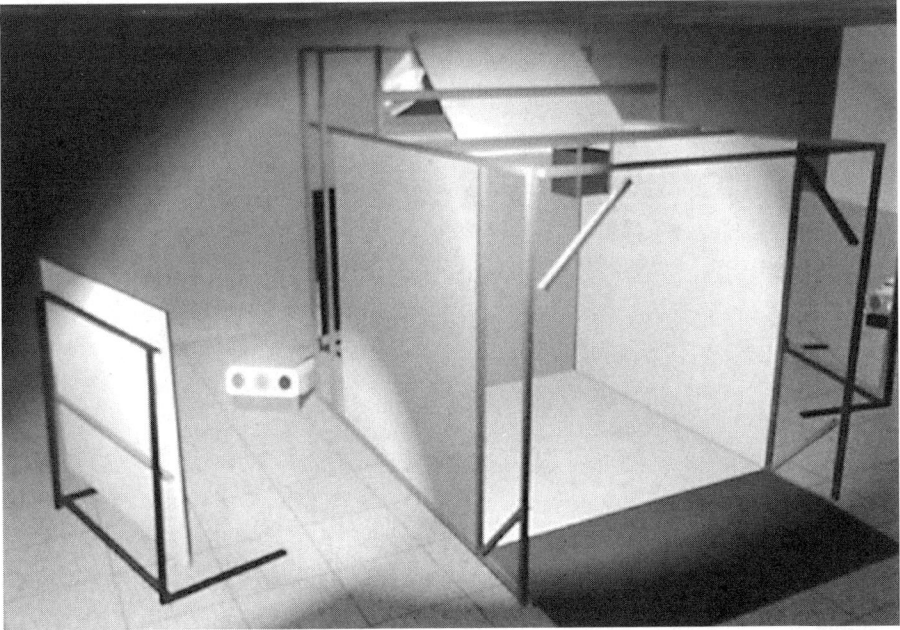

Figure 1.2 CAVE is a multiperson room-sized high-resolution three-dimensional video and audio environment. Graphics are projected in stereo onto three walls and the floor, and viewed with stereo glasses. As a viewer wearing a location sensor moves within its display boundaries, the correct perspective and stereo projections of the environment are constantly updated, so the image moves with and surrounds the viewer to achieve immersion. *Source:* picture courtesy of Electronic Visualization Lab., University of Illinois at Chicago.

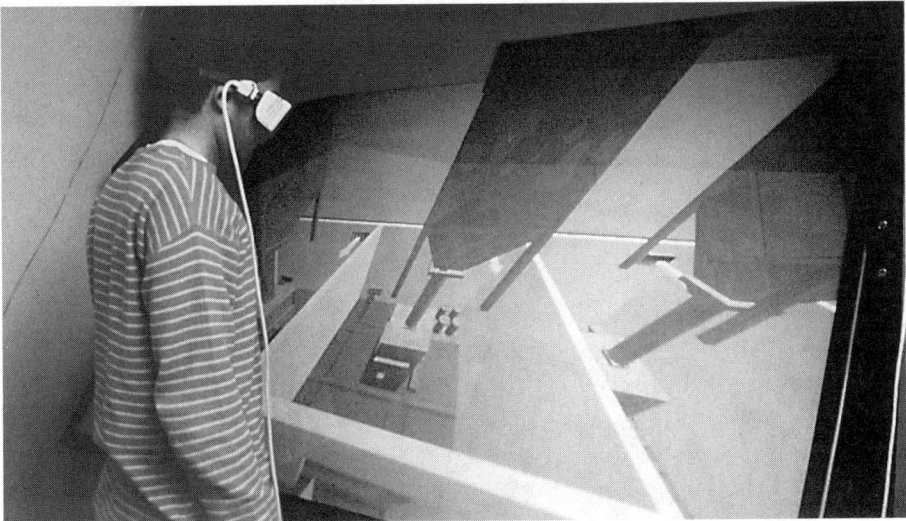

Figure 1.3 ImmersaDesk is a drafting table–format version of CAVE. When folded up, it fits through a standard institutional door and deploys into a 6-by 8-foot footprint. It requires a single graphics engine of the SGI Onyx or Octane class, one projector, and no architectural modifications to the working space.

Figure 1.4 The PowerWall achieves very high display resolution through parallelism, building up a single image from an array of display panels projected from the rear onto a single screen. High-speed playback of previously rendered images is possible by attaching extremely fast disk subsystems, accessed in parallel, to an Onyx.

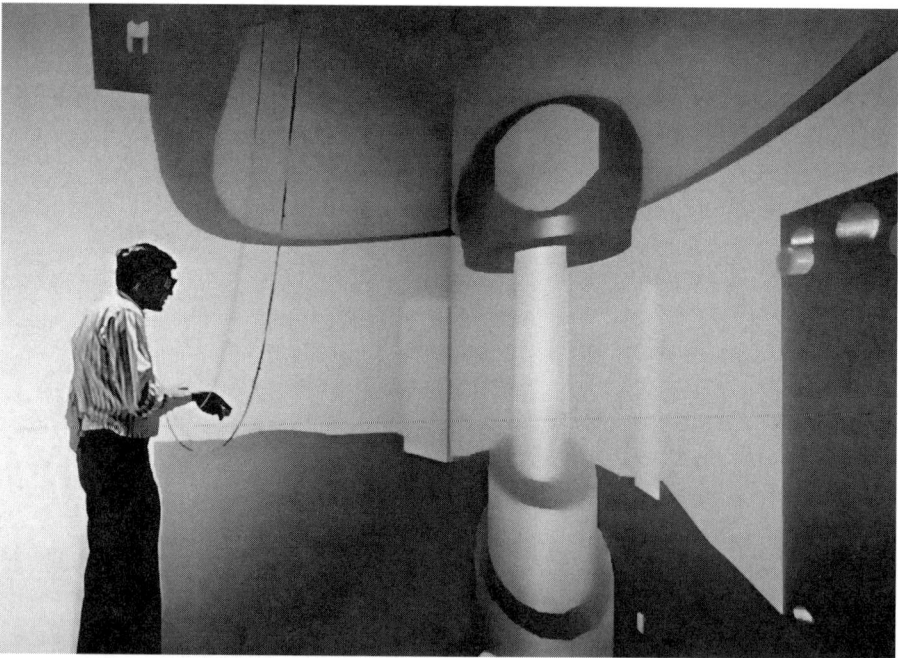

Figure 1.5 View of CAVE from the interior.

3. *Retinal display.* Such a display is based on laser microscanner technology; it will use tiny solid-state lasers to scan color images directly onto the retina. The laser microscanner display, however, still faces substantial technical obstacles. It aims to provide high resolution (1000 × 1000 pixels) through a lightweight device.

There are two distinct groups in VR, based on the means used. The first group is *immersive VR,* which is based on immersive display technologies such as HMDs or stereo projection. The other group is *nonimmersive desktop monitor VR,* which has emerged from animation of computer-aided drafting (CAD).

Other Terms

Historically, a number of researchers termed related phenomena differently; for example, Myron Krueger coined the term *artificial reality,* William Gibson coined *cyberspace,* and Jaron Lanier coined *virtual reality.* There are many terms that are somewhat similar in scope to VR or VE, such as *synthetic environments, artificial reality, virtual world,* and *augmented reality.* For example, instead of completely immersing a user in a simulated environment, an augmented reality system can combine computer-generated imagery with a view of the real world. A typical application would be to overlay information on real-world objects, such as showing the location of a component on the inside of a machine instead of navigating inside a machine in an immersive environment. Terms such as *virtual reality, cyberspace, virtual environments, teleoperation, telerobotics, augmented reality,* and *synthetic environments* have a common feature in that in all such systems the basic components are a human operator, a machine, and a human–machine interface linking the operator to the machine.

Comparison with Related Phenomena

1. *Augmented reality system.* In an augmented reality system, the operator's interaction with the real world (either directly or via a teleoperator system) is enhanced by overlaying the associated real-world information with information stored in the computer (generated from models, derived previously from other sensing systems, etc.).
2. *Traditional simulators (e.g., flight simulators).* Virtual environment systems differ from traditional simulator systems in that they rely much less on physical mock-ups for simulating objects within reach of the operator and are much more flexible and reconfigurable.
3. *Merging of traditional simulator and VR.* Since traditional simulators came earlier and VR later, it is natural to see the two technologies starting to merge. Here is an interesting incident that illustrates this trend. Recently, a motorcycle simulator technology was introduced in Japan in which Honda Motor Company and Salt Lake City, Utah–based Evans and Sutherland Computer Corporation collaborated on the develop-

ment of new simulation technology that all new motorcyclists in Japan will have to use before they are given a license. Some 15,000 motorcyclists were killed in accidents in Japan between 1988 and 1995, and the Japanese government has passed a law dictating that all new riders be trained on a simulator. Evans and Sutherland shipped 300 visual systems in the first phase, based on its Liberty image generation technology. The systems have been engineered so they simulate motion, sound, and vision, as well as traffic flow control.

1.3 REPRESENTATIVE APPLICATIONS

Representative applications of virtual reality technology are presented in a number of areas. Applications in manufacturing or pointers to it have been emphasized particularly. Immersive display technology can be used for creating virtual prototypes of products and processes. The user can then be exposed to an environment that is next best only to an actual product or process. Examples from the product standpoint include virtual prototyping of a product, such as earthmoving equipment, instead of expensive physical prototyping. From the process standpoint, such examples include detailed layout design involving hard-to-quantify factors such as adequate illumination, sources of distractions for operators caused by heavy goods, and personnel movement in the aisles.

The issues here are concerned with CAD model portability among systems, trade-offs between highly-detailed models and real-time interaction and display, rapid prototyping, collaborative design using VR over distance, use of the World Wide Web for virtual manufacturing in small and medium-sized business, using qualitative information (illumination, sound levels, ease of supervision, handicap accessibility) to design manufacturing systems, use of intelligent and autonomous agents in virtual environments, and determining the validity of VR versus reality (quantitative testing of virtual versus real assemblies/equipment).

A number of initiatives in this area have been undertaken at the National Institute of Standards and Technology (NIST). Engineering tool kit environments are needed that integrate clusters of functions that manufacturing engineers need in order to perform related sets of tasks. Integrated production system engineering environments would provide functions to specify, design, engineer, simulate, analyze, and evaluate a production system. Some examples of the functions that might be included in an integrated production system engineering environment are:

- Identification of product specifications and production system requirements
- Producibility analysis for individual products

- Modeling and specification of manufacturing processes
- Measurement and analysis of process capabilities
- Modification of product designs to address manufacturability issues
- Plant layout and facilities planning
- Simulation and analysis of system performance
- Consideration of various economic/cost trade-offs of different manufacturing processes, systems, tools, and materials
- Analysis supporting selection of systems/vendors
- Procurement of manufacturing equipment and support systems
- Specification of interfaces and the integration of information systems
- Task and workplace design
- Management, scheduling, and tracking of projects.

The interoperability of the commercial engineering tools that are available today is extremely limited, so as users move back and forth between different software applications carrying out the engineering process, they must reenter data. Examples of production systems that may eventually be engineered using this type of integrated environment include transfer lines, group technology cells, automated or manually operated workstations, customized multipurpose equipment, and entire plants.

Virtual Collaborative Environments

The term *global virtual manufacturing* (GVM) extends the definition of VM to include, and emphasize, the use of Internet/intranet global communications networks for virtual component sourcing, and multisite multiorganization virtual collaborative design and testing environments. Companies that commit to GVM may be able to dramatically shorten the time to market for new products, cut the cost of prototyping and preproduction engineering, enable many more variations to be tried out before committing to manufacture, and increase the range and effectiveness of quality assurance testing. Virtual prototypes can be virtually assembled, tested, and inspected as part of production planning and operative training procedures: They can be demonstrated, market tested, used to brief and train sales and customer staff, transmitted instantly from site to site via communications links, and modified and recycled rapidly in response to feedback.

Manufacturers and their worldwide subcontractors and main suppliers can establish agile manufacturing teams that will work together on the design, virtual prototyping, and simulated assembly of a particular product while establishing confidence in the virtual supply chain. Using the most advanced VR systems, geographically remote members of the team can meet together in the same virtual design environment to discuss and implement changes to virtual prototypes. Examples of recent developments in virtual collaborative environments include projection of gestures and movements of multiple remote designers as voice-activated avatars to help explain the intention of the designer to others in real time using high-speed ATM networks.

1.4 IMPORTANT OUTCOME OF VIRTUAL MANUFACTURING: VIRTUAL FACTORIES OF THE FUTURE

The Need
When a single factory may cost over a billion dollars (as is the case in the semiconductor industry), it is evident that manufacturing decision makers need tools that support good decision making about their design, deployment, and operation. However, in the case of manufacturing models, there is usually no testbed but the factory itself; development of models of manufacturing operations is very likely to disrupt factory operations while the models are being developed and tested.

Virtual Factories
Sophisticated computer simulations — what might be called *virtual factories* — call for a distributed, integrated, computer-based composite model of a total manufacturing environment, incorporating all the tasks and resources necessary to accomplish the operation of designing, producing, and delivering a product. With virtual factories capable of accurately simulating factory operations over time scales of months, managers would be able to explore many potential production configurations and schedules or different control and organizational schemes at significant savings of cost and time to determine how best to improve performance.

Simulation
Since a factory model running in simulation mode would run thousands of times faster than real factory operations and would probably cost much less as well, managers would have a rapid, nondisruptive methodology for testing various manufacturing strategies. Improvements suggested by real operations could be tested without risk in the simulation. Simulations could also assist in training tool operators and floor managers, who would be able to use factory models in simulation mode much as pilots use simulators to gain experience in flying real airplanes, especially under stressful or unusual conditions.

Control
Computer-based factory models might also be coupled to real factories in what could be called *control mode,* in which the factory model would actually control and run the operation of the real factory through manipulation of the objects in the virtual factory. Operating procedures and scheduling protocols would be validated in the virtual factory and then applied in or transferred to the real production facility. Control mode would enable direct electronic transfer of modularized capabilities from computer simulation to production line.

Self Diagnosis

Coupled to appropriate computer-based reasoning and decision-support tools, a virtual factory operating in control mode would be capable of a significant amount of self-diagnosis. Driven by data from the real factory, the virtual factory would be able to analyze the performance of the entire factory continuously to determine the potential for optimizing operations to reduce costs, reduce production time, improve quality, or reuse materials. For example, the virtual factory would be able to use the data collected by a factory monitoring system, analyze potential and actual failures, and identify the cause of a problem. Such a system assumes the availability of a knowledge base for every piece of equipment in the factory that, given certain monitored data, can be used in conjunction with a diagnostic system and reasoning and decision-support tools to identify the source of a problem.

Toward Monitoring and Control for Complex Manufacturing Systems

For monitoring and control of complex manufacturing systems, four dimensions can be conceived to express complexity:

1. *Space* permits us to examine the physical location, layout, and flow issues critical in all manufacturing operations.
2. *Time* permits us to address facility life-cycle and operational dynamic issues, beginning with concurrent engineering of the production process and testing facilities during product design, extending through production and decline of the initial generation product(s), cycling through the same process for future-generation products.
3. *Process* allows us to study the coherent integration of engineering, management, and manufacturing processes. It permits examination of the important, yet intricate interplay of relationships between classically isolated functions. As examples, consider relationships between production planning and purchasing, production control and marketing, quality and maintenance, and design and manufacturing. Processes involve decisions ranging from long-range operational planning to machine/device-level short-term planning and control. The integration between various levels of aggregation is essential.
4. *Network* deals with organization and infrastructure integration. Whereas the third dimension focuses on the actions, this dimension concentrates on the actors and their needs and responsibilities. Clearly including personnel, the set of actors also includes all devices, equipment, and workstations; all organizational units, be they cells, teams, departments, or factories; and all external interactors, such as customers, vendors, subcontractors, and partners. Issues such as contrasting hierarchically controlled networks with heterarchical, autonomous agent networks must be addressed.

Virtual manufacturing techniques enhance our ability to understand the four dimensions described above by addressing issues such as designing products that can be evaluated and tested for structural properties, ergonomic functionality, and reliability, without having to build actual scale models; designing products for aesthetic value, meeting individual customer preferences; ensuring facility and equipment compliance with various federally mandated standards, facilitating remote operation and control of equipment (telemanufacturing and telerobotics); developing processes to ensure manufacturability without having to manufacture the product (e.g. avoiding destructive testing); developing production plans and schedules and simulating their correctness; and educating employees on advanced manufacturing techniques, worldwide, with emphasis on safety.

1.5 CONSIDERATIONS IN THE DEVELOPMENT OF VIRTUAL MANUFACTURING MODELS

Learning from Past Effort

- The initial models were too poor in detail to provide answers that satisfied factory demands. Even small events may produce large fluctuations in factory operations, and adequate models must be capable of reflecting these subtle influences.
- The user interfaces were so complicated and/or incomprehensible that they were unusable. The most common user of information is a human being, who can be overwhelmed by either the number or the complexity of the user interfaces required to access or disseminate information. For example, senior factory managers are best able to comprehend results that are explicitly tied to financial metrics of performance; results tied to metrics relevant at lower levels in the hierarchy will be less helpful to them.
- There were insufficient skilled personnel to understand and apply the models intelligently. Use of models is not inherently easy. It requires skills that enable using models and simulations, as well as understanding and analyzing the results. Learning these skills requires education and training.
- There were sufficient skilled experts on the factory floor to manage operations, so that modeling and simulation were considered unnecessary. When things are going relatively smoothly, new tools such as modeling and simulation are felt to be superfluous; current skill sets are thought to be sufficient to do the job.
- Factories themselves constantly change (e.g., new or modified manufacturing processes may be installed), and a lack of synchronization between a model and what is actually being done on the shop floor may invalidate the model.

For such reasons, factory-level modeling has never succeeded in capturing the attention of senior manufacturing management in the way that process and product models have. However, recent advances in information technology make the idea of realistic simulations of factory operations much more feasible than they have been in the past.

Determining the Requirements for Effective Factory Models

A virtual factory model will involve a comprehensive model or structure for integrating a set of heterogeneous and hierarchical submodels at various levels of abstraction. Each submodel will be designed for a specific purpose, but together they will operate from a common source of data or knowledge base and will be able to deal with the task at hand without expensive or time-consuming hand-tailoring of interfaces for a particular user's needs. To a very high degree, the software used to control actual factory operations will also drive the operation of the virtual factory model, although it will do so very rapidly, so that simulations can be run on a timely basis.

An ultimate goal would be the creation of a demonstration platform that would compare the results of real factory operations with the results of simulated factory operations using information technology applications such as those discussed above. This demonstration platform would use a computer-based model of an existing factory and would compare its performance with that of a similarly equipped factory running the same product line but using, for example, a new layout of equipment, a better scheduling system, a paperless product and process description, or fewer or more human operators. The entire factory would have to be represented in sufficient detail so that any model user, from factory manager to equipment operator, would be able to extract useful results.

There are two broad areas of need:

(1) hardware and software technology to handle sophisticated graphics and data-oriented models in a useful and timely manner, and
(2) representation of manufacturing expertise in models in such a way that the results of model operation satisfy manufacturing experts' needs for accurate responses.

Research Considerations Related to Virtual Modeling Technology

- *User interfaces.* Since factory personnel in the twenty-first century will be interacting with many applications, it will not suffice for each application to have its own set of interfaces, no matter how good any one application is. Much thought has to be given both to the nature of the specific interfaces and to the integration of interfaces in a system designed so as not to confuse the user. Interface tools that allow the user to filter and abstract large volumes of data will be particularly important.

- *Model consistency.* Models will be used to perform a variety of geographically dispersed functions over a short time scale during which the model information must be globally accurate. In addition, pieces of the model may themselves be widely distributed. As a result, a method must be devised to ensure model consistency and concurrency, perhaps for extended time periods. The accuracy of models used concurrently for different purposes is a key determinant of the benefits of using such models.
- *Testing and validation of model concepts.* Because a major use of models is to make predictions about matters that are not intuitively obvious to decision makers, testing and validation of models and their use are very difficult. For factory operations and design alike, there are many potential "right answers" to important questions, and none of these is "provably correct" (e.g., what is the "right" schedule?). As a result, models have to be validated by being tested against understandable conditions, and in many cases, common sense must be used to judge if a model is "correct." Because models must be tested under stochastic factory conditions, which are hard to duplicate or emulate outside the factory environment, an important area for research involves developing tools for use in both testing and validating model operation and behavior. Tools for automating sensitivity analysis in the testing of simulation models would help to overcome model validation problems inherent in a stochastic environment.

FURTHER READING

For further material on some of the topics in this chapter, the reader is referred to ITM (1995), Banerjee and Kesavadas (1999), McLean (1998), and Vince (1998).

REFERENCES

Banerjee, P., and Kesavadas, T., eds., *Industrial Virtual Reality,* MH-Vol. 5 / MED-Vol. 9, ASME, New York, 1999.

ITM, *Information Technology for Manufacturing: A Research Agenda,* National Academy Press, Washington, DC, 1995.

McLean, C., "Modeling Production Systems Using Virtual Reality Techniques," *Proceedings of the IEEE Systems, Man and Cybernetics Conference,* San Diego, CA, pp. 344–347, 1998.

Vince, J., *Essential Virtual Reality — Fast,* Springer Verlag, New York, 1998.

CHAPTER

2

PRINCIPLES OF THREE-DIMENSIONAL COMPUTER GRAPHICS AND GEOMETRICAL TRANSFORMATIONS

2.1 INTRODUCTION

To develop virtual reality applications systematically, a brief review of three-dimensional computer graphics and geometric transformations is needed.

The purpose of three-dimensional computer graphics in virtual reality is to support realism. The issue of realism arises because on a graphics display, just as on a painter's canvas, it is impractical to produce an image that is a perfectly realistic representation of an actual scene. A compromise has to be made among a number of factors, such as the different kinds of realism needed by applications, the amount of processing required to generate the image, the capabilities of the display hardware, the level of detail recorded in the model of the scene, and the perceptual effects of the image on the observer. The basic problem addressed by visualization techniques is sometimes referred to as *depth cueing*. When a three-dimensional scene is projected on a two-dimensional display screen, information about the depth of objects in the images tends to be reduced or lost and techniques that provide depth cues are designed to restore or enhance the communication of depth to the observer. Techniques to communicate the depth cues include *perspective projection,* in which distant objects are made smaller than near ones; *stereoscopic views,* in which one image is generated for the location of the left eye and a slightly rotated image is generated for the location of the right eye and the two images are fused; and *hidden line* or *surface elimination* and *rendering.* We look next at some of these concepts in detail.

2.2 VIRTUAL WORLD AND OBSERVER SPACE

A right-handed Cartesian coordinate system with *X, Y,* and *Z* axes is used. As a convention a right-hand system (Figure 2.1) assumes that when using one's right hand, the outstretched thumb, first, and middle fingers align with the *X, Y,* and *Z* axes, respectively. The *Y* axis is considered to be the vertical axis. If the existence of a virtual world constructed out of Cartesian coordinates is assumed, the next step is to consider how it can be viewed from any point in space. The term *virtual observer* is used to describe this point in space.

Figure 2.1 Right-handed coordinate system with our convention.

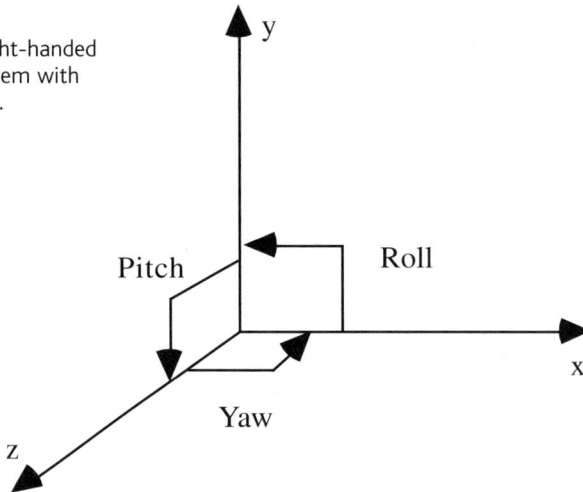

2.2.1 Positioning the Virtual Observer

There are many conventions for computing the virtual observer's (VO's) position, such as direction cosines, *XYZ* fixed angles, *XYZ* Euler angles, and quaternions. For example, in an implementation of *XYZ* Euler angles in IRIS Performer™, the term *static coordinate system* (SCS) is used to denote a fixed frame of reference. The term *dynamic coordinate system* (DCS) is used for a local, rotating frame of reference. In DCS, three separate rotations are pro-

Figure 2.2 Different software packages assume different conventions. (The one in this figure has been used in IRIS Performer.)

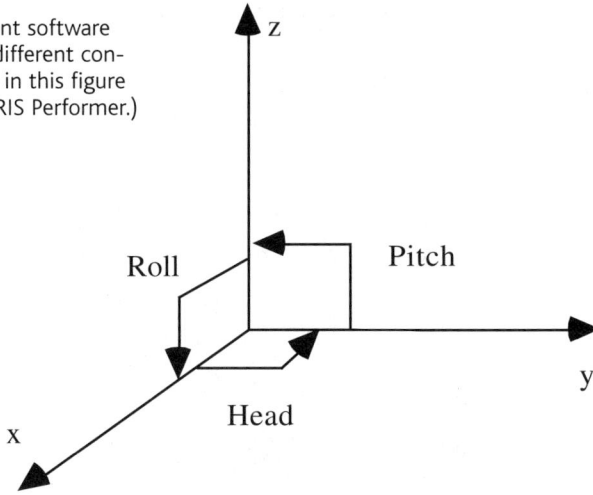

vided for, referred to as pitch, roll, and yaw. The conventions and terminology vary; the convention of one of the versions of IRIS Performer is shown in Figure 2.2. In the convention that we shall follow, *roll* is the angle of rotation about the Z axis, *pitch* is the angle of rotation about the X axis, and *yaw* is the angle of rotation about the Y axis (Figure 2.1).

2.2.2 *XYZ* Fixed Angles

XYZ is the virtual environment (VE) frame of reference, and *uvw* is the virtual observer (VO) frame of reference (Figure 2.3). It is easy to verify that for a point P in space, the following coordinate transformations apply for a rotation angle "roll" about the Z axis, separating the VO frame of reference from the VE frame of reference:

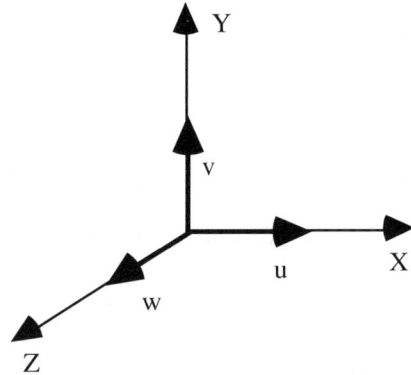

Figure 2.3 *XYZ* is the virtual environment coordinate system (VECS), *uvw* is the virtual observer coordinate system (VOCS).

$$\begin{bmatrix} X_P \\ Y_P \\ Z_P \end{bmatrix} = \begin{bmatrix} \cos roll & -\sin roll & 0 \\ \sin roll & \cos roll & 0 \\ 0 & 0 & 1 \end{bmatrix} \begin{bmatrix} u_P \\ v_P \\ w_P \end{bmatrix} \qquad (2.1)$$

Let us use an abbreviated notation for the transformation above:

$$\begin{bmatrix} X_P \\ Y_P \\ Z_P \end{bmatrix} = [roll] \begin{bmatrix} u_P \\ v_P \\ w_P \end{bmatrix} \qquad (2.2)$$

The following transformations apply for a rotation angle pitch about the X axis and a rotation angle yaw about the Y axis, separating the VO and VE frames of reference:

$$\begin{bmatrix} X_P \\ Y_P \\ Z_P \end{bmatrix} = \begin{bmatrix} 1 & 0 & 0 \\ 0 & \cos pitch & -\sin pitch \\ 0 & \sin pitch & \cos pitch \end{bmatrix} \begin{bmatrix} u_P \\ v_P \\ w_P \end{bmatrix} \quad \text{or} \quad \begin{bmatrix} X_P \\ Y_P \\ Z_P \end{bmatrix} = [pitch] \begin{bmatrix} u_P \\ v_P \\ w_P \end{bmatrix} \qquad (2.3)$$

$$\begin{bmatrix} X_P \\ Y_P \\ Z_P \end{bmatrix} = \begin{bmatrix} \cos yaw & 0 & \sin yaw \\ 0 & 1 & 0 \\ -\sin yaw & 0 & \cos yaw \end{bmatrix} \begin{bmatrix} u_P \\ v_P \\ w_P \end{bmatrix} \quad \text{or} \quad \begin{bmatrix} X_P \\ Y_P \\ Z_P \end{bmatrix} = [yaw] \begin{bmatrix} u_P \\ v_P \\ w_P \end{bmatrix} \qquad (2.4)$$

The transformations above apply for viewing the VO frame of reference from the VE frame of reference. We are interested in viewing the VE from the VO frame of reference when using devices such as head trackers and wands. The angles *roll, pitch,* and *yaw* get reversed and can be represented by $-roll$, $-pitch$, and $-yaw$, respectively. The following are the reverse transformations.

For $-roll$:

$$\begin{bmatrix} u_P \\ v_P \\ w_P \end{bmatrix} = \begin{bmatrix} \cos roll & \sin roll & 0 \\ -\sin roll & \cos roll & 0 \\ 0 & 0 & 1 \end{bmatrix} \begin{bmatrix} X_P \\ Y_P \\ Z_P \end{bmatrix} \qquad (2.5)$$

Let us abbreviate this as

$$\begin{bmatrix} u_P \\ v_P \\ w_P \end{bmatrix} = [-roll] \begin{bmatrix} X_P \\ Y_P \\ Z_P \end{bmatrix} \qquad (2.6)$$

For $-pitch$:

$$\begin{bmatrix} u_P \\ v_P \\ w_P \end{bmatrix} = \begin{bmatrix} 1 & 0 & 0 \\ 0 & \cos pitch & \sin pitch \\ 0 & -\sin pitch & \cos pitch \end{bmatrix} \begin{bmatrix} X_P \\ Y_P \\ Z_P \end{bmatrix} \quad \text{or} \quad \begin{bmatrix} u_P \\ v_P \\ w_P \end{bmatrix} = [-pitch] \begin{bmatrix} X_P \\ Y_P \\ Z_P \end{bmatrix} \qquad (2.7)$$

For $-yaw$:

$$\begin{bmatrix} u_P \\ v_P \\ w_P \end{bmatrix} = \begin{bmatrix} \cos yaw & 0 & -\sin yaw \\ 0 & 1 & 0 \\ \sin yaw & 0 & \cos yaw \end{bmatrix} \begin{bmatrix} X_P \\ Y_P \\ Z_P \end{bmatrix} \quad \text{or} \quad \begin{bmatrix} u_P \\ v_P \\ w_P \end{bmatrix} = [-yaw] \begin{bmatrix} X_P \\ Y_P \\ Z_P \end{bmatrix} \qquad (2.8)$$

Notice that because of special structure of the matrices above, the following equalities hold:

$$
\begin{aligned}
[-roll] &= [roll]^T = [roll]^{-1} \\
[-pitch] &= [pitch]^T = [pitch]^{-1} \\
[-yaw] &= [yaw]^T = [yaw]^{-1}
\end{aligned}
\tag{2.9}
$$

In addition to the angular transformations above, there is a translation of the VO by (t_x, t_y, t_z) with respect to the VE frame of reference. This can be expressed as

$$
\begin{bmatrix} X_P \\ Y_P \\ Z_P \\ 1 \end{bmatrix} =
\begin{bmatrix} 1 & 0 & 0 & t_x \\ 0 & 1 & 0 & t_y \\ 0 & 0 & 1 & t_z \\ 0 & 0 & 0 & 1 \end{bmatrix}
\begin{bmatrix} u_P \\ v_P \\ w_P \\ 1 \end{bmatrix}
$$

Let us abbreviate the above by
$$
\begin{bmatrix} X_P \\ Y_P \\ Z_P \\ 1 \end{bmatrix} = [translate] \begin{bmatrix} u_P \\ v_P \\ w_P \\ 1 \end{bmatrix}
\tag{2.10}
$$

So far we have modeled the transformations individually. In practice, these transformations have to be combined. This can be achieved by applying the transformations in sequence. A popular convention of sequencing the transformations is roll, pitch, yaw and finally, translation to locate VO in the VE frame of reference. This can be expressed as

$$
\begin{bmatrix} X_P \\ Y_P \\ Z_P \\ 1 \end{bmatrix} = [translate][yaw][pitch][roll] \begin{bmatrix} u_P \\ v_P \\ w_P \\ 1 \end{bmatrix}
\tag{2.11}
$$

To locate VE in the VO frame of reference, the following reverse transformations are obtained:

$$
\begin{bmatrix} u_P \\ v_P \\ w_P \\ 1 \end{bmatrix} = [roll]^{-1}[pitch]^{-1}[yaw]^{-1}[translate]^{-1} \begin{bmatrix} X_P \\ Y_P \\ Z_P \\ 1 \end{bmatrix}
\tag{2.12}
$$

Using the equalities above this can be rewritten as:

$$
\begin{bmatrix} u_P \\ v_P \\ w_P \\ 1 \end{bmatrix} = [-roll][-pitch][-yaw][-translate] \begin{bmatrix} X_P \\ Y_P \\ Z_P \\ 1 \end{bmatrix}
\tag{2.13}
$$

The above can be expressed by a single homogeneous transformation matrix:

$$
\begin{bmatrix} u_P \\ v_P \\ w_P \\ 1 \end{bmatrix} = \begin{bmatrix} T_{11} & T_{12} & T_{13} & T_{14} \\ T_{21} & T_{22} & T_{23} & T_{24} \\ T_{31} & T_{32} & T_{33} & T_{34} \\ T_{41} & T_{42} & T_{43} & T_{44} \end{bmatrix} \begin{bmatrix} X_P \\ Y_P \\ Z_P \\ 1 \end{bmatrix}
\tag{2.14}
$$

where

$T_{11} = \cos yaw \cos roll + \sin yaw \sin pitch \sin roll$

$T_{12} = \cos pitch \sin roll$

$T_{13} = -\sin yaw \cos roll + \cos yaw \sin pitch \sin roll$

$T_{14} = -(t_x T_{11} + t_y T_{12} + t_z T_{13})$

$T_{21} = -\cos yaw \sin roll + \sin yaw \sin pitch \cos roll$

$T_{22} = \cos pitch \cos roll$

$T_{23} = \sin yaw \sin roll + \cos yaw \sin pitch \cos roll$

$T_{24} = -(t_x T_{21} + t_y T_{22} + t_z T_{23})$

$T_{31} = \sin yaw \cos pitch$

$T_{32} = -\sin pitch$

$T_{33} = \cos yaw \cos pitch$

$T_{34} = -(t_x T_{31} + t_y T_{32} + t_z T_{33})$

$T_{41} = 0$

$T_{42} = 0$

$T_{43} = 0$

$T_{44} = 1$

Note that in many software packages the order of rotation is predefined. In Performer v. 1.2 it is first *head* [with respect to (w.r.t.) Z], then *pitch* (w.r.t. Y), and finally, *roll* (w.r.t. X). In Softimage any order is possible. In much of the robotics literature the coordinates are realigned to perform rotation around only the Z axis. We have also been noticing changes in some conventions from one software version release to another (e.g., Performer v. 2.0 underwent some changes in coordinate system conventions). Hence different software packages (and sometimes even versions!) define their own conventions.

2.2.3 *XYZ* Euler Angles

XYZ fixed angles are relative to a fixed frame of reference, whereas *XYZ* Euler angles are relative to the local rotating frame of reference. For example, *XYZ* fixed angles can be conveniently expressed by attaching them to the VE coordinate system because it is generally fixed, whereas *XYZ* Euler angles can be expressed more conveniently by attaching them to the VO coordinate system because it is generally moving.

To illustrate the difference between *XYZ* fixed angles and *XYZ* Euler angles; let us consider an example where we have a pitch rotation and then a yaw rotation relative to a rotating frame of reference, let us call it the VO reference frame, *uvw*. Figure 2.3 shows the VE reference frame *XYZ* and the VO reference frame *uvw* mutually aligned. Figure 2.4 shows the orientation of *uvw* after it is subjected to a pitch rotation of 90° about *u*. Figure 2.5 shows the new orientation after *uvw* is subjected to a yaw rotation of 90° about *v*. The new ori-

Figure 2.4 Orientation of *uvw* after pitch rotation of 90° about *u*.

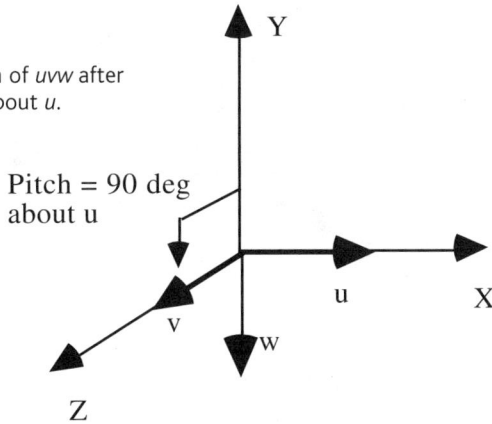

Figure 2.5 *uvw* orientation after yaw rotation of 90° about *v*.

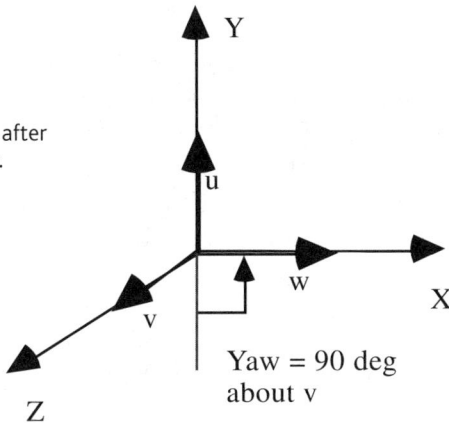

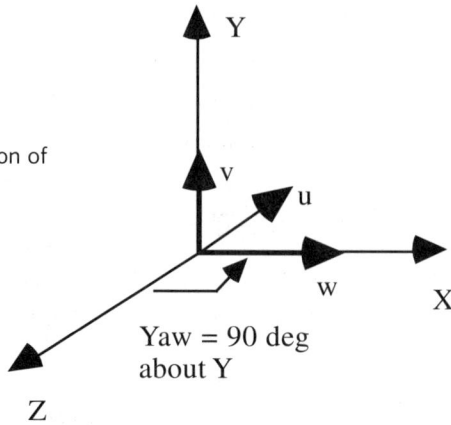

Figure 2.6 Yaw orientation of 90° about Y.

Yaw = 90 deg
about Y

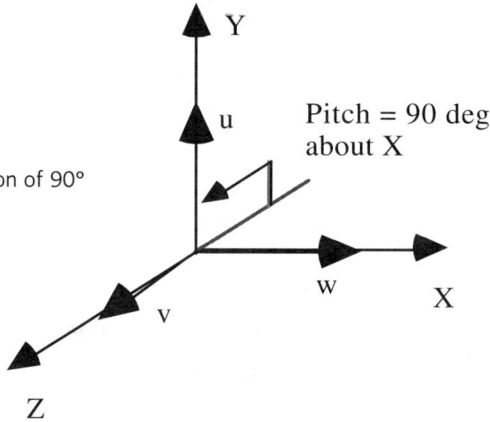

Figure 2.7 Pitch rotation of 90° about X.

Pitch = 90 deg
about X

entation of *uvw* could have been achieved using *XYZ* fixed angles by first performing a yaw rotation of 90° (Figure 2.6), followed by a pitch rotation of 90° (Figure 2.7).

To summarize, in Figures 2.3 and 2.4, one is considering the transformations with respect to a moving frame of reference *(uvw)*. This represents the Euler mode. In Figures 2.6 and 2.7, one is considering the transformations with respect to a fixed frame of reference *(XYZ)*. This represents the fixed mode. We observe that the order of rotational transformations in the Euler mode is in the reverse order compared to that in the fixed mode. This happens to be true in general for all rotational transformations.

Without developing the matrices for roll, pitch, yaw, and translate again for the Euler mode, we can state that if a VO is located in the VE using *XYZ* Euler angles, then any point (X_P, Y_P, Z_P) in the VE can be related to (u_P, v_P, w_P) for the VO by the following equation:

$$\begin{bmatrix} u_P \\ v_P \\ w_P \\ 1 \end{bmatrix} = [-yaw][-pitch][-roll][-translate] \begin{bmatrix} X_P \\ Y_P \\ Z_P \\ 1 \end{bmatrix} \qquad (2.15)$$

Note that the only difference between equation (2.15) and that used for XYZ fixed angles is that the roll, pitch, and yaw matrices have been reversed. This, too, can be represented by a single homogeneous matrix operation:

$$\begin{bmatrix} u_P \\ v_P \\ w_P \\ 1 \end{bmatrix} = \begin{bmatrix} T_{11} & T_{12} & T_{13} & T_{14} \\ T_{21} & T_{22} & T_{23} & T_{24} \\ T_{31} & T_{32} & T_{33} & T_{34} \\ T_{41} & T_{42} & T_{43} & T_{44} \end{bmatrix} \begin{bmatrix} X_P \\ Y_P \\ Z_P \\ 1 \end{bmatrix} \qquad (2.16)$$

where
$T_{11} = \cos yaw \cos roll - \sin yaw \sin pitch \sin roll$
$T_{12} = \cos yaw \sin roll + \sin yaw \sin pitch \cos roll$
$T_{13} = -\sin yaw \cos pitch$
$T_{14} = -(t_x T_{11} + t_y T_{12} + t_z T_{13})$
$T_{21} = -\cos pitch \sin roll$
$T_{22} = \cos pitch \cos roll$
$T_{23} = \sin pitch$
$T_{24} = -(t_x T_{21} + t_y T_{22} + t_z T_{23})$
$T_{31} = \sin yaw \cos roll + \cos yaw \sin pitch \sin roll$
$T_{32} = \sin yaw \sin roll - \cos yaw \sin pitch \sin roll$
$T_{33} = \cos yaw \cos pitch$
$T_{34} = -(t_x T_{31} + t_y T_{32} + t_z T_{33})$
$T_{41} = 0$
$T_{42} = 0$
$T_{43} = 0$
$T_{44} = 1$

2.2.4 Quaternions

Quaternions are useful for representing rotations about any arbitrary axis and are hence useful for modeling motion in virtual manufacturing and automation. Quaternions were discovered by Hamilton around 1860 as a generalization of complex numbers. Just as a complex number has the form $x + iy$, where $i^2 = -1$, a *quaternion* **q** is defined as a 4-tuple

$$\mathbf{q} = [s + xi + yj + zk]$$

where: $i^2 = j^2 = k^2 = ijk = -1$, and s, x, y, and z are real numbers.

A quaternion can be abbreviated as

$$\mathbf{q} = [s, \mathbf{v}]$$

where s is a scalar and $\mathbf{v}$ is a three-dimensional vector.

A *unit quaternion* is defined as a quaternion whose norm, $x^2 + y^2 + z^2 + s^2 = 1$. Two quaternions are equal if, and only if, their corresponding terms are equal. They can be added and subtracted as follows:

$$\mathbf{q}_1 \pm \mathbf{q}_2 = [(s_1 \pm s_2) + (x_1 \pm x_2)i + (y_1 \pm y_2)j + (z_1 \pm z_2)k] \tag{2.17}$$

The product $\mathbf{q}_1 \mathbf{q}_2$ is given by

$$\mathbf{q}_1 \mathbf{q}_2 = [(s_1 s_2 - x_1 x_2 - y_1 y_2 - z_1 z_2) + (s_1 x_2 + s_2 x_1 + y_1 z_2 - y_2 z_1)i + (s_1 y_2 + s_2 y_1 + z_1 x_2 - z_2 x_1)j + (s_1 z_2 + s_2 z_1 + x_1 y_2 - x_2 y_1)k] \tag{2.18}$$

The inverse $\mathbf{q}^{-1}$ is given by

$$\mathbf{q}^{-1} = \frac{[s - xi - yj - zk}{|\mathbf{q}|^2}$$

where $|\mathbf{q}|$ is the magnitude or modulus of $\mathbf{q}$ and is given by

$$|\mathbf{q}| = \sqrt{s^2 + x^2 + y^2 + z^2}$$

Furthermore, it can be shown that: $\mathbf{q}\mathbf{q}^{-1} = \mathbf{q}^{-1}\mathbf{q} = [1 + 0i + 0j + 0k] = 1$ Any position vector can be represented as a quaternion with a zero scalar term. For example, a point $P(x,y,z)$ is represented in quaternion form by: $P = [0 + xi + yj + zk]$. This form is used for rotating individual vertices about an axis.

Returning to the original application for quaternions, that is, to rotate vectors, let us consider how this is achieved. It can be shown that a position vector $\mathbf{u}$ can be rotated about an axis by some angle using the following operation:

$$\mathbf{u}9 = \mathbf{q}\mathbf{u}\mathbf{q}^{-1} \tag{2.19}$$

where the axis and angle of rotation are encoded within the unit quaternion $\mathbf{q}$, whose modulus is 1, and $\mathbf{u}9$ is the rotated vector. For example, say that we needed to rotate a vector through an angle u about an axis defined by $\mathbf{u}$. The quaternion for achieving this is given by

$$\mathbf{q} = [\cos u/2, \sin u/2\ \mathbf{u}] \tag{2.20}$$

where $\mathbf{u}$ is a unit vector. Another way of stating this is as follows:

$$s = \cos[u/2],\ x = s_1 \sin[u/2],\ y = s_2 \sin[u/2],\ z = s_3 \sin[u/2]$$

This represents a rotation by an angle u about the axis defined by the unit vector $\mathbf{u} = (s_1, s_2, s_3)$. Specifically, the point $P = xi + yj + zk$ is rotated to the point $P9 = x9\,i + y9\,j + z9\,k$ as $P9 = \mathbf{q}P\mathbf{q}^{-1}$, where $\mathbf{q}^{-1}$ is the conjugate or inverse of unit quaternion $\mathbf{q}$ and is given by $\mathbf{q}^{-1} = s - xi - yj - zk$.

A simple application of this notation is found in roll, pitch, and yaw rotations. For instance, a roll rotation about the Z axis is given by the quaternion:

$$\mathbf{q}_{\text{roll}} = [\cos u/2, \sin u/2\ [0,0,1]]$$

while a pitch rotation about the x axis is given by

$\mathbf{q}_{pitch} = [\cos\theta/2, \sin\theta/2\ [1,0,0]]$

and a yaw rotation about the y axis is given by

$\mathbf{q}_{yaw} = [\cos\theta/2, \sin\theta/2\ [0,1,0]]$

where θ is the angle of rotation.

Note that all of the above are unit quaternions, where norm $s^2+x^2+y^2+z^2=1$. If we define the three quaternions above to represent the rotations of roll, pitch, and yaw, then by multiplying them together we can arrive at a single quaternion representing the compound rotation:

$\mathbf{q}_{yaw}\mathbf{q}_{pitch}\mathbf{q}_{roll} = [s + xi + yj + zk]$

where

$s =$ cos[yaw/2] cos[$pitch$/2] cos[$roll$/2] + sin[yaw/2] sin[$pitch$/2] sin[$roll$/2]
$x =$ cos[yaw/2] sin[$pitch$/2] cos[$roll$/2] + sin[yaw/2] cos[$pitch$/2] sin[$roll$/2]
$y =$ sin[yaw/2] cos[$pitch$/2] cos[$roll$/2] - cos[yaw/2] sin[$pitch$/2] sin[$roll$/2]
$z =$ cos[yaw/2] cos[$pitch$/2] sin[$roll$/2] - sin[yaw/2] sin[$pitch$/2] cos [$roll$/2]

Notice that this compound quaternion is equivalent to XYZ fixed angles. If XYZ Euler angles were required, the quaternion sequence would have to be reversed as follows:

$\mathbf{q} = \mathbf{q}_{roll}\mathbf{q}_{pitch}\mathbf{q}_{yaw}$

A quaternion can also be represented as an equivalent matrix notation. Converting from a rotation matrix $\mathbf{M}$ to a quaternion representation, we have

$\mathbf{q} = [s,(x, y, z)]$

where

$$s = \pm\tfrac{1}{2}\sqrt{M_{11} + M_{22} + M_{33} + M_{44}}$$

$$x = \frac{M_{32} - M_{23}}{4s}$$

$$y = \frac{M_{13} - M_{31}}{4s}$$

$$z = \frac{M_{21} - M_{12}}{4s}$$

where M_{ij}, $i = 1,2,3; j = 1,2,3$ are the elements of matrix $\mathbf{M}$.

Disadvantages of Euler Angles and Matrices

Traditionally, homogeneous matrices have been used to represent Euler angles because the basic rotation matrices for rotation about the u, v, and w axes are simple and well known. There are 12 different conventions that you can use to represent rotations using Euler angles, since you can use any combination of axes to represent rotations. Six combinations are for rotation about three axes:

uvw, uwv, vuw, wvu, vwu, and *wuv.* Six combinations are for rotations about 2 axes: *vwv, wvw, uwu, wuw, uvu,* and *vuv.* Note that *vwv* can also be represented as *vvw* or simply *vw.* It is easy to verify this noting the fact that the coordinates are moving with the object. The first convention *(uvw)* is used most frequently. Euler angle representation is very efficient because it uses only three variables to represent the three degrees of freedom (DOF). However, there is no easy way to represent a single rotation with Euler angles that corresponds to a series of concatenated rotations. The following disadvantages are apparent:

1. *Lack of intuition.* Describing a general rotation as rotations about the three axes is not natural for an animation designer. If, for instance, the animator wants to rotate an object 30° about a rotation axis given by the vector (1,1,1), it is quite tedious to derive the corresponding Euler angles about the three basic axes.

2. *Importance of the order of rotation axes.* The user of a graphical system must express rotations with respect to a certain convention that defines the order in which the three basic rotations are applied. Different conventions yield different results.

3. *Gimbal lock.* It is possible to create a series of rotations in which 1 DOF in the rotation is lost. This situation is called gimbal lock. Gimbal lock happens when a series of rotations at 90° is performed; suddenly, the rotation does not occur due to the alignment of the axes. For example, imagine that a series of rotations is to be performed by a flight simulator. The first rotation is Q_1 around the *u* axis, the second rotation is 90° around the *v* axis, and Q_3 is the rotation around the *w* axis. If one performs the specified rotations in succession, one discovers that Q_3 rotation around the *w* axis has the same effect as the rotation around the initial *u* axis (see Figure 2.8). The *v* axis rotation has caused the *u* and *w* axes to get aligned, and 1 DOF is lost since the last rotation is not unique; it can be expressed by a combination of previous rotations. Getting an intuitive understanding of how rotation matrices work is

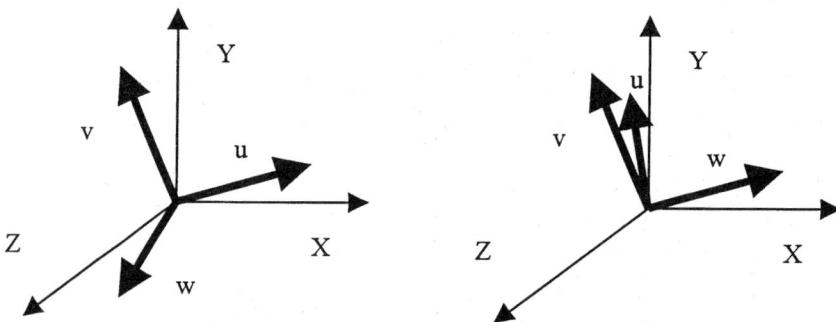

Figure 2.8 Example of gimbal lock.

quite difficult. In particular, it is difficult to predict how successive rotations about the basic axes affect each other.

4. *Ambiguous correspondence to rotations.* Given a rotation matrix, it is difficult to solve the inverse problem: What are the original rotations about the basic axes? In general, there is an ambiguous solution to this problem.

5. *Unapparent result of composition.* In *XYZ* Euler coordinate system, two successive rotations can be expressed as one. For composition, the two rotation matrices must be multiplied followed by extraction of the resulting rotation. Determination of this rotation is tedious and in general not possible.

6. *Redundancy of representation.* Homogeneous matrices contain expendable information. If the matrices are to be used exclusively for rotation, the matrices will have zeros for indices $(4, i)$ and $(i, 4)$, $\forall i \in \{1, 2, 3\}$. In addition to this, the matrix uses nine places for the 4 DOF that are necessary to describe a rotation according to *XYZ* Euler system. On top of this, numerical inaccuracies can be problematic. Since rotation matrices must be orthonormal, there are six constraints that must be maintained during the computations (each row must be a unit vector and the columns must be mutually orthogonal; that is, $i \perp j, j \perp k, k \perp i, |i| = |j| = |k| = 1$.

Advantages of Quaternions

1. *Obvious geometrical interpretation.* Quaternions express rotation as a rotation angle about a rotation axis. This is a more natural way to perceive rotation than is the *XYZ* Euler system. The mapping between rotations and quaternions is unambiguous, with the exception that every rotation can be represented by two quaternions. That **q** and −**q** correspond to the same rotation is mathematically pleasing. This is because rotations themselves come in pairs. Given a rotation, the same rotation is obtained by rotating in the opposite direction about the opposite axis.

2. *Coordinate system independence.* Quaternion rotation is not influenced by the choice of coordinate system. It is not necessary to have any convention as to the order of rotation about explicit axes.

3. *Compact representation.* The representation of rotation using quaternions is compact in the sense that it is four-dimensional and thus contains only the 4 DOF required according to *XYZ* Euler system. In theory, all nonzero quaternions can be used for rotation. In practical applications only unit quaternions will be used. Thus only one constraint on the representation must be upheld during computation compared to the six constraints on rotation matrices.

4. *No gimbal lock.* Since gimbal lock is innate to the matrix representation of *XYZ* Euler system, this problem does not appear in the quaternion representation.

5. *Simple composition.* Rotations are easily composed when using quaternions. The composition corresponds to multiplication of the quaternions involved. Rotation with $\mathbf{q}_1$ followed by rotation with $\mathbf{q}_2$ is achieved by rotating with the quaternion $\mathbf{q}_2\mathbf{q}_1$.
6. *Simple and smoother interpolation methods.* Achieving a smooth interpolation is simpler and more accurate using quaternions than using the *XYZ* Euler system.

Smooth Interpolation of Orientations Using Quaternions

The problem of using spline curves to smoothly interpolate mathematical quantities in Euclidean spaces is common. Many quantities, however, such as rotations, lie in non-Euclidean spaces. In 1992, a method was introduced to smoothly interpolate orientations given *N* rotational key frames of an object along a trajectory. This method allows the user to impose constraints on the rotational path, such as the angular velocity at the endpoints of the trajectory. The rotations are converted to quaternions and then to spline in that non-Euclidean space.

Advantages of the Smooth Interpolation Technique

1. The paths generated through rotation space are very smooth.
2. The smooth interpolation technique allows the user to specify arbitrarily large initial and final angular velocities of a rotating body; by assigning large angular velocities, a user can make an object tumble several full turns between successive keypoints.
3. The technique permits fairly easy addition of additional constraints.
4. The technique generalizes to interpolations of other quantities in non-Euclidean spaces.
5. The technique is sufficiently fast to use in experimentation, taking only a few minutes per interpolation.

Various Methods of Interpolating Quaternions.
There are three common methods:

1. Linear quaternion interpolation (Lerp)
2. Spherical linear quaternion interpolation (Slerp)
3. Spherical and quadrangle quaternion interpolation (Squad)

Linear Quaternion Interpolation (Lerp)

Let *H* be the quaternion space, *h* be a weighting factor, and $\mathbf{p}$ and $\mathbf{q}$ be quaternions. For $\mathbf{p},\mathbf{q} \in H$ and $h \in [0,1]$, the interpolation curve can be stated:

$$Lerp\,(\mathbf{p},\mathbf{q},h) = \mathbf{p}\,(1 - h) + \mathbf{q}h \qquad (2.21)$$

The curve for linear interpolation between quaternions gives a straight line in quaternion space. However, the distribution of interpolation points is sparse in the middle of the curve and dense at the ends. This is not a desired property since the interpolation is not uniform. This property is also expressed by the fact that Lerp does not yield unit quaternions. Another explanation of a unit quaternion is that it lies on the surface of a four-dimensional equivalent (called a hypersphere) of a three-dimensional sphere, whereas a nonunit quaternion cuts through the hypersphere.

Spherical Linear Quaternion Interpolation (Slerp)

Simple linear quaternion interpolation yields a secant between the two quaternions, meaning that it cuts through the hypersphere. An obvious idea is to define an interpolation method yielding the same interpolation curve using unit quaternions as the interpolated quaternions. Instead of performing simple linear interpolation, the curve should follow a great arc on the quaternion unit sphere from one key frame to the other. This is called *great arc interpolation* or *spherical linear interpolation*. Given $\mathbf{p},\mathbf{q} \in H$ and $h \in [0,1]$, the Slerp interpolation curve can be stated:

$$Slerp\ (\mathbf{p},\mathbf{q},h) = \frac{\mathbf{p} \sin[(1 - h)\ \Omega] + \mathbf{q} \sin h\Omega}{\sin \Omega} \qquad (2.22)$$

$$\text{where } \cos \Omega = \mathbf{p} \cdot \mathbf{q}$$

Spherical and Quadrangle Quaternion Interpolation (Squad)

For interpolation between two rotations, Slerp is a good solution. But when we have to interpolate among a series of rotations, some problems emerge at control points where (1) the curve is not smooth (see Figure 2.9), (2) the angular velocity is not constant, or (3) the angular velocity is not continuous. A better idea is to apply a Bezier curve to interpolate between the control points. Such an interpolation curve is called *Squad* ("spherical and quadrangle") and is defined as (with $h \in [0,1]$)

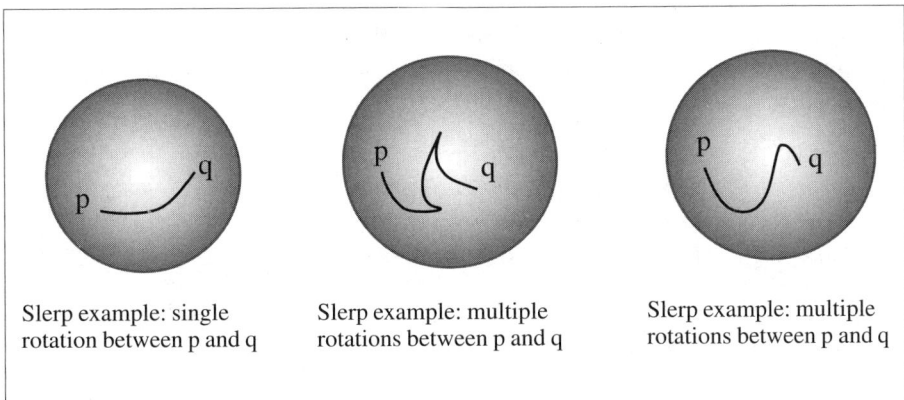

Slerp example: single rotation between p and q	Slerp example: multiple rotations between p and q	Slerp example: multiple rotations between p and q

Figure 2.9 Slerp and Squad concept.

$$Squad(\mathbf{q}_i,\mathbf{q}_{i+1},\mathbf{s}_i,\mathbf{s}_{i+1},h) = Slerp\ (Slerp(\mathbf{q}_i,\mathbf{q}_{i+1},h),\ Slerp(\mathbf{s}_i,\mathbf{s}_{i+1},h),\ 2h(1-h)) \qquad (2.23)$$

$$\mathbf{s}_i = \mathbf{q}_i \exp\left[\frac{-\log(\mathbf{q}_i^{-1}\mathbf{q}_{i+1})}{4} + \log(\mathbf{q}_i^{-1}\mathbf{q}_{i-1})\right] \qquad (2.24)$$

Numerical Example

$Q_1 = X\,\text{Quat}[300°]$
$Q_2 = X\,\text{Quat}[20°]$
$Q_3 = X\,\text{Quat}[250°]$
$Q_4 = X\,\text{Quat}[20°]$
$Q_5 = X\,\text{Quat}[180°]$

We use the following code, written in Mathematica:

```
For [h = 0, h <= 1,
                    Print    ["h=", h] ;
Print["Lerp:      ", N[Lerp[Q2, Q3, h]], "  ",   "θ=",
N[AngLerp[Q2, Q3, h]]];
Print["Slerp:       ", N[Slerp[Q2, Q3, h]], "  ",   "θ=",
N[AngSlerp[Q2, Q3, h]]];
Print [
    "Squad: ", N[Squad[Q1, Q2, Q3, Q4, h]], "  ",   "θ=",
N[AngSquad[Q1, Q2, Q3, Q4, h]]];
Print["  "];
h += 0.25;
]
```

For the segment Q_2–Q_3, the values are as follows:

$h = 0.0$
Lerp: Quat[0.984808,0.173648,0,0] $\theta = 20$
Slerp: Quat[0.984808, 0.173648, 0, 0] $\theta = 20$
Squad: Quat[0.984808, 0.173648, 0, 0] $\theta = 20$

$h = 0.25$
Lerp: Quat[0.871439,0.490503,0,0] $\theta = 58.7473$
Slerp: Quat[0.779884, 0.625923, 0, 0] $\theta = 77.5$
Squad: Quat[0.468388, 0.883523, 0, 0] $\theta = 124.141$

$h = 0.5$
Lerp: Quat[0.382683,0.92388,0,0] $\theta = 135$
Slerp: Quat[0.382683, 0.92388, 0, 0] $\theta = 135$
Squad: Quat[-0.135932, 0.990718, 0, 0] $\theta = 195.625$

$h = 0.75$
Lerp: Quat[-0.269363,0.963039,0,0] $\theta = 211.253$
Slerp: Quat[0.108867, 0.994056, 0, 0] $\theta = 192.5$
Squad: Quat[-0.4756, 0.879662, 0, 0] $\theta = 236.797$

$h = 1$
Lerp: Quat[-0.573576,0.819152,0,0] $\theta = 250$
Slerp: Quat[-0.573576, 0.819152, 0, 0] $\theta = 250$
Squad: Quat[-0.573576, 0.819152, 0, 0] $\theta = 250$

Figure 2.10 illustrates the x component of the Lerp interpolation, and Figure 2.11 illustrates the x component of the Squad interpolation. The Slerp interpolation is not shown, it is linear and hence is easy to understand. It can be seen that the change in x component of θ with respect to h increases in the middle segment in Lerp because the interpolation points are concentrated at the two ends. The change in θ is gradual for Squad (and constant for Slerp). Also note that in Squad the transition of θ is smooth at the ends, which enables smooth interpolation and is one of the main strengths of Squad, as mentioned before. Similar behavior is observed for the y and z components of θ.

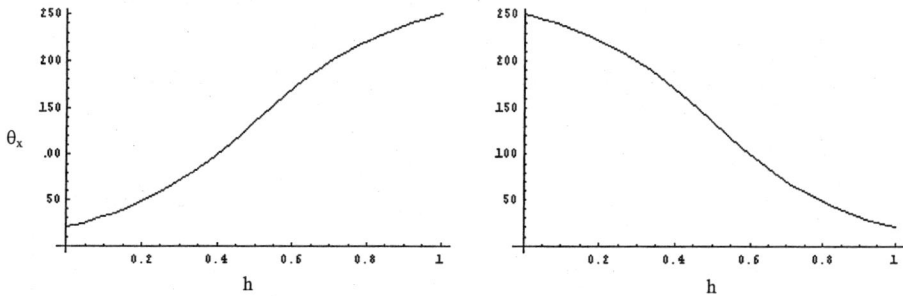

Figure 2.10 Lerp example.

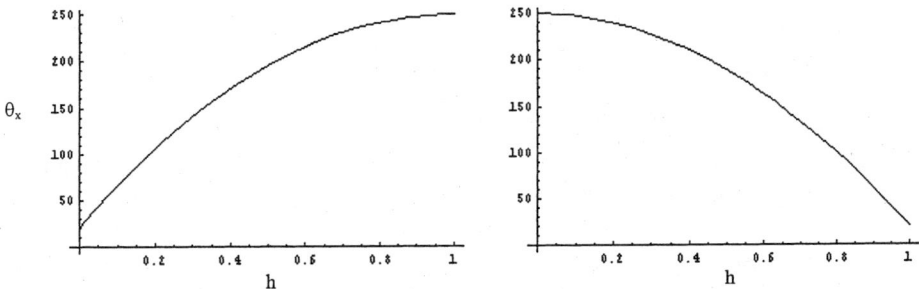

Figure 2.11 Squad example.

Dual Quaternions to Represent Both Rotations and Translations

Given a set of spatial displacements (locations and orientations), the problem of motion interpolation deals with computation of a continuous set of intermediate displacements. Most current approaches separate the interpolations of rotations and translations. The problem with separating translation from rotation is that the resulting motion depends on the choice of the origin of the moving coordinate frame and is thus not unique. Also, separating translation from rotation tends to result in motions with larger variations than necessary in velocity and acceleration. There is an additional problem of coordinating rotations with translations to interpolate the key frames properly. Dual quaternions have been proposed as a possible method to combine translations with rotations.

By approximating three-dimensional displacements with small rotations of a sufficiently large four-dimensional sphere, a double quaternion formulation provides a natural and geometrically intuitive way of handling the trade-off between robustness and accuracy. A dual quaternion belongs to the same family as double quaternions but has a slightly different interpretation. The dual quaternion approach provides a natural way of measuring the spatial distance between two displacements. This is because in the space of dual quaternions, spatial distance is characterized by angular measures that can naturally be combined. On the contrary, in quaternion–translation representation the spatial distance between two displacements is characterized by an angular distance and a linear distance, which do not have any geometrically intuitive way of combining.

The dual quaternion interpolation can be handled by interpolating each of the two quaternions separately. All existing implementations for quaternion interpolation can be easily modified to handle dual quaternions. Given a normalized rotation axis $\mathbf{R} = \{x,y,z\}$, an angle θ, and a three-dimensional translational vector $\mathbf{T} = \{dx,dy,dz\}$, the function $\mathbf{DQ}$ returns the dual quaternion as a vector in $\mathbf{R}^8$ space:

$$\mathbf{DQ}\,(\mathbf{R},\theta,\mathbf{T}) = \{x \cdot s, y \cdot s, z \cdot s, c, vx \cdot c + (vy \cdot z - vz \cdot y)\,s, vy \cdot c + (vz \cdot x - vx \cdot z)$$
$$s, vz \cdot c + (vx \cdot y - vy \cdot x)\,s, -(x \cdot vx + y \cdot vy + z \cdot vz)\,s\}$$

where

$s = \sin[\theta/2]$, $c = \cos[\theta/2]$, $vx = dx/2$, $vy = dy/2$, and $vz = dz/2$.

To obtain the original rotation axis $\mathbf{R}$, the angle θ, and the translational vector $\mathbf{T}$ from a normalized dual quaternion $\mathbf{DQ}$, it is first necessary to determine a dual quaternion representing the rotation only, and then to apply its inverse (conjugate) to extract another dual quaternion representing the translation. Given $\mathbf{DQ} = \{q_0,q_1,q_2,q_3,q_4,q_5,q_6,q_7\}$, the conjugate of the rotation part is $\mathbf{QR}^{-1} = \{-q_0,-q_1,-q_2,q_3,0,0,0,0\}$ and the translational part is the product: $\mathbf{QT} = \mathbf{DQ} \cdot \mathbf{QR}^{-1}$.

Then the original data is defined as

$$\mathbf{R} = \{\frac{q_0}{L}, \frac{q_1}{L}, \frac{q_2}{L}\}, \quad \text{where} \quad L = \sqrt{q_0^2 + q_1^2 + q_2^2 + q_3^2}$$

$$\theta = 2 \text{ Arccos}[q_3]$$

$$\mathbf{T} = \{2\mathbf{QT}[4], 2\mathbf{QT}[5], 2\mathbf{QT}[6]\}$$

Let us now discuss the conversion to dual quaternion from Euler angles and a translational vector. Given the three Euler angles (pitch, yaw, and roll) and a three-dimensional translational vector, it is possible to obtain the corresponding dual quaternion. First, pure rotations in each of the three axes are defined as

$$Q_X = \mathbf{DQ}\,(X, \text{pitch}, \varnothing)$$
$$Q_Y = \mathbf{DQ}\,(Y, \text{yaw}, \varnothing)$$
$$Q_Z = \mathbf{DQ}\,(Z, \text{roll}, \varnothing)$$

where $X = \{1,0,0\}$, $Y = \{0,1,0\}$, $Z = \{0,0,1\}$, and $\varnothing = \{0,0,0\}$. Supposing that the order of the desired rotation is roll, then yaw, and finally, pitch, the product $\mathbf{Q}$ of these three dual quaternions represents the orientation of the body:

$$\mathbf{QR} = Q_X \cdot Q_Y \cdot Q_Z$$

To translate the body to the position $\mathbf{T} = \{dx, dy, dz\}$, the dual quaternion QT is defined as

$$\mathbf{QT} = \mathbf{DQ}\,(\varnothing, 1, \mathbf{T})$$

Note that here the angle is 1 in order to get a normalized dual quaternion, but because of the null axis $\varnothing$, no rotation is performed. Finally, the dual quaternion $\mathbf{Q}$ represents both the rotation and the translation and is given by

$$\mathbf{Q} = \mathbf{QT} \cdot \mathbf{QR}$$

Example: Animating Articulated Figures in Virtual Reality

The animation of articulated figures is quite popular because of the desire to use human beings as avatars in virtual environments. An articulated figure, also called *kinematic linkage,* consists of a series of rigid links, which are connected at joints. In this case a configuration of the linkage is described by a series of rotation angles, which are commonly called *joint angles.* Roughly speaking, each of these angles describes the rotation angle between consecutive links.

There are basically two tasks that have to be solved to animate these structures: forward and inverse kinematics. In *forward kinematics,* motion is defined explicitly by prescribing the set of joint angles needed to get the position and orientation of the last link, also called the *end-effector*. In *inverse kinematics,* the animator defines the position and orientation of the end-effector only and the kinematics gives the corresponding set of joint angles. The inverse kinematics problem is intrinsically not unique in terms of its solutions. Given the end-effector position and orientation, generally one of several sets of joint angles could be used. It might even happen that there exist an infinite number of solutions when the linkage has more than 6 DOF. In this case, the design of the linkage (e.g., the human arm) is called *redundant*. On the other hand, it is also very likely that there exists no solution at all. This is the case if the prescribed position of the end-effector is out of reach. Note that this does not necessarily mean that the position is too far away. It is also possible that the linkage is restricted in its motion capabilities, due to a poor design. For example, a linkage where all joint axes are parallel is planar and will never be able to get out of this plane.

A common approach to solving the inverse kinematics problem is based on the *manipulator Jacobian,* which is, roughly speaking, a matrix that describes the relationship between joint and end-effector velocities. Given the velocity of the end-effector, the Jacobian allows recalculation of the corresponding joint velocities simply by solving a system of linear equations. This can be used to deduce a gradient-based iterative technique for solving the inverse kinematics. Since we can apply this method to solve redundant linkages, we pursue this approach in our example. Because of their simplicity we use dual quaternions for these computations.

A simple model of the human torso and arm is shown in Figure 2.12. To track the position of a person in the virtual environment (e.g., CAVE™), we nor-

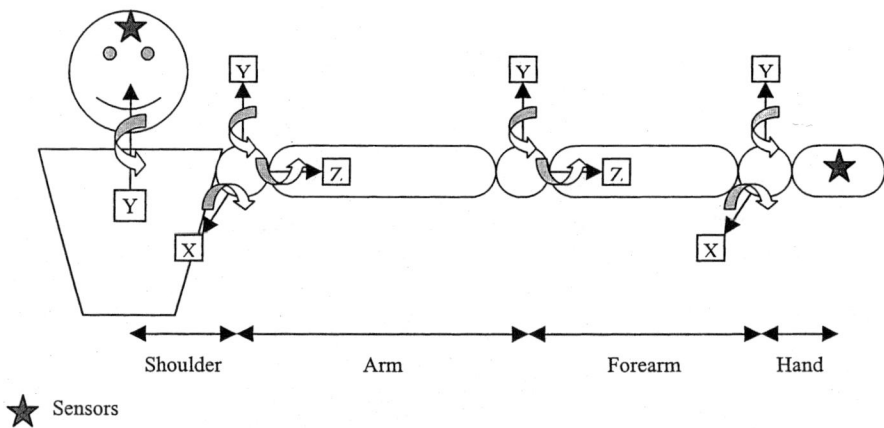

Figure 2.12 Simple model of human torso and arm in a virtual environment, with a sensor attached to the head (e.g., shutter glasses) and another sensor attached to the hand (e.g., wand).

mally have two sensors, one attached to the head and the other to the hand (represented by stars in the figure). We will apply inverse kinematics to calculate the relative angles of the "missing parts" (torso, arm, and forearm) to represent his or her complete avatar.

The human torso and arm can be modeled as an 8-DOF mechanism. This admittedly simple model neglects some movements of the torso and the scapula but is adequate for many applications. Placing a fixed coordinate system at the middle axis of torso [0] and moving coordinate systems at each joint [i], we define eight dual quaternions Q_i to represent the transformations from frame [$i - 1$] to [i] as a function of joint variable θ_i and the corresponding translations.

- *Torso:* Q_1 = Rotation(0,θ_1,0) Translation(0,0,Shoulder). Note that the shoulder translation gives the torso exterior (where the arm is attached) with respect to (w.r.t.) the torso middle axis, and this is followed by the rotation θ_1 w.r.t. the torso middle axis to indicate torso movement w.r.t. the middle axis.
- *Shoulder:* Q_2 = Rotation(θ_2,0,0), Q_3 = Rotation(0,θ_3,0), Q_4 = Rotation(0,0,θ_4) Translation(0,0,Arm).
- *Elbow:* Q_5 = Rotation(0,θ_5,0), Q_6 = Rotation(0,0,θ_6) Translation(0,0,Forearm).
- *Wrist:* Q_7 = Rotation(θ_7,0,0), Q_8 = Rotation(0,θ_8,0) Translation(0,0,Hand).

Manipulator Jacobian.
Given a desired differential change in position and orientation of the end-effector, we need to know the required changes in joint coordinates. The differential change in position and orientation of the end-effector as a function of all joint coordinates is written as an $M \times 6$ matrix consisting of differential rotation and translation vector (where M is the number of DOFs of the kinematic structure). In this matrix, known as the Jacobian matrix, each column consists of the differential translation and rotation vector corresponding to differential changes of each of the joint coordinates:
Differential changes of the end-effector = Jacobian · joint velocities

Applying dual quaternion algebra, we can calculate the columns of the Jacobian:

$$J_i = Q_i \cdot \mathbf{Axis}_i \cdot Q_i^c$$

where

J_i = column i of the Jacobian matrix
Q_i = dual quaternion representing the movement of the joint θ_i
$\mathbf{Axis}_i$ = dual quaternion representing the rotation axis of the joint θ_i:
- Dual Quaternion[1,0,0,0,0,0,0,0] for rotation about X
- Dual Quaternion[0,1,0,0,0,0,0,0] for rotation about Y

- Dual Quaternion[0,0,1,0,0,0,0,0] for rotation about Z

Q_i^c = conjugate of Q_i

The Jacobian matrix can be used to solve the inverse kinematics problem. The idea is to push the manipulator toward its desired position. We therefore first estimate the differential changes of the end-effector such that the arm will move toward the desired position. Then we recalculate the joint velocities necessary to realize this push. Afterward we apply the push and continue on this motion for a specified amount of time:

joint velocities = Jacobian^{-1} · differential changes in the end-effector

Although we pushed directly toward the position desired, the manipulator will follow only approximately. The manipulator will move into a position that is closer to the position desired than was the previous one. To solve the inverse kinematics problem, we therefore iterate this process. Solving the joint velocities means inverting the manipulator Jacobian, but due to the fact that in our case this is a rectangular matrix (8 × 6), we cannot calculate the inverse. Instead of this, we apply the pseudoinverse of the Jacobian matrix to solve the system of linear equations:

joint velocities = pseudoinverse(Jacobian) · differential changes in the end-effector

2.3 PERSPECTIVE PROJECTION

In two dimesions we specify a window on the two-dimensional world and a viewport on the corresponding view surface. Conceptually, objects in the world are clipped against the window and are then transformed into the viewport for display.

In general, *projections* transform points in a coordinate system of dimension n into points in a coordinate system of dimension less than n. Here we limit ourselves to the projection from three-dimensions to two dimensions. The projection of a three-dimensional object is defined by straight projection rays (called *projectors*) emanating from a center of projection, passing through each point of the object, and intersecting a projection plane to form the projection. The class of projections we deal with here is known as *planar geometric projections* because the projection is onto a plane rather than a curved surface and uses straight rather than curved projectors. Many cartographic projections are either nonplanar or nongeometric. Similarly, the Omnimax film format requires a nongeometric projection.

In Section 2.2 we developed a matrix to change the coordinate values of points in the VE's frame of reference to the VO's frame of reference. After this operation, the virtual objects and the VO (who is effectively located at the origin of this local frame of reference and is gazing along the positive *w* axis) share a common space (meaning that the VO and VE frames are merged, for convenience, into one frame, say the VO frame). (X_{VO}, Y_{VO}, and Z_{VO}) are abbreviated as *(u,v,w)* using our convention defined earlier. This is shown in Figure 2.13.

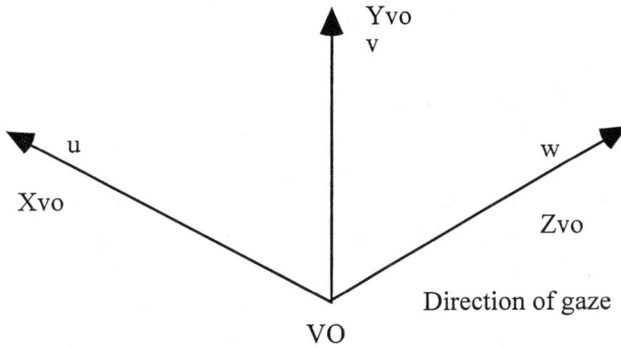

Figure 2.13 The virtual observer (VO) is located at the origin of its local frame of reference and is gazing along the positive *w* axis.

Note that the origin of the VO frame of reference is on the projection plane, not at the VO (Figure 2.14). VO is the center of projection.

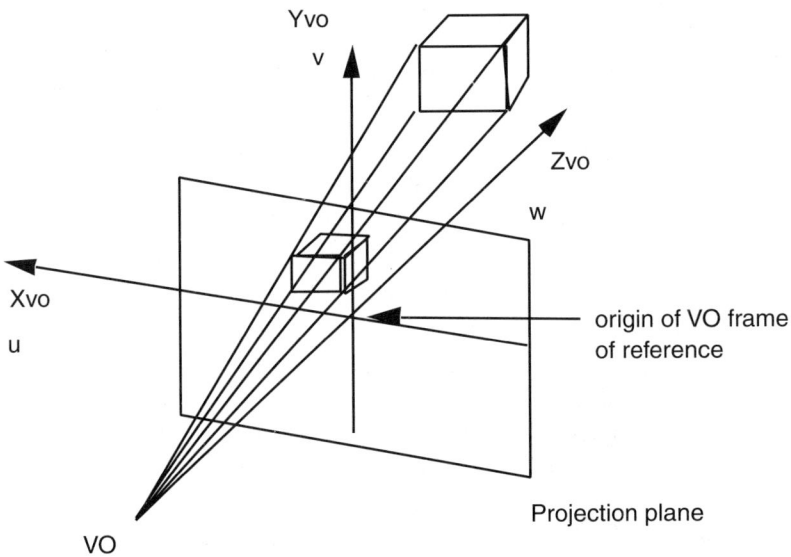

Figure 2.14 VO gazing at an object to form a perspective projection.

Planar projections can be divided into two basic classes: perspective and parallel. The distinction is in the relation of the center of projection (VO location) to the projection plane. If the distance from the one to the other is finite, the projection is *perspective;* if the distance is infinite, the projection is *parallel.* The visual effect of a perspective projection is similar to that of photographic systems and of the human visual system. The size of the perspective projection of an object varies inversely with the distance of that object from the center of projection. Thus, although the perspective projections of objects tend to look realistic, it is not particularly useful for recording the exact shape and measurements of the objects: Distances cannot be taken from the projection, angles are preserved only on those faces of the object parallel to the projection plane, and parallel lines do not in general project as parallel lines. The parallel projection is a less realistic view, although the projection can be used for exact measurements and parallel lines do remain parallel. As with the perspective projection, angles are preserved only on faces of the object parallel to the projection plane. Since the purpose of our discussion here is primarily to produce visual effects, we are concentrating on perspective projection. For other design purposes these can be converted to parallel projections.

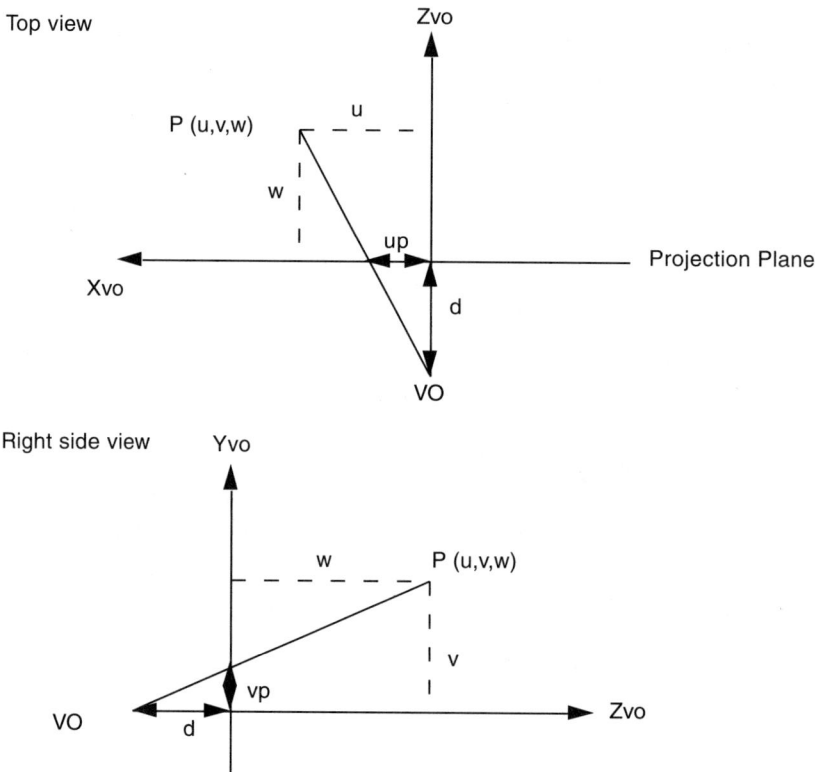

Figure 2.15 Geometry of perspective projection.

Coming back to perspective projections, we are interested in determining the dimensions of an actual object (box of Figure 2.14, in this case) on a projection plane to give an appropriate perspective view. For this let us choose any point P on the actual object and let its coordinates in VO frame of reference be (u,v,w), as shown in Figure 2.15. Let (u_p, v_p, w_p) be the perspective view in a two-dimensional projection of the three-dimensional object on the (X_{VO}, Y_{VO}) plane (i.e., projection plane). The Z_{VO} values are all constant ($w_p = 0$ in this case, based on the way the VO axes are defined). Hence we only need to find u_p and v_p. Using geometry of similar triangles in Figure 2.15, we have

$$\frac{u_p}{d} = \frac{u}{(w+d)} \quad \text{or} \quad u_p = \frac{ud}{(w+d)} \tag{2.25}$$

$$\frac{v_p}{d} = \frac{v}{(w+d)} \quad \text{or} \quad v_p = \frac{vd}{(w+d)} \tag{2.26}$$

where d is the distance between VO and projection plane, as shown in Figure 2.15. The action of d is one of scaling: By increasing its value, the size of the perspective image increases and vice versa. For example, if d is increased, the projection plane can be moved closer and closer to the actual object in Figure 2.14 and the perspective image will increase. If the projection plane is moved beyond the actual object, the perspective image will appear larger than the object.

2.3.1 Perspective Projection and Field of View

So far the equations were derived based on the location of a point P. Let us now extend the concept to a line (e.g., screen size s) instead of a point because the perspective projection applies to all points on the line (Figure 2.16). When a set of uv coordinates are divided individually by relatively large w values in equations (2.25) and (2.26), the numerical results are consistently small and

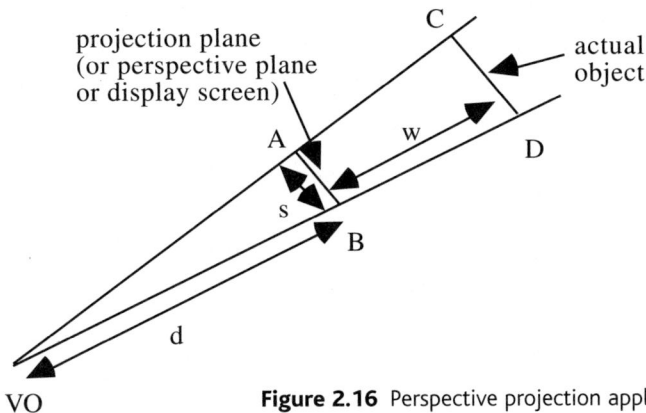

Figure 2.16 Perspective projection applied to a line.

little perspective is detected in the resulting image. But when the same set of *uv* coordinates are individually divided by relatively small *w* values, the values are much larger and have a wide variation, thus resulting in dramatic perspective characteristics. Hence the values of *w* relative to *u* and *v* are useful in controlling the change in perspective. Controlling the values of *u* and *v* visible to the VO leads to what is known as *field of view* (FOV). Points inside the FOV range are visible and points outside are invisible.

So far, the impact of *w* on FOV has been considered. Let us see the impact of *d* on FOV. Let *s* be the screen size (or any other standardizing measure). The ratio *d/s* can be viewed as a measure of FOV. The following characteristics of this ratio influence FOV:

- A small ratio means a broad FOV. A wide-angle lens is needed to capture it. Such a lens has a short focal length and a wide aperture.
- A large ratio, on the other hand, implies a narrow FOV. A telephoto lens is needed here. It has a large focal length and a narrow aperture.

The ratio *d/s* can also be used to make the perspective image dimensionless (e.g., u_p/s and v_p/s represent measures on a dimensionless coordinate system):

$$\frac{u_p}{s} = \frac{ud}{(w+d)s} \qquad \frac{v_p}{s} = \frac{vd}{(w+d)s} \tag{2.27}$$

2.3.2 Mapping to the Display Device

The perspective plane above has to be mapped to a display device for displaying purposes. We assume the existence of a memory device called a *frame store* (or *frame buffer*) that will store the perspective image for display purposes. The spatial resolution of the frame store will match that of the display device. The perspective projection creates a planar coordinate description of the scene formed on the projection plane. This can be expressed as fractional values of the screen space, which enables the projection to be mapped to different resolution displays. For instance, let the perspective domain be bounded by the rectangular limits (0,0) to (1,1), and the screen space domain be bounded by pixel addresses ranging from (0,0) to (colmax,rowmax). A point (X_{persp}, Y_{persp}) is mapped into (X_{pixel}, Y_{pixel}) as follows:

$$X_{pixel} = colmax \cdot X_{persp}$$
$$Y_{pixel} = rowmax \cdot Y_{persp}$$

This assumes that the *aspect ratio* (length/width ratio) of the perspective projection matches that of the display device. If this were not so, the image would appear distorted. Note that in our discussion on perspective projection we considered a parameter *s* for perspective plane size. Here colmax and rowmax are related to *s*.

2.4 VIEWING FRUSTUM, FIELD OF VIEW, AND CLIPPING PLANES

Near and far *clipping planes* are set to specify the distance along the line of sight that is to be considered for display. The frustum formed between the near and far clipping planes is the *viewing frustum* (Figure 2.17). For example, in Performer, the API pfFOV(chan,horiz,vert) is used to set up the channel, horizontal dimensions, and vertical dimensions of the FOV, respectively (the concept of channel is explained in Section 2.9). The FOV is the angular width of view measured in degrees. The API pfNearFar(chan,near,far) is used to specify the distance along the line of sight from the viewpoint (eye) to near and far clipping planes along the line of sight. The planes are set perpendicular to the line of sight.

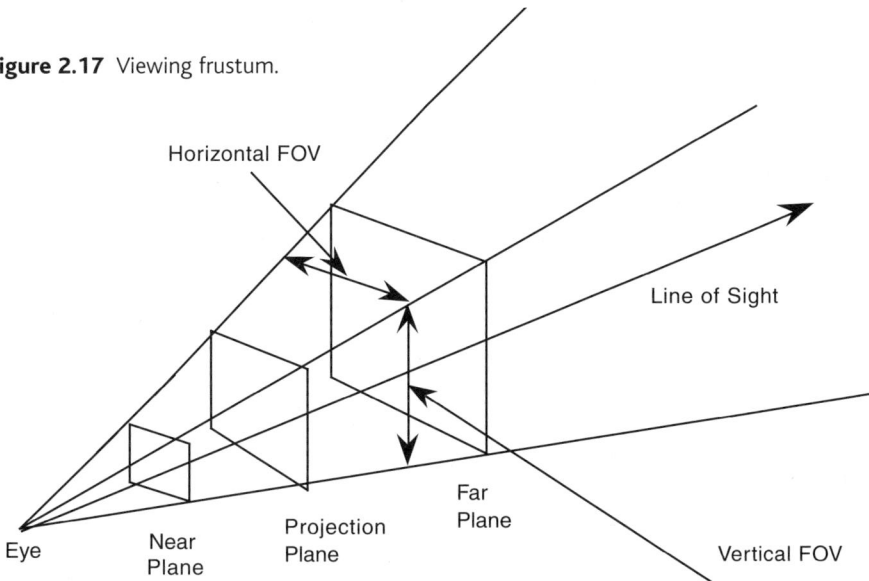

Figure 2.17 Viewing frustum.

2.5 Z BUFFER FOR HIDDEN SURFACE REMOVAL

Z buffer is the most frequently used mechanism for hidden surface removal in sophisticated graphics systems. A buffer stores the depth or distance from the eye point of the closest surface "seen" at that pixel. When a new surface is scan converted, the depth at each pixel is computed. If the new depth at a given pixel is closer to the eye point than the depth currently stored in the z buffer at that pixel, the new depth and intensity information are written into both the z buffer and the frame buffer. Otherwise, the new information is discarded and the next pixel is examined. In this way, nearer objects always overwrite more distant objects, and when every object has been scan converted, all surfaces have been correctly ordered in depth. Back face removal or back face culling is based on a related concept. Culling is discussed in more detail later in this chapter and in Chapter 3.

2.6 ILLUMINATION MODELS

Two approaches are available to create a colored view of a three-dimensional scene. The first, which is not very flexible, simply assigns a fixed color to every surface of an object. No matter how the object is viewed, its colors remain constant. The second approach attempts to simulate the interaction of light sources with colored surfaces, which is the technique now used in most computer graphics systems. Illumination models are useful for the second approach.

2.6.1. Point Light Sources

A *point light source* is the simplest light source to model. The intensity of light is generally specified in terms of three additive primary color components: red, green, and blue. The light source requires a location in space (L_x, L_y, L_z) and a luminous value $(I_{red}, I_{green}, I_{blue})$. The components of vector **S** are as follows (see Figure 2.18):

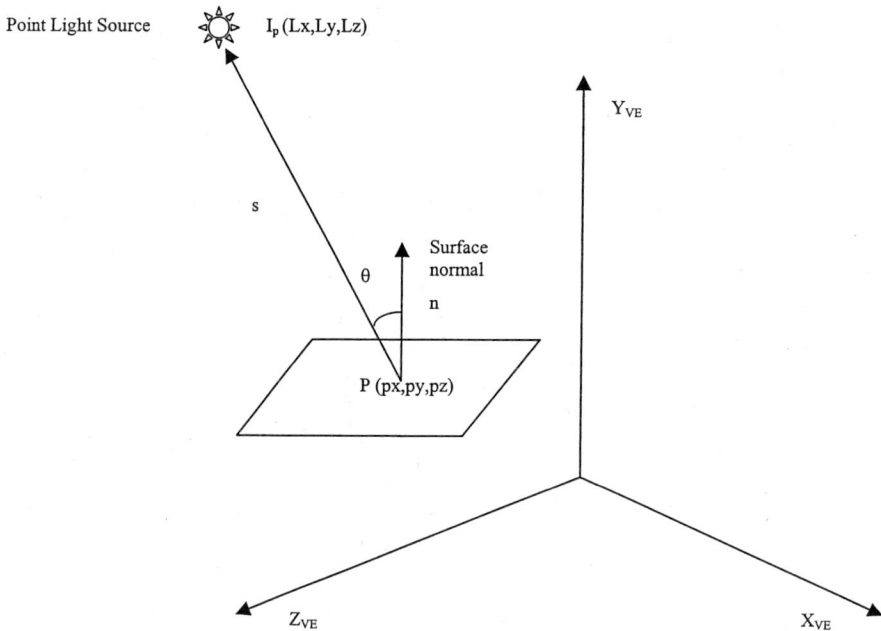

Figure 2.18 Illumination model for point light source.

$$S = \begin{bmatrix} s_1 \\ s_2 \\ s_3 \end{bmatrix} = \begin{bmatrix} L_x - p_x \\ L_y - p_y \\ L_z - p_z \end{bmatrix} \qquad (2.28)$$

The location of a point P is (p_x, p_y, p_z). Calculating the level of light incident on a surface is a function of the angle between the surface normal and an incident ray. Ignoring the falloff in light intensity with distance, the incident flux is given by

$$I_i = I_p \cos \theta = \frac{I_p (\mathbf{s} \cdot \mathbf{n})}{|\mathbf{s}| \cdot |\mathbf{n}|} = \frac{I_p (s_1 n_1 + s_2 n_2 + s_3 n_3)}{|\mathbf{s}| \cdot |\mathbf{n}|} \qquad (2.29)$$

where I_p is the intensity of point light source and θ is the angle between incident ray and surface normal.

Ambient Light
Unless we allow for light to be reflected from one surface to another, there is a very good chance that some surfaces will not receive any illumination at all. Consequently, when these surfaces are rendered, they will appear black and unnatural. In anticipation of this happening, illumination schemes allow the existence of some level of background light level called the *ambient light*. This is incorporated as a constant within the lighting calculations and is set to some convenient value; typically, it accounts for 20 to 25 % of the total illumination.

Transparency
Transparency is an important attribute that needs to be incorporated in the model, otherwise, it will be impossible to simulate the effects of glass and other transparent media. Normally, it is implemented by associating with a polygon or surface patch a transparency coefficient that varies between 0 and 1.

2.6.2 Multiple Light Sources

More computational power is needed for real-time calculations with multiple light sources. Several light sources introduce the problem of light-balancing, which prevents some surfaces from being overilluminated, while others remain in shadow. Let us move to another topic. So far, light intensity reduction with distance was ignored. For intensity falloff with distance, there are many ways of modeling such situations; two are given below.

- *Alternative 1*. Intensity of light radiation falls off as a function of the inverse square of the propagation distance.
- *Alternative 2*. Simple tests have shown that good results can be obtained by dividing the light source intensity by d + k, where d is the distance of the VO from the illuminated surface and k is some system-defined constant.

2.7 REFLECTION MODELS

Taking a simplistic approach to light reflection, we can express the behavior of light in terms of *diffuse* and *specular reflection*. Light reflected by a diffuse surface is radiated equally in all directions and is therefore independent of the observer's position, whereas light reflected by a polished surface creates a specular highlight for certain viewpoints. In addition, there is a third type, known as *ambient reflection*.

2.7.1 Diffuse Reflection

Rough surfaces such as carpets, textiles, and some papers exhibit diffuse reflection properties. To compute the underlying diffuse component, we need to know the reflection coefficient of the surface, K_d, the intensity of the light source, I_i, and the cosine of the angle subtended by the incident light and the surface normal. The diffuse term I_d for one light source is expressed as

$$I_d = I_i K_d \cos \theta \qquad (2.30)$$

In general, there will be three such expressions associated with the red, green, and blue components of the light model.

2.7.2 Specular Reflection

Smooth or polished surfaces can be simulated by combining a diffuse reflection with a specular highlight. This extra spot of light represents the reflection of the light source into the observer's field of view, and is a powerful visual cue for distinguishing between matte and polished surfaces.

When an incident ray represented by L is reflected, the angle of reflection equals the angle of incidence θ. However, if the observer is displaced by an angle ϕ, the intensity of the specular highlight will be attenuated accordingly. This can be accounted for by modulating the intensity by a $\cos^g \phi$ term, where g is a gloss parameter. The complete expression is

$$I_s = I_i K_s \cos^g \phi \qquad (2.31)$$

where I_i is the light source intensity, K_s a color-independent specular coefficient, ϕ the error angle, and g the gloss parameter (Figure 2.19).

2.7.3 Ambient Reflection

Just like ambient illumination, there is an ambient reflection associated with objects. This simple reflection model has three elements, ambient, diffuse, and

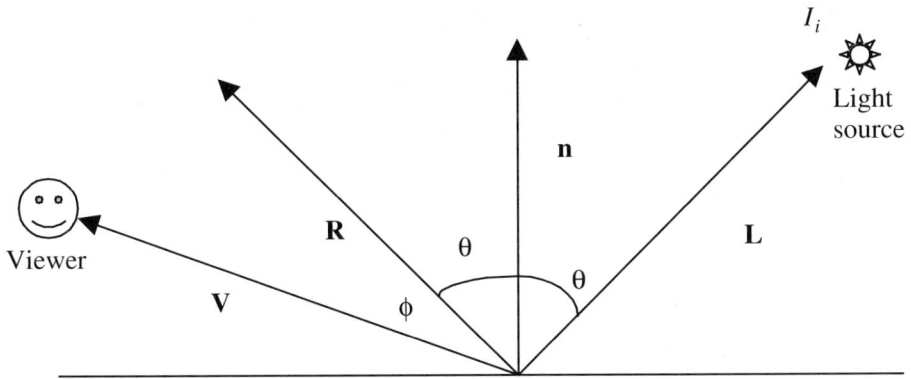

Figure 2.19 Specular reflection model.

specular, for each of which there will be a red, a green, and a blue component. For one light source this can be summarized as

$$I = I_{ambient} + [I_{diffuse} + I_{specular}]$$

$$I = I_i K_a + [I_i K_d (\mathbf{L} \cdot \mathbf{n}) + I_i K_s \cos^g \phi]$$

(2.32)

where the term in brackets has to be computed for every light source. For an example, see the semi truck example described in Chapter 3.

2.8 COLOR MODELS

Two popular models to manipulate color in VR displays are:

1. *Red–green–blue (RGB) color model.* The RGB model is implemented in a scale of 0 to 1, as shown by the color cube in Figure 2.20. The color cube can be interpreted in a number of ways, for example, as shown in Figure 2.20, where three colors formed by various combinations of red, green, and blue are black (0,0,0); gray (0.5,0.5,0.5), and white (1,1,1).

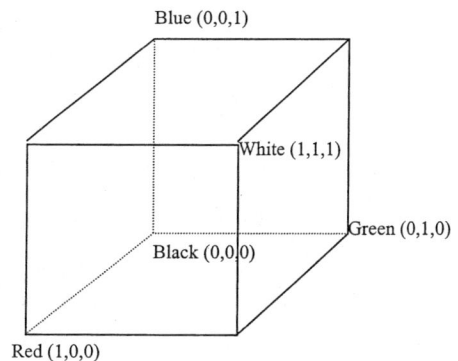

Figure 2.20 RGB color cube.

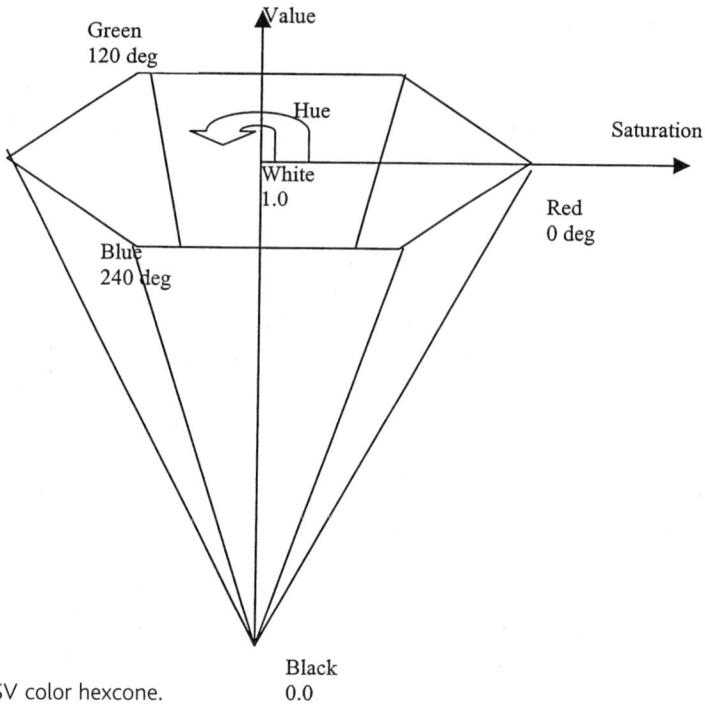

Figure 2.21 HSV color hexcone.

2. *Hue–saturation–value (HSV) color model.* This represents another color model. In the HSV color space, *hue* identifies the underlying color, *saturation* controls the level of white light desaturating the color, and *value* is a measure of the color's darkness or lightness. Figure 2.21 shows a spatial interpretation of this space that is organized as a hexcone as derived from a diagonal projection through the RGB color cube. Notice that the cone results from the fact that saturation options keep progressively increasing from dark to white, being nonexistent at dark. Conversion algorithms exist between RGB and HSV (see, e.g., Foley et al., 1990).

2.9 RENDERING

The perspective projection of an object needs to be rendered using the illumination models. *Rendering* means forming a color pixel–based image from the geometric database of the VE using the illumination models. Rendering can be quite complex. A large database may include several thousand objects, each of which contains dozens of triangles. Not all of them may be visible, but the renderer may still have to access them all to identify those that are seen.

Rendering Example: IRIS Performer™

It is convenient to explain the rendering concept by using a library of software constructs. These constructs are used to facilitate rendering a visual database. The software constructs used here for illustration purposes are from IRIS Performer. Performer provides a pipelined multiprocessing model for implementing visual simulation. Pipelines can be split into separate processes (one to three) to tailor the application to the number of available central processing units (CPUs). The Performer APIs begin with the prefix *pf*. In Performer, a pipeline is implemented as pfPipe, and it is a software rendering pipeline that renders one or more channels (pfChannels) into a graphics window. A Performer channel (pfChannel) is a view into a visual scene graph or scene database (pfScene). Alternatively, a channel can be thought of as a rendering viewport into a pipe. Scene graphs are addressed in more detail in Chapter 3.

There are three functional stages of a graphics pipeline.

1. *Application.* This stage refers to the application being addressed: for example, doing requisite processing for the visual simulation application, including reading input from control devices, simulating the dynamics of moving models, updating the visual database, and interacting with other networked simulation stations.
2. *Cull.* The tasks in this stage include traversing the visual database and determining which portions of it are potentially visible, performing level-of-detail selection for models with multiple representations, and building a sorted, optimized display list for the draw stage.
3. *Draw.* In this stage, graphics library commands are issued to a geometry pipeline to create an image for subsequent display.

Figure 2.22 shows the process flow for a single-pipe system. The application constructs and modifies the scene definition (a pfScene) associated with a channel. The traversal process associated with that channel's pfPipe then traverses the scene graph, building an IRIS Performer display list. As shown in the figure, this display list is used as input to the draw process that performs the actual graphics library actions required to draw the image. The *frame buffer* is a dedicated block of memory that holds intensity and other information for

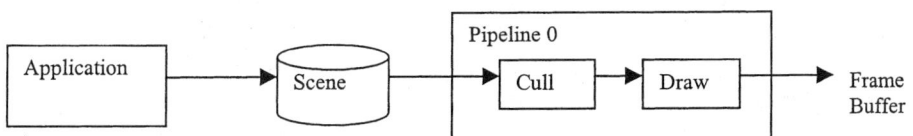

Figure 2.22 Single graphics pipeline.

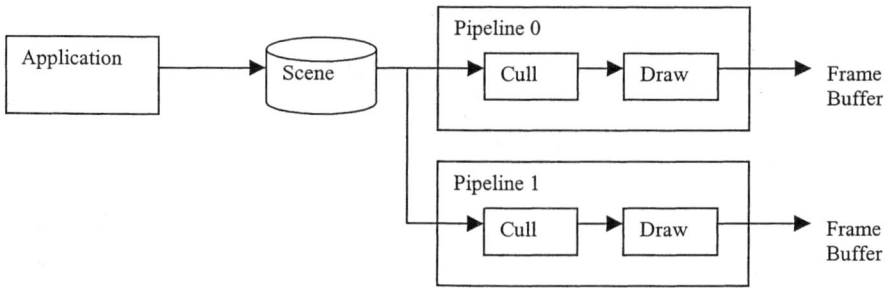

Figure 2.23 Dual graphics pipeline.

every pixel on the display surface. The frame buffer is scanned repeatedly by the display hardware to generate visual imagery.

Applications can be rendered into multiple windows, each of which is connected to a single geometry pipeline through a pfPipe rendering pipeline. Figure 2.23 shows the process flow for a dual-pipe system. Notice both the differences and similarities between these two figures. Each pipeline (pfPipe) is independent in multiple-pipe configurations; the cull and draw tasks are separate, as are the frame buffers. In contrast, these pfPipes are controlled by the same application process, and in many situations access the same shared scene definition. Each of these stages can be combined into a single IRIX™ process or split into multiple processes (pfMultiprocess) for enhanced performance on multiple CPU systems.

2.10 ANTIALIASING

The pixel nature of the image creates problems if one is attempting to display very small features or straight edges. Certain small features may be smaller than the spatial resolution of a pixel and will result in some form of visual approximation. An object containing straight edges will also cause problems when it is converted into a pixel description, and will result in jagged pixel edges.

Consider using a midpoint algorithm to draw a one-pixel-thick black line, with slope between 0 and 1, on a white background. In each column through which the line passes, the algorithm sets the color of the pixel that is closest to the line. Each time the line moves between columns in which the pixels closest to the line are not in the same row, there is a sharp jag in the line drawn into the canvas, as is clear in Figure 2.24a. The same is true for other scan-converted primitives that can assign only one of two intensity values to pixels.

Suppose we now use a display device with twice the horizontal and vertical resolution. As shown in Figure 2.24b, the line passes through twice as many columns and therefore has twice as many jags, but each jag is half as

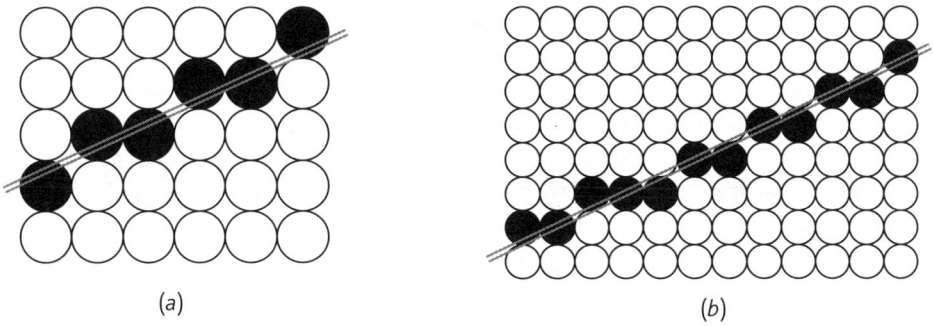

Figure 2.24 Antialiasing to smooth out jagged edges.

large in x and in y. Although the resulting picture looks better, the improvement comes at the price of quadrupling the memory cost, memory bandwidth, and scan-conversion time. Increasing resolution is an expensive solution that only diminishes the problem of jaggies — it does not eliminate the problem.

Such sampling artifacts are referred to as *aliasing*. However, they can be minimized by implementing various antialiasing algorithms, which are generally computationally expensive. Some modern image generators do implement antialiasing strategies in hardware and can therefore work in real time.

2.11 GEOMETRIC TRANSFORMATIONS FOR OBJECTS

The change-of-coordinate-system point of view is useful when information for subobjects is specified in the subobject's own local coordinate systems. For example, if the front wheel of the bicycle in Figure 2.25 is made to rotate about its Z_{wh} coordinate, all wheels must be rotated appropriately, and we need to know how the bicycle as a whole moves in the world-coordinate system. This problem is complex because several successive changes of coordinate systems

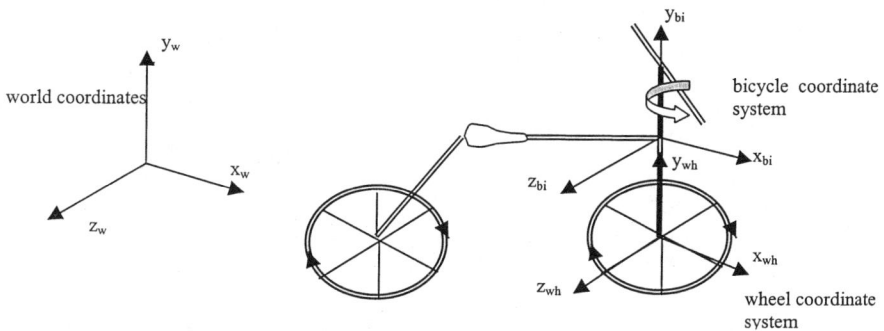

Figure 2.25 Example of local subobject coordinate system.

occur. First, the bicycle and front-wheel coordinate systems have initial positions in the world-coordinate system. As the bike moves forward, the front wheel rotates about the z axis of the wheel-coordinate system, and simultaneously the wheel- and bicycle-coordinate systems move relative to the world-coordinate system. The wheel and bicycle-coordinate systems are related to the world-coordinate system by time-varying translations in x and z plus a rotation about y. The bicycle- and wheel-coordinate systems are related to each other by a time-varying rotation about y as the handle is turned. (The bicycle-coordinate system is fixed to the frame, not to the handle.)

To make the problem a bit easier, we assume that the wheel and bicycle axes are parallel to the world-coordinate axes and that the wheel moves in a straight line parallel to the world-coordinate x axis. As the wheel rotates by an angle α, a point P on the wheel rotates through the distance αr, where r is the radius of the wheel. Since the wheel is on the ground, the bicycle moves forward αr units. Therefore, the rim point P on the wheel moves and rotates with respect to the initial wheel-coordinate system, with a net effect of translation by αr and rotation by α. Its new coordinates, P', in the original wheel-coordinate system are thus

$$P'_{wh} = T(\alpha r, 0, 0) \cdot Rz(\alpha) \bullet P_{wh}$$

and its coordinates in the new (translated) wheel-coordinate system are given by just the rotation $P'_{wh} = Rz(\alpha) \cdot P_{wh}$. To find the points P_W and P'_W in the world-coordinate system, we transform from the wheel- to the world-coordinate system:

$$P_w = Mw_{wh} \bullet P_{wh} = Mw_{bi} \bullet Mbi_{wh} \bullet P_{wh}$$

where $Mw(bi)$ and $Mbi(wh)$ are translations given by the initial positions of the bicycle and wheel. P'_W is defined as

$$P'_w = Mw_{wh} \bullet P'_{wh} = Mw_{wh} \bullet T(\alpha r, 0, 0) \bullet Rz(\alpha) \bullet P_{wh}$$

Thus it is natural to think of each object as being defined in its own coordinate system and then being scaled, rotated, and translated by redefinition of its coordinates in the new world-coordinate system. In this point of view, one thinks naturally of separate files, each with an object in it, being shrunk or stretched, rotated or placed on the world-coordinate plane. This concept is developed further next.

Frames of Reference

Objects are first modeled in the object-coordinate system (OCS). Objects are then visualized to be consistent with the VE-coordinate system (VECS), also referred to as the world-coordinate system (WCS), which represents the VE (Figure 2.26). Various modeling transformations are then used to locate

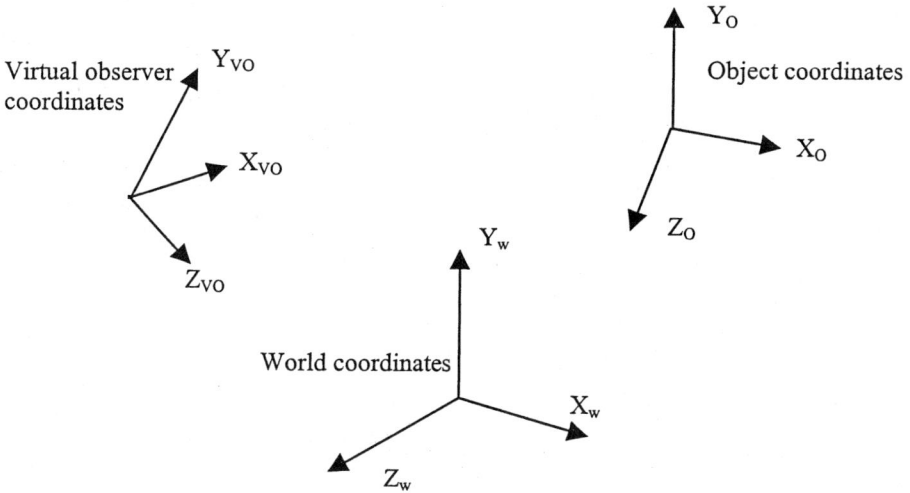

Figure 2.26 Frames of reference in virtual modeling.

objects in the VE. The VO's position is determined from the three-dimensional head or hand tracking system.

Transformations for Object Manipulation in OCS.

$$
\begin{bmatrix} x' \\ y' \\ z' \\ 1 \end{bmatrix} = \begin{bmatrix} 1 & 0 & 0 & t_x \\ 0 & 1 & 0 & t_y \\ 0 & 0 & 1 & t_z \\ 0 & 0 & 0 & 1 \end{bmatrix} \begin{bmatrix} x \\ y \\ z \\ 1 \end{bmatrix}
\tag{2.33}
$$

Any point on the object (x,y,z) can be translated to new point (x',y',z'), as follows:

$$
\begin{bmatrix} x' \\ y' \\ z' \\ 1 \end{bmatrix} = \begin{bmatrix} S_x & 0 & 0 & 0 \\ 0 & S_y & 0 & 0 \\ 0 & 0 & S_z & 0 \\ 0 & 0 & 0 & 1 \end{bmatrix} \begin{bmatrix} x \\ y \\ z \\ 1 \end{bmatrix}
\tag{2.34}
$$

Scaling.
When objects offset from the origin are scaled, their offsets are also scaled.

Reflection.
Reflection is a form of scaling where one or more of the scaling factors are negative.

Rotation.
The equations developed earlier for *XYZ* fixed angles, and so on, can be used to model object rotations.

FURTHER READING

For more details about quaternions, a good reference is Kuipers (1999). An excellent book in computer graphics is that of Foley et al. (1990).

REFERENCES

Foley, J. D., A. van Dam, S. K. Feiner, J. F. and Hughes, *Computer Graphics,* 2nd ed., Addison-Wesley, Reading, MA, 1990.
Kuipers, J. B., *Quaternions and Rotation Sequences,* Princeton University Press, Princeton, NJ, 1999.

EXERCISES

2.1.(a) Consider a fire safety evacuation drill simulation using VR. For simplicity, suppose that there is a fire in the corner of a room as indicated in Figure 2.27. Assume that the location of a person, the fire, and an exit are approximated by the respective room corner points as indicated. The room is cubical, with the length of each side being 5 m. The coordinate system associated with the person and exit are shown. Indicate which coordinate system is VOCS and which is VECS in this case. Also find the fire location in each coordi-

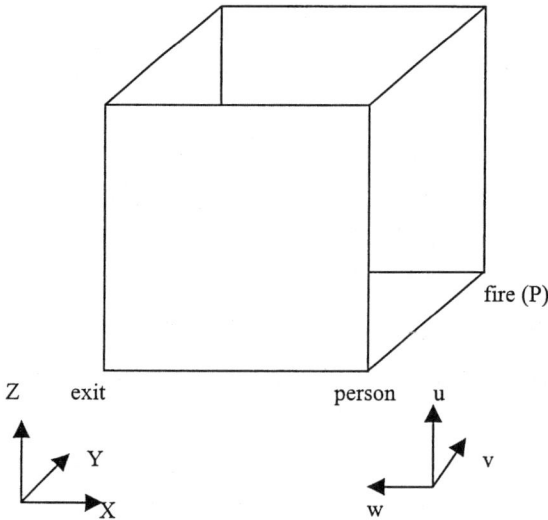

Figure 2.27 Simple fire safety evacuation drill simulation in VR.

P1(3,4,1)

P2(7,3,0)

Motorcyclist

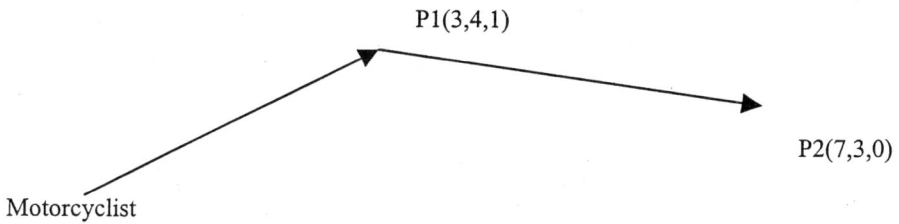

Figure 2.28 Motorcyclist training example.

nate system and find the coordinate transformation from the VE to the VO coordinate system in this case.

(b) Suppose that the person runs to the exit. Find the transformations needed to align the person and exit coordinate systems assuming **(i)** a fixed coordinate system and **(ii)** an Euler coordinate system. Which of the coordinate systems above is a better choice in this case? Why?

2.2. Suppose that you want to use VR for training motorcyclists. The motorcyclist (approximated by a point for simplicity) starts at the location shown in Figure 2.28 and the path passes through points P_1 and P_2, as shown (the P_1 and P_2 locations shown are w.r.t. VECS). Assume that the motorcyclist is represented in the VOCS as in Exercise 2.1, and points P_1 and P_2 use the same VECS in Exercise 2.1. Also, the VOCS origin is located at the motorcyclist location shown at the beginning of the path. The motorcyclist makes a "roll" turn of $-30°$ at point P_1 w.r.t. VECS and keeps driving along a straight line thereafter. How much deviation from the path P_1-P_2 will he make? What is one way of getting back to the path? Consider using fixed coordinates, Euler coordinates, and quaternions. Which coordinate system(s) will you recommend for solving this problem most effectively? Why? [*Hint:* There can be two interpretations for this problem: (1) the motorcyclist's original location is at (0,0,0) w.r.t. VECS, and (2) the motorcyclist's original location is the same as that of the person in Exercise 2.1 w.r.t. VECS. Solve for both cases.]

3

PRINCIPLES OF VIRTUAL REALITY

3.1 STEREO PERSPECTIVE PROJECTION

The viewing frustum introduced in Chapter 2 can be expanded to the frustum shown in Figure 3.1. It shows two frustums, one for the left eye and the other for the right eye. Stereoscopic vision helps us estimate depth and works up to distances of approximately 30 meters (m); beyond this, the disparity between the images captured by the left and right eyes is insignificant. Assuming that

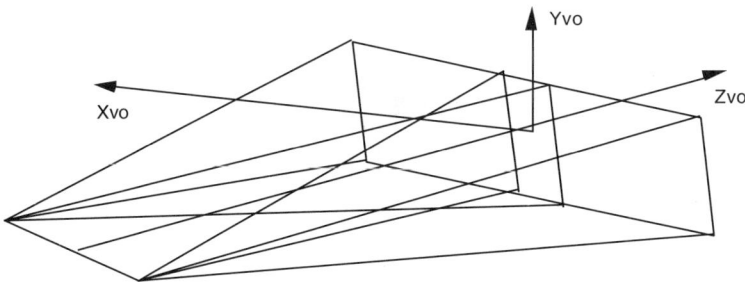

Figure 3.1 Viewing frustum in stereo.

the user is gazing at a distant horizon, the user has two parallel lines of sight aligned with the z axis. Figure 3.2 illustrates these two lines of sight, with an interocular distance (the distance between the two pupils) of S_e, which is approximately 6.5 centimeters (cm).

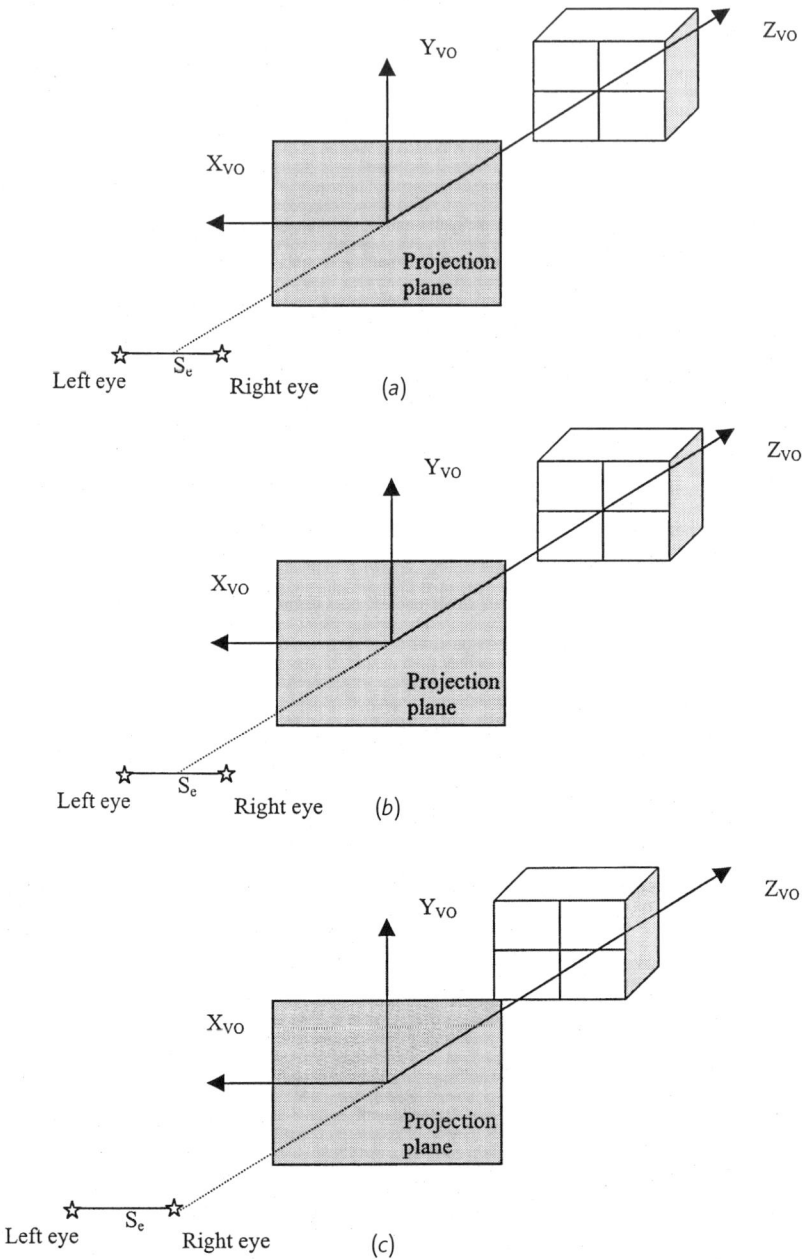

Figure 3.2 Viewer gazing at a box in stereo.

In Figure 3.2*a* we see a pair of eyes gazing toward a box. To obtain the view from the left eye we position this eye at the origin, shift the box to the right (Figure 3.2*b*) by an amount of $S_e/2$ in the negative X_{VO} direction. Gaze is now with the left eye passing through the origin. Now $X_{new} = X_{old} - S_e/2$. To obtain the view from the right eye, we position this eye at the origin and shift the original box by a corresponding distance $S_e/2$ to the left (Figure 3.2*c*) (i.e., in the positive X_{VE} direction). Gaze is now with the right eye passing through the origin and $X_{new} = X_{old} + S_e/2$.

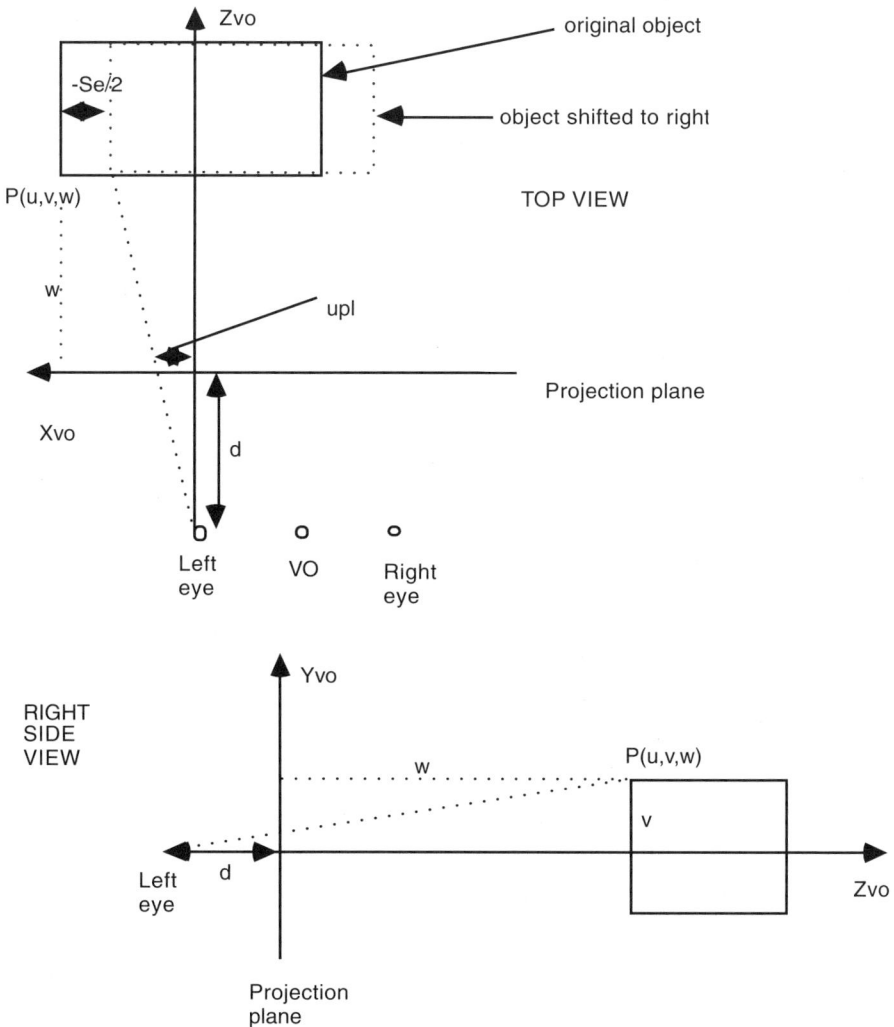

Figure 3.3 Geometry of stereo perspective projection.

Let $(u_{pl}, v_{pl}, 0)$ be the point at which $P(u,v,w)$ intersects the projection plane after the point P on the object is shifted by $-S_e/2$ along X_{VO} axis (refer to Figure 3.3). Using the geometry of similar triangles, we have

$$\frac{u_{pl}}{d} = \frac{u - S_e/2}{w + d} \quad \text{or} \quad u_{pl} = \frac{d(u - S_e/2)}{w + d}$$

$$\frac{v_{pl}}{d} = \frac{v}{w + d} \quad \text{or} \quad v_{pl} = \frac{dv}{w + d}$$

We can do a similar exercise for the right eye by moving the object by $S_e/2$ units along the X_{VO} axis. Let $(u_{pr}, v_{pr}, 0)$ be the point of intersection of $P(u,v,w)$ after moving to the right with the projection plane. The coordinates u_{pr} and v_{pr} are as follows:

$$u_{pr} = \frac{d(u + S_e/2)}{w + d} \quad \text{and} \quad v_{pr} = \frac{dv}{w + d}$$

Example

Consider the case of an observer with an interocular distance of 6.5 cm looking toward a cube with 20-cm sides. The cube's center intersects the z axis as shown in Figure 3.4 and the nearest side of the cube is 80 cm from the observer. Substituting these values into the perspective projection formulas we obtain values for u_{pl}, v_{pl}, u_{pr}, and v_{pr}, as shown in Table 3.1. The distance of the projection plane d is 40 cm.

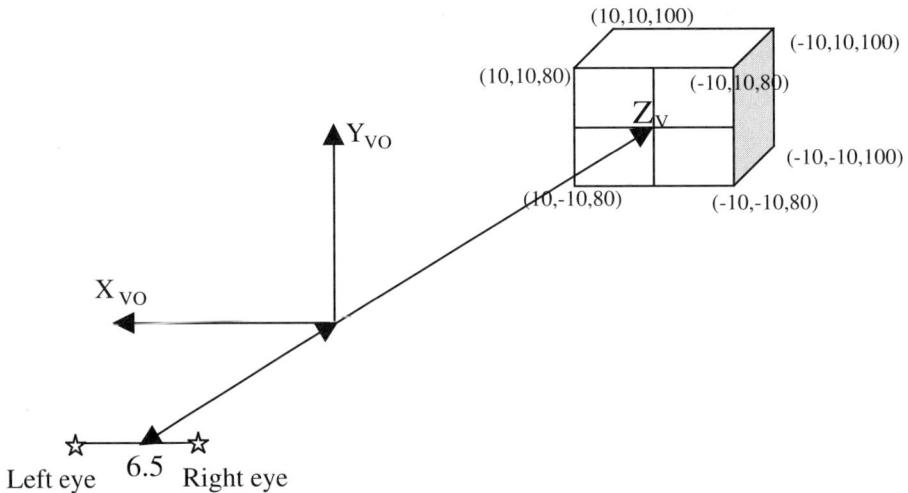

Figure 3.4 Stereo perspective numerical example.

TABLE 3.1

u	v	w	u_{pl}	v_{pl}	u_{pr}	v_{pr}
10	10	80	2.25	3.33	4.42	3.33
10	−10	80	2.25	−3.33	4.42	−3.33
−10	10	80	−4.42	3.33	−2.25	3.33
−10	−10	80	−4.42	−3.33	−2.25	−3.33
10	10	100	1.93	2.86	3.79	2.86
10	−10	100	1.93	−2.86	3.79	−2.86
− 10	10	100	−3.79	2.86	−1.93	2.86
−10	−10	100	−3.79	−2.86	−1.93	−2.86

3.2 SIMPLE THREE-DIMENSIONAL MODELING

Euler's rule for a polyhedron without holes has the following formula:

$$\text{no. edges} = \text{no. faces} + \text{no. vertices} - 2 \qquad (3.1)$$

Thus, if two of these quantities are known, the third can be calculated.

3.2.1 Polygonal Mesh

A *polygonal mesh* is a collection of edges, vertices, and polygons connected such that each edge is shared by at most two polygons. An edge connects two vertices, and a polygon is a closed sequence of edges. A polygonal mesh is a set of connected polygonally bounded planar surfaces. Volumes bounded by planar surfaces, such as boxes, cabinets, and building exteriors, can be represented easily and naturally by polygonal meshes. Polygon meshes can be used for approximate representation of objects with curved surfaces, as shown in Figure 3.5. It shows the cross–section of a curve and the polygon mesh representing it. The errors in the representation can be made arbitrarily small by using more and more polygons to create a better piecewise linear approxima-

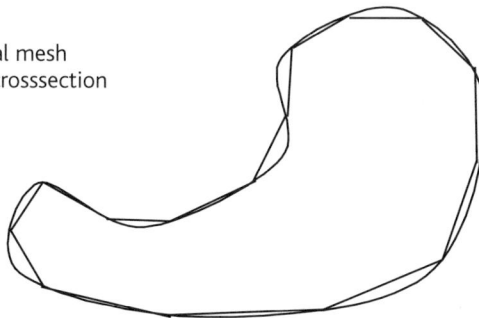

Figure 3.5 Polygonal mesh representation of a crosssection of a curved object.

tion, but this increases space requirements and the execution time of algorithms processing the representation.

Representation of a Polygon

In an explicit representation, each polygon is represented by a list of vertex coordinates:

$$P = ((X_1,Y_1,Z_1), (X_2,Y_2,Z_2), \ldots, (X_n,Y_n,Z_n)) \tag{3.2}$$

The vertices are stored in the order in which they would be encountered traveling around the polygon. For a single polygon, this is space-efficient; for a polygonal mesh, however, much space is lost because the coordinates of shared vertices are duplicated. Polygons defined with pointers to a vertex list have each vertex in the polygon mesh stored just once, in the vertex list $V = ((X_1,Y_1,Z_1), (X_2,Y_2,Z_2), \ldots, (X_n,Y_n,Z_n))$. A polygon is defined by a list of indices (or pointers) into the vertex list. A polygon made up of vertices 3, 5, 7, and 10 in the vertex list would thus be represented as $P = (3, 5, 7, 10)$. An example is shown in Figure 3.6, where polygon P_1 is represented by vertices 1, 2, and 4 and polygon P_2 is represented by vertices 2, 3, and 4.

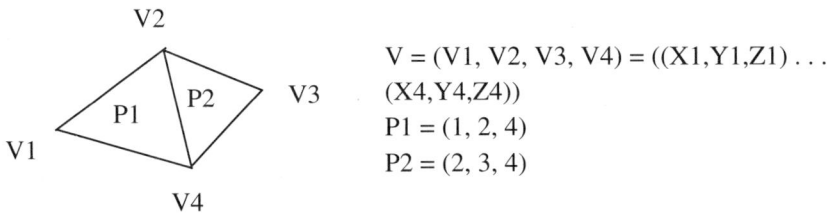

$$V = (V1, V2, V3, V4) = ((X1,Y1,Z1) \ldots (X4,Y4,Z4))$$
$$P1 = (1, 2, 4)$$
$$P2 = (2, 3, 4)$$

Figure 3.6 Polygons represented by indices or pointers to vertex list.

Simple Object Display

Just as with a vertex list (or table; Table 3.2), one can prepare an edge table. However, to draw objects, one needs to know faces (e.g., for rendering) in addition to edges and vertices (Table 3.3). To reduce duplication (e.g., Euler's rule) and for convenience, a popular convention is to use pointers to vertices and faces and use surface normals to specify the inside or outside nature of a surface from the visibility standpoint. Note that all the vertex sequences have been specified taking a clockwise journey around the boundary, as viewed from the outside of the cube (Figure 3.7). Such an arrangement is often very important to rendering programs that have to compute surface normal vectors. The clockwise sense in itself is not important, as a counterclockwise sequence would work just as well, but a consistent approach to definition is normally required.

TABLE 3.2 VERTEX TABLE

No.	X	Y	Z
1	0	0	0
2	0	1	0
3	0	1	1
4	0	0	1
5	1	0	0
6	1	0	1
7	1	1	1
8	1	1	0

TABLE 3.3 FACE TABLE

No.	Vertices			
1	1	2	3	4
2	4	3	7	6
3	6	7	8	5
4	5	8	2	1
5	2	8	7	3
6	1	4	6	5

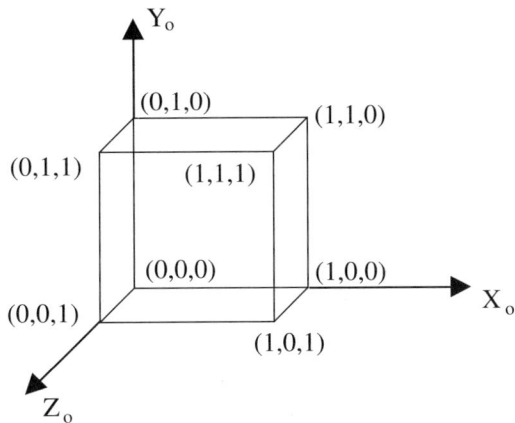

Figure 3.7 Example of clockwise convention when viewing an object from outside.

Note that a popular format considers a counterclockwise convention when viewing an object from outside. The format is Wavefront Technology's .obj and a small example is shown below.

```
v 2.828425 0.000000 4.898977   (represents vertex A)
v 2.828425 0.000000 -4.898977  (vertex B)
v 0.000002 7.999997 0.000000   (vertex C)
v -5.656852 0.000003 0.000000  (vertex D)
# 4 vertices
vn -0.471404 0.333333 -0.816497
vn -0.471404 0.333333 0.816497
vn 0.000000 - 1.000000 0.000000
vn 0.942809 0.333333 0.000000
# 4 normals
# 0 texture vertices
usemtl light_steel_blue
f \
4//1 3//1 2//1
f \
```

```
3//2 4//2 1//2
f \
2//3 1//3 4//3
f \
1//4 2//4 3//4
# 4 elements
```

In the foregoing example of the .obj format, f denotes faces, v denotes vertices, and vn denotes vertex or face normal. The vertices are expressed in the order 1,2,3,4, or A,B,C,D as shown in Figure 3.8. The vertices and face normals are both indexed as 1,2,3,4. The faces are expressed as vertex index/texture index # (currently empty)/face normal index.

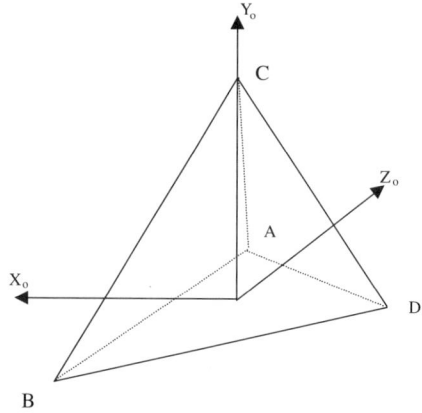

Figure 3.8 Example of counterclockwise convention when viewing an object from outside.

Face Normals

As shown in the preceding section, face normals can be computed by taking cross products of two edges on a face (e.g., form two vectors $\mathbf{v}_1$ and $\mathbf{v}_2$ as shown in Figure 3.9).

$$\mathbf{v}_1 = [(0 - 0), (1 - 0), (0 - 0)] = [0,1,0]$$
$$\mathbf{v}_2 = [(0 - 0), (1 - 1), (1 - 0)] = [0,0,1]$$

The cross product $\mathbf{v}_1 \times \mathbf{v}_2$ produces [1,0,0], which is a unit vector directed along the positive X axis. This is a valid surface normal, but it is pointing toward the interior of the object. If we reverse $\mathbf{v}_2$ and perform the cross product again, we have

$$\mathbf{v}_2 = [(0 - 0), (1 - 1), (0 - 1)] = [0,0,-1]$$

and the cross product $\mathbf{v}_1 \times \mathbf{v}_2$ produces [-1,0,0], which is now directed along the negative X axis, thus pointing outside the object.

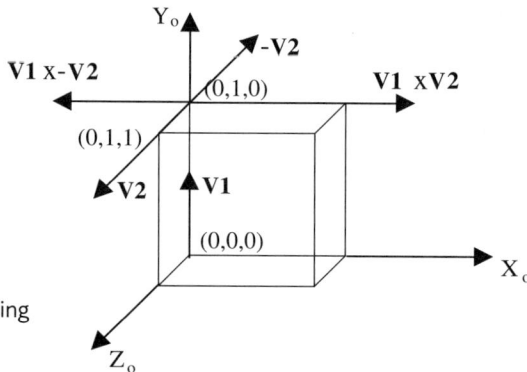

Figure 3.9 Computing face normal.

Mesh Handling

Definition.

Polygonal mesh is a collection of nonintersecting polygons, possibly joined along common edges or at vertices. *Triangle mesh* is a polygonal mesh whose polygons are triangles. Meshes may be nonmanifold. In a nonmanifold mesh, there exist edges in which more than two faces are incident (Figure 3.10).

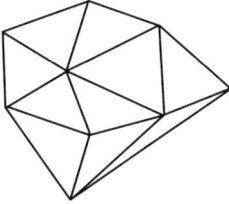

Triangular mesh

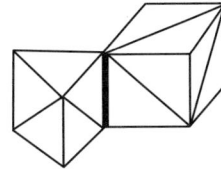

Non Manifold mesh (common edge of a box and a prism)

Figure 3.10 Triangular and nonmanifold meshes.

Mesh Generation/Tessellation.

A *triangulation* is a connectivity scheme composed of triangular elements which tessellate a planar region or perhaps a three-dimensional surface. A *tessellation* is a general subdivision of a spatial domain, with no implied maximum dimension. A *Delaunay triangulation* enforces the additional constraint that the circumcircle uniquely defined by the three points of a face will contain no other points in the domain. Thus a Delaunay triangulation of an arbitrary point set guarantees a unique connectivity scheme, assuming that four or more points are not co-circular. (Delaunay triangulation is discussed in more detail in Chapter 5.) Similarly, there may be other schemes for mesh generation or tessellation.

Motivation for Reducing Triangles.

Triangle meshes are large: for example, tessellation algorithms often deal with the order of 10^4 polygons, digital elevation data often contain on the order of 10^5 to 10^7 polygons, three-dimensional digitizers often deal with 10^5 to 10^6 polygons, and iso-surface generation schemes often handle 10^6 triangles. Large meshes mean slower rendering speeds, large memory requirements, and more expensive analysis. Another major source of triangle accumulation is error or noise in data sets. Hence there is a need for triangle reduction.

Methods for Triangle Reduction.

Many CAD packages nowadays provide triangle reduction options. Some of the basic principles include (Figure 3.11):

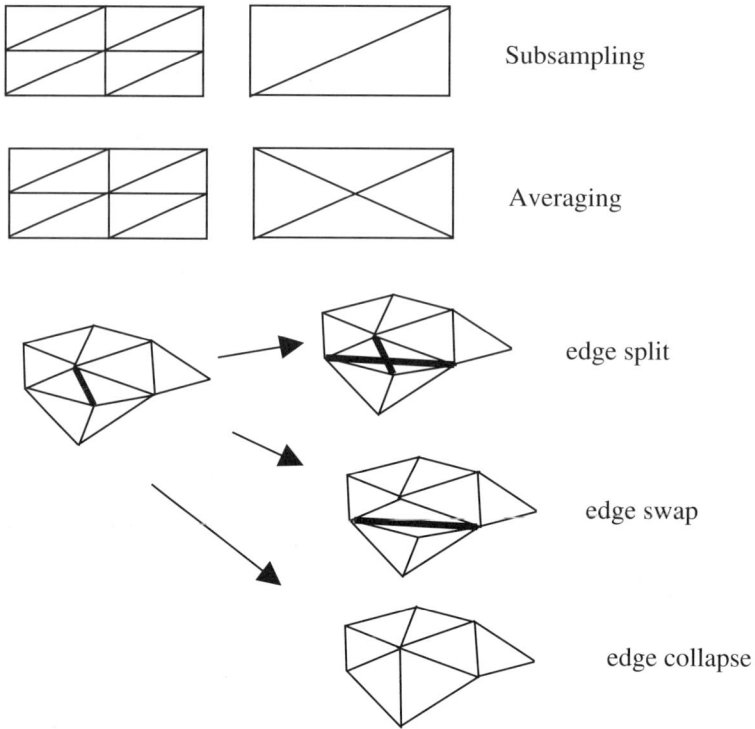

Figure 3.11 Triangle reduction principles.

- *Subsampling.* Gross triangles sample a chosen region.
- *Averaging.* Triangles are reduced by averaging triangles in a region.
- *Edge collapse.* Vertices are removed and edges collapsed.
- *Edge split.* Edges are split to add more triangles at places as an intermediate step in hopes of achieving a better overall triangle count.
- *Edge swap.* Edges are swapped between vertices.

Figure 3.11 illustrates some of these techniques. Some of the other techniques involve retriangulation using a constrained Delaunay technique and evaluating local topology to use local knowledge to reduce triangles (e.g., planarity of region, mesh mismatches due to errors).

3.2.2 Useful Model Building Techniques

A couple of useful model building techniques are illustrated next. A data extrusion example is shown in Figure 3.12, in which a surface is extruded along a path to create a volume quickly. The figure also illustrates a sweeping technique by which a surface is rotated to create a volume quickly. These techniques are often used to create CAD or VR models.

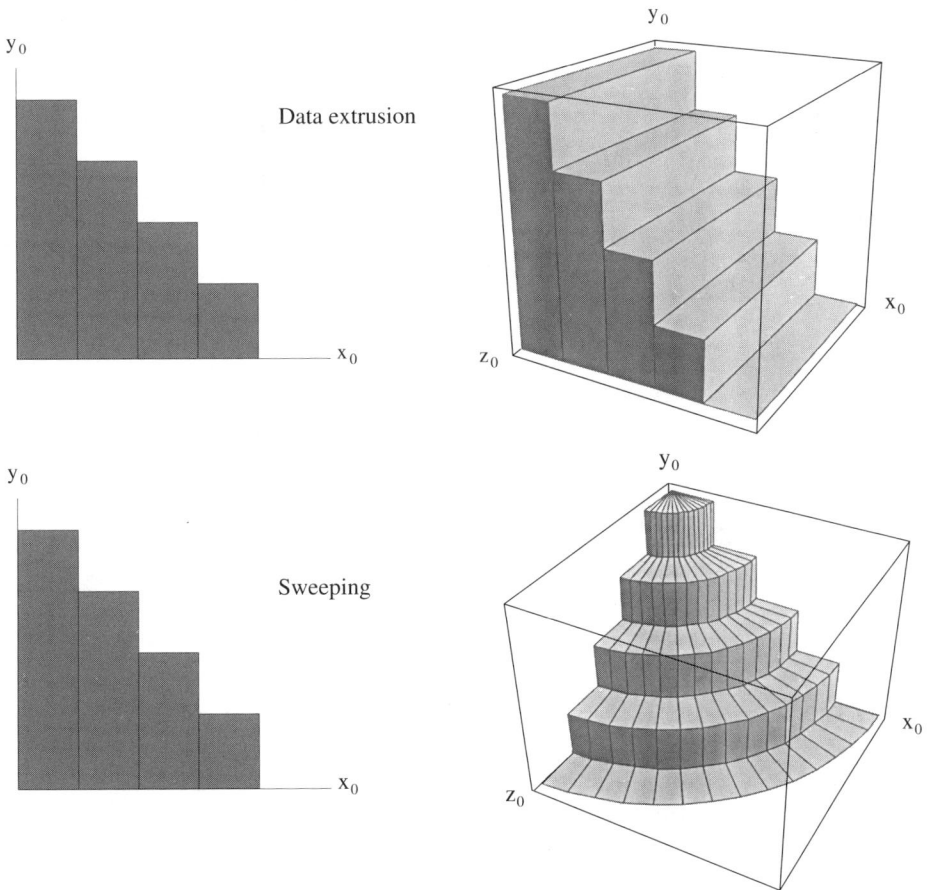

Figure 3.12 Model building techniques.

3.2.3 Useful Model Assembly Techniques

Once the CAD models are built, they often have to be assembled. A useful assembly technique adopted by CAD packages, such as ProEngineer™, is based on placement constraints between surfaces built out of primitives, such as box, sphere, torus, and so on. A number of placement constraints are illustrated in Figure 3.13. They include:

- *Align:* ensures that two surfaces are aligned
- *Mate:* ensures that two surfaces touch each other
- *Offset:* ensures a certain offset
- *Orient:* ensures that two surfaces are oriented in the same direction

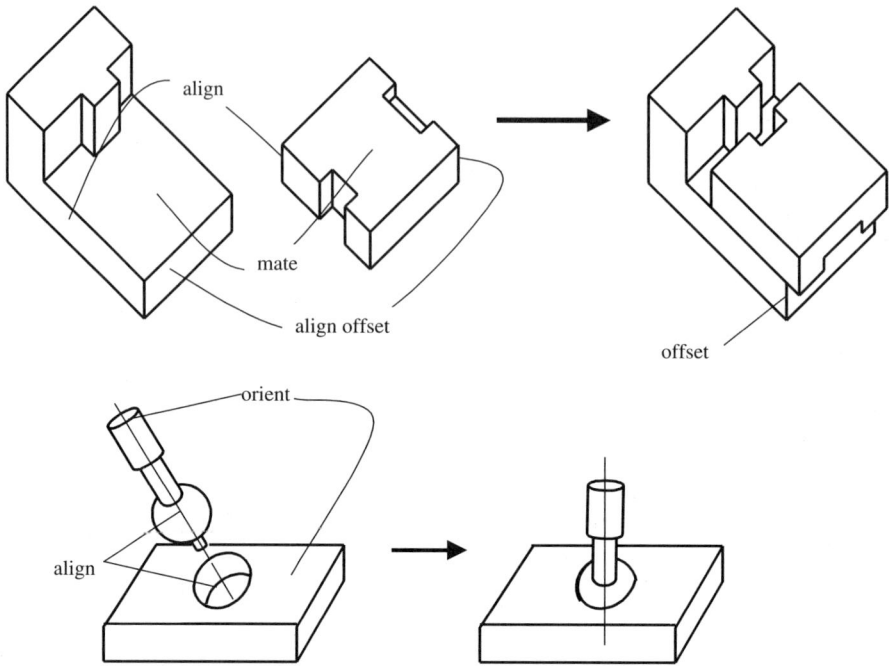

Figure 3.13 Model assembly techniques.

3.2.4 Model Preparation, Validation, and Repair: Rapid Prototyping Example

Introduction to Rapid Prototyping

From a systems perspective, a rapid prototyping (RP) machine transforms an input geometric model into a corresponding physical artifact. Once material/process parameters are set, the fabrication is driven primarily by the input geometry. Consequently, much of the current software associated with RP machines deals with geometry. Actually, the geometry and materials/process characteristics are not as independent as implied above. Just consider distortion compensation, where the input geometric shape must be altered according to a predictive materials/process model to make up for shrinkage. This coupling will take an even more fundamental role in the future, when RP systems will be able to blend different materials and trade off geometric properties with material properties.

Three-Dimensional Solids

Three-dimensional solid modeling gained acceptance in engineering practice in the early 1980s with the availability of cost-effective computational power. Design engineers embraced solid modeling, but manufacturing engineers by and large resisted the change and continued to rely on drawings. The reason is primarily that in addition to geometry, a drawing contains many nongeometric elements, such as tolerances, annotations, notes, and procedures meaningful in production. A three-dimensional solid model, on the other hand, simply describes three-dimensional geometric shape.

The unique advantage of three-dimensional solid modeling in engineering is the ability to compute geometry without the need for human interpretation. To enable this, three-dimensional solid models must be defined explicitly, precisely, and completely. As a result, their representation is verbose compared to the minimal data and computation requirements for wireframe models and two-dimensional drawings. For example, screw threads are generally expressed in a two-dimensional drawing by a note specifying the type of thread. Three-dimensional solids require an explicit and detailed geometric model of the full thread.

Interface Formats.
For reasons of competitiveness, targeted applications, and system performance, CAD vendors utilize a variety of proprietary internal geometric mathematical forms for defining three-dimensional solids. Typical boundary surface forms include planes, conics, Bezier surfaces, Coons-type patches, and nonuniform rational B-spline (NURBS) surfaces, while their corresponding geometry manipulation routines include interpolation, blending, intersecting, and trimming. RP vendors accommodate this diversity by basically supporting a single common mathematical surface form and shift the burden of providing translators (postprocessors) to CAD vendors. Technically, the translation approximates all surfaces by a network of surface elements of a single mathematical form. The RP *common surface form* is the simplest and lowest common denominator of all the surface forms, namely, surfaces approximated by triangular meshes. This representation, properly ordered, along with its data types, delimiters, and other file information, is the well-known de facto standard STL format, which is similar in many respects to other formats, such as wavefront (OBJ) and Inventor (IV).

Three-Dimensional Formats.
Although many agree that the STL format leaves much to be desired, there is no unanimity on its replacement. Solid Concepts Inc. introduced its own file format [SFX (Solid File exchange)], which removes the redundancy in STL and includes topology. VRML 2.0 or VRML 97 is another promising data format. It has been approved as an International Standards Organization (ISO) standard. It accommodates STL files, AutoCAD™ DXF, and parts of the Initial Graphics Exchange Specification (IGES). VRML certainly makes RP more accessible globally, especially for applications dealing with the creation of artifacts for viewing. Yet RP applications dealing with tooling and functional parts in manufacturing are too demanding for this level of modeling capability.

STEP Standard.

Since there is no single all-encompassing general mathematics form that includes all other practical geometric surfaces as special cases, it makes sense to focus on those math forms that support a large application base. The STEP standard is based on NURBS geometry. The CAD community has been developing standards for the exchange of product data since 1979. IGES became a U.S. national standard in 1981. IGES 5.2 version includes three-dimensional solid models, transcending its early manifestation as strictly a standard for exchanging two-dimensional graphical drawings. Similar efforts in Europe led to the German DIN 66301 standard, *Verband der Automobilindustrie* (VDAFS) in 1987, and the French AFNORZ68-300 standard, *Standard d'Echange et de Transfert* (SET). Some RP software packages allow the import–export of IGES, STL, DXF, ASCII, VRML, SLC, and three-dimensional measurement data, among many other useful operations for verification, editing, CAD-like operations, and tooling operations. These past efforts are now being superseded by STEP, a much more extensive data exchange standard that focuses on full product data (materials, physical behavior, bills of materials, data management) and will eventually extend beyond product data and include process and resource data. One can think of IGES as specifying a neutral flat file geometry format, while STEP specifies the associativities of a product-neutral database. The first phase of STEP was approved as an international standard in late 1994 as ISO 10303. Many CAD vendors are committed to supporting STEP. The STEP standard provides extensibility to many application domains through a structure called *application protocols.* Each application domain has its own application protocol, with substantial sharing of common core functions.

Model Preparation

A designer creating a part model at a CAD station is generally absorbed in the functionality of his design. Once the model is transferred to the RP machine, the constraints of the RP physical process on the model become the dominant consideration. In this process planning step, issues such as part size, orientation for optimum build, whether portions are made solid or hollow, whether several parts are nested or subdivided to fit into one build volume, compensating for distortion and shrinkage, and generating supports for the part during build need to be considered. Software for preparing the model for fabrication is generally provided by the design and system houses rather than RP machine vendors. Such software contains many of the same surface and solid modeling operations found in CAD systems (Booleans, offsetting), except that the operations are done on tessellated models and stored in the STL format. The trend in RP software is clearly toward providing a complete geometric manipulation environment for facet solid models that can interface with future analysis capabilities. This entire RP process planning operation is predicated on having a valid tessellated solid model represented in the STL format (or an equivalent model in the case of two-dimensional contours.)

Model Validation and Repair

It is a fact of life that input STL files are seldom free of errors. Three-dimensional solid models, composed of intricate blends, intersections, trims, and offsets of many complex surface shapes, cause many geometrical special cases that must be accounted for during tessellation (Figure 3.14). In addition, many numerical idiosyncrasies are usually present in data files consisting of tens or hundreds of thousands of very small and ill-conditioned triangles. Figure 3.14 illustrates how triangular gaps appear during tessellation. Surface 1, being relatively flat, is approximated by large triangles, which results in a set of widely

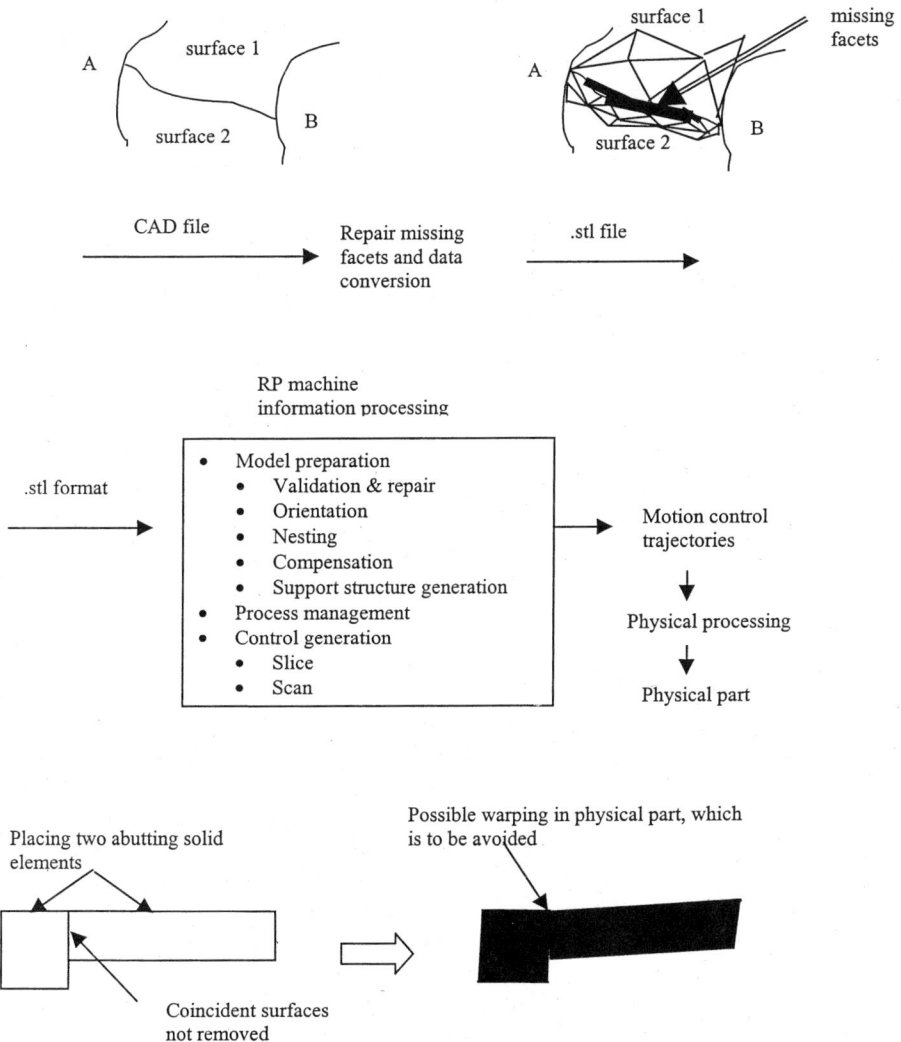

Figure 3.14 Rapid prototyping example for model preparation, validation, and repair.

spaced points along the intersection curve with surface 2. Surface 2, having high curvature, must be approximated by small triangles, resulting in a set of closely spaced points along the common intersection curve. When each respective surface's set of distinct points along the common intersection curve are connected by straight lines to create triangles, triangular gaps appear (shaded).

Numerical pathologies occur in STL data files, for example, when vertices of triangles, which are supposed to be coincident, have different values. Commercial software packages are available that preprocess STL files and add topology adjacency information, *a posteriori*, by testing edges and vertices for sameness and adding pointers in the file.

Finally, anomalies can occur because the geometric elements unnecessarily constrain the physical fabrication. As shown in Figure 3.14, if the coincident surface between two abutting solids is not removed, certain fabrication processes will solidify each portion of the solid independently, possibly leading to internal stress buildup and ultimate distortion of the physical part.

Software for three-dimensional viewing and editing of STL files typically supports the following functions: highlighting in color or by other means the display of inverted triangles and improper edges, monitoring the number of errors and the status of repair, automatically or interactively inverting reversed normals, stitching edges to eliminate gaps, deleting double surfaces, removing penetrating volumes, filling open volumes, and cleaning up intersecting triangles and topological inconsistencies. Errors corrected interactively provide picking, deleting, and adding of triangles. Some packages reduce the number of triangles in a file to within a specific tolerance. Major deficiencies, such as gaps consisting of many missing contiguous triangles, require manual repair because a significant number of sampled surface values needed to make a decision are lost. One may have to extrapolate the local surface curvature from known neighboring vertices to solve the problem.

3.3 REAL-TIME IMAGE GENERATION

As already introduced in Chapter 2, for real-time image generation three principal kinds of database traversals are normally used:

1. *Application.* For example, do requisite processing for the visual simulation application, including reading input from control devices, simulating the dynamics of moving models, updating the visual database, and interacting with other networked simulation stations.
2. *Cull.* Traverse the visual database and determine which portions of it are potentially visible, perform level-of-detail selection for models with multiple representations, and build sorted, optimized display lists for the draw stage.

3. *Draw (and render).* The draw traversal then runs through the display list and draws the geometries. After the geometries are drawn, the rendering phase fills up the objects where needed to give the feeling of surface color and quality (e.g., shininess, dull, transparent, translucent). Usually, rendering is considered as a part of the drawing phase.

Frame

A frame is the period of time in which all visual processing must be completed. In many implementations, the frame time is usually a positive integer, a multiple of the video refresh time span. For example, if the video scan rate is 60 hertz (Hz) and the frame rate is 20 Hz, the frame time [$\frac{1}{20}$ second (s)] is three times the video refresh time ($\frac{1}{60}$ s) (note that the standard TV refresh rate is 60 Hz and the standard computer screen refresh rate is 72 Hz, the latter being adjustable).

Frame Synchronization

To avoid difficulties such as motion sickness to participants, the frame refresh rate needs to be synchronized with the video rate, especially in view of situations where computation of the new frame takes longer than the allotted time (a situation referred to as *overrun*).

3.3.1 Delays and Frame Rates

The feeling of motion is achieved by updating picture frames very rapidly (e.g., 10 to 120 times a second) with objects separated by small displacements. The update rate is measured in frames per second. Although delays are clearly related to frame rates, they are not the same: A system may have a high frame rate, but the image being displayed or the computational result being presented may be several frames old. Research has shown that delays of longer than a few milliseconds can measurably affect user performance, whereas delays of longer than a tenth of a second can have a severe impact. Relatively static environments with slowly moving objects are usable with frame rates as low as 8 to 10 per second and delays of up to 0.1 s.

Environments with objects exhibiting high frequencies of motion (such as virtual racquetball or skiing) will require very high frame rates (> 60 Hz) and very short delays. In all cases, however, if the frame rate falls below 8 frames per second, the sense of an animated three-dimensional environment begins to fail, and if delays become greater than 0.1 s, manipulation of the environment becomes very difficult.

Current VR technology uses a technique known as *stereoscopy,* which creates a three-dimensional sensation of depth by creating two distinct views of an object — one for the left eye and the other for the right eye — in every two consecutive frames. VR has normally half the refresh rate of animation, because in an immersive environment almost twice the number of images need to be created, one for the left eye and one for the right eye. Because of this, VR can achieve the same level of modeling effect as animation only if it can achieve twice the frame update rate of animation. Typical refresh rates range from 10 frames/s (Hz) to 120 Hz. Whereas animation is essentially a two-dimensional model of the motion parameters, virtual reality (VR) extends this concept to three dimensions. *VR can thus be viewed as a complete means of simulation in space because it captures all three dimensions.*

To achieve high frame update rates, we need efficiency in many respects. Although many of these measures are embedded in current sophisticated graphics hardware, there are two important requirements from application software to effectively drive simulation in space. One of them is efficient data structures for object representation, and the other is efficient algorithms for motion representation.

3.4 LEVEL OF DETAIL

For real-time image generation, one has to deal with frame update time on the order of $1/20$ sec. This means that there is very little time available for updating one frame to the next. One important technique in updating frames with complex displays is by controlling the *level of detail* (LOD) by appropriately managing the culling and drawing operations. Controlling LOD is achieved by organizing the display information in a more detailed format for closer objects and in a more aggregate format for objects far away from the observer.

A number of examples are provided at this point to illustrate use of the concepts mentioned so far. Although many of the examples have been provided within the framework of a number of programming languages to provide realism, we do not expect readers to be familiar with these languages. Much of the discussion focuses around pseudocode.

Example 1: Scene Graph

VR programming languages such as Inventor™, VRML, Java3D, and Performer™, among others, focus on creating and/or handling three-dimensional objects. All information about these objects — their shape, size, coloring, surface texture, location in three-dimensional space — is stored in a scene database (also known as a scene diagram or graph). These languages have functions, or methods and descriptions, analogous to those in most other graphics libraries for modeling and displaying three-dimensional scenes. An orthogonal or perspective camera can be defined, with various properties, such as front and back clipping plane, angle of view, and so on, to create a viewing frustum. Polygonal objects can be defined as primitives, or using polygon lists or patches. Various kinds of lights, texture mappings, and so on, can be created, and rendering methods from wireframe to Gouraud shading can be bound to a scene. Note that all features may not be available in all languages.

Most languages use a right-handed coordinate system. Polygons are defined counterclockwise in some languages, whereas in others they are defined clockwise.

Scene database.

The node is the basic building block used to create three-dimensional scene databases. The style and contents of nodes vary with different languages. Each node holds a piece of information, such as a surface material, shape description, geometric transformation, light, or camera. All three-dimensional shapes, attributes, cameras, and light sources present in a scene are represented as nodes. An ordered collection of nodes is referred to as a *scenegraph* or *diagram*. This scenegraph is stored in the database. The language takes care of storing and managing the scenegraph in the database. The database can contain more than one scenegraph. After you have constructed a scenegraph, you can apply a number of operations or actions to it, including rendering, picking, searching, computing a bounding box, and writing to a file.

Classes of database primitives include shape nodes (e.g., sphere, cube, cylinder, quad mesh), property nodes (e.g., material, lighting model, textures, environment), and group nodes (e.g., separator, level of detail, and switch; Figure 3.15). Paths are used to isolate particular objects in a scenegraph. Other special database primitives are engines and sensors. Engines are objects that

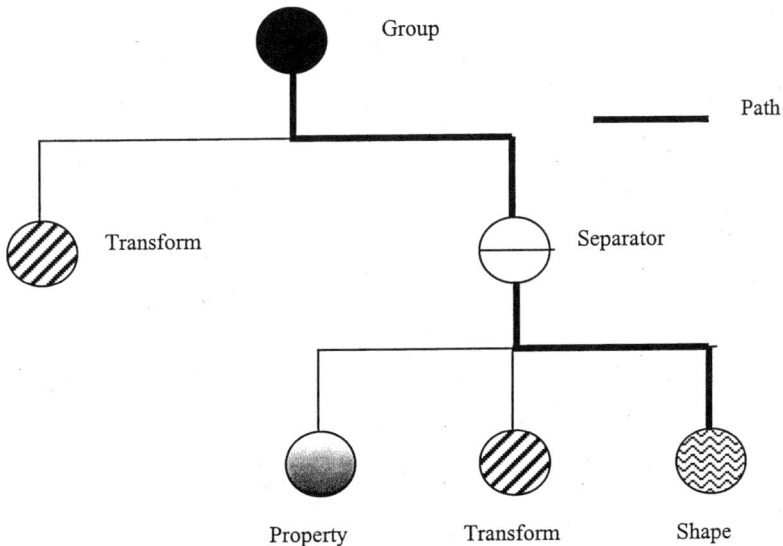

Figure 3.15 Example of a scene graph.

can be connected to other objects in the scenegraph and used to animate parts of the scene or constrain certain parts of the scene in relation to other parts. A *sensor* is an object that detects when something in the database changes and calls a function supplied by the application. Sensors can respond to specified timing requirements (e.g., "Do this every *n* seconds") or to changes in the scenegraph data.

Most VR languages are structured using the pattern of C + + classes, which are organized hierarchically in a tree. There are *base* (or *root*) classes, and there are *derived classes*. Based on the C + + principle, classes down the hierarchy in the tree inherit the fields and methods of the classes they are derived from. One of the most important aspects of these languages is the ability to program new objects and operations as extensions to the existing classes. One way to extend the set of objects is to derive new classes from existing ones. Another way to include new features is by using *callback functions*, which provide an easy mechanism for introducing specialized behavior into a scene graph or prototyping new nodes without subclassing. A callback function is a user-written function that is called under certain conditions.

Node manipulation: inventor example.
To isolate the effects of nodes in a group, a *Separator node* is used. Before traversing its children, a Separator node saves the current traversal state. When it has finished traversing its children, the Separator restores the previous traversal state. Nodes within a Separator thus do not affect anything above or to the right (meaning going above and then traversing down to the right) in the graph.

Figure 3.16 shows the body and head for a very simple robot shown in Figure 3.17. The body group, a separator, contains a *Transform node* (marked

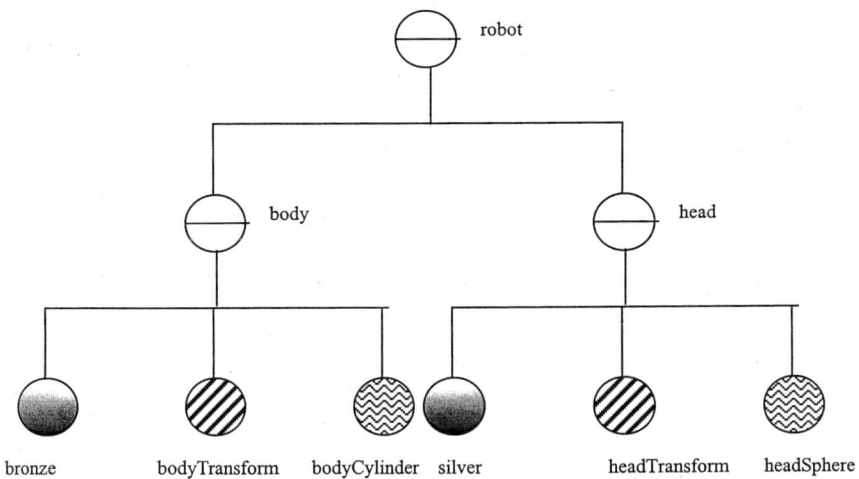

Figure 3.16 Example of separator node.

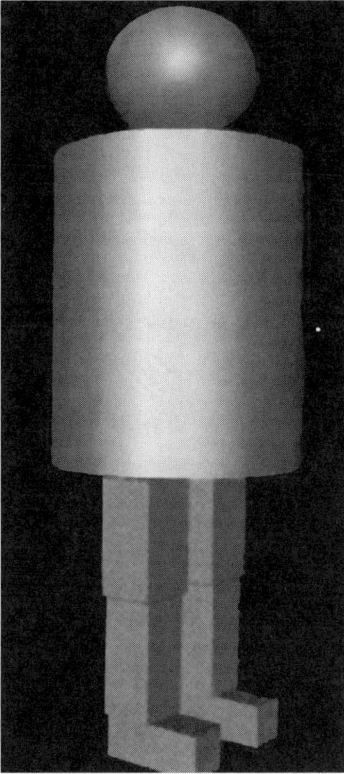

Figure 3.17 Simple robot example.

"bodyTransform") and a *Material node* (marked "bronze") that affect the traversal state used by the cylinder in that group. These values are restored when all children in the body group have been visited, so the head group is unaffected by the body-group nodes. Because the head group is also a separator group, the traversal state is again saved when group traversal begins and restored when group traversal finishes.

Scene graph concepts.

Often, the data file format mirrors the scene graph. It is essentially a description of the entire scenegraph or a portion of it. It is very important to note that the scene database is traversed from left to right and from top to bottom. For example, a transformation node or property node affects everything to the right and below it. Unlike transformation nodes, a property node can be effectively replaced by new property nodes farther downstream. With transformation nodes, new transformations are postmultiplied (or premultiplied if the transformations are expressed in reverse notation in the graphics library: for example, $\mathbf{T}_{xy}$ vs. $\mathbf{T}_{yx}$ — if one travels down from node u to x to y to z, premultiplying would give $P_u = P_z\mathbf{T}_{zy}\mathbf{T}_{yx}\mathbf{T}_{xu}$, whereas postmultiplying would give $P_u = \mathbf{T}_{ux}\mathbf{T}_{xy}\mathbf{T}_{yz}P_z$) onto the current or existing transformation matrix. This is commonly the way graphics libraries maintain the current transformation matrix. As a result of this, it often makes more sense to read multiple transformations from bottom to top/right to left.

Example 2: Simple Robot

Paths are used to isolate particular objects in the scenegraph. Suppose that you want to refer to the left foot of the robot. Which node in previous graph represents the left foot? You cannot refer simply to the foot node, since that node is used for both the left and right feet. The answer is that the left foot is represented by the path, or chain, starting at the robot node (the root), and leading all the way down the graph to the foot node. Figure 3.18 indicates the path for the left foot node. A path contains references to a chain of nodes, each of which is a child of the preceding node. A path represents a scenegraph or subgraph (part of a scenegraph). The path is represented here by a heavy line that connects the chain of nodes.

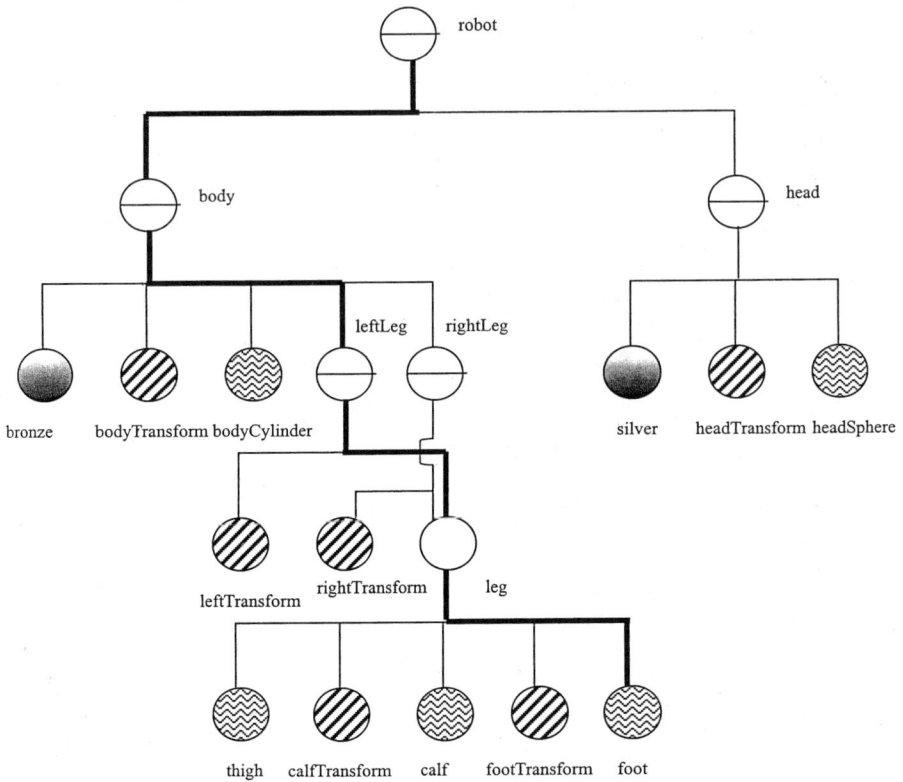

Figure 3.18 Robot scene graph with a path for the left foot.

Example 3: Culling

If a given Transform or Group node contains objects on opposite sides of the room (or worse, on opposite sides of the world), the browser won't be able to cut out part of the scene. *View culling* is the term used to describe this process of selecting objects to be discarded or ignored. If the north wall of the room is under one Transform, for example, and the south wall is under a different Transform, half of the scene can be culled at any given time, depending on which wall the user is viewing.

Here is a detailed example of how you might model the room whose shape is shown in Figure 3.19 (Indexed Face Set is explained in Appendix A5.) At first, you might be tempted to create the walls as a single IndexedFaceSet. If you do, the browser tries to draw every polygon in the set since some portion of the room is always visible. It is far more efficient to break up the walls into several different sets based on location. For example, you might break them up into 20 indexed face sets, as shown in Figure 3.20. Group the indexed face

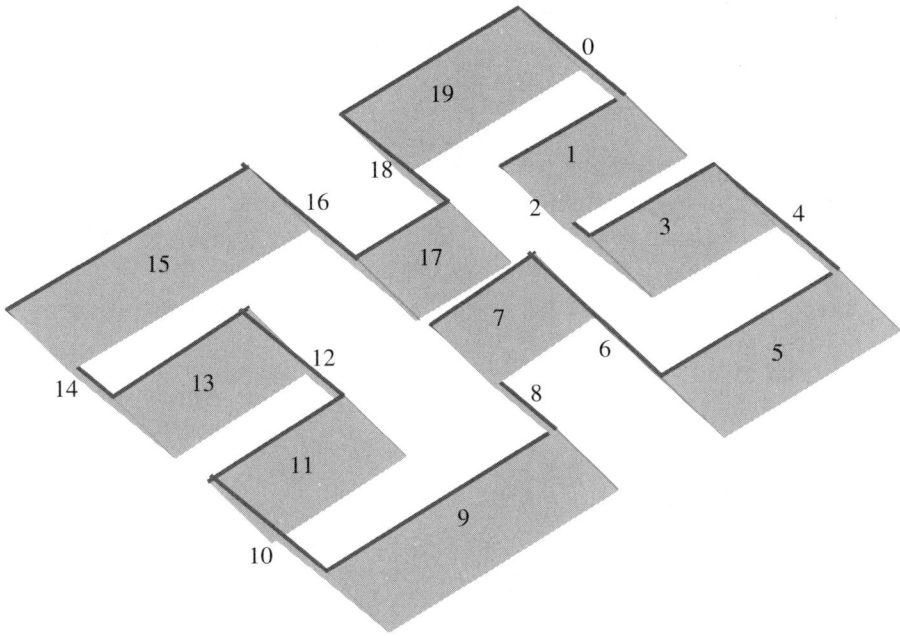

Figure 3.19 Division of a large object into smaller objects for culling.

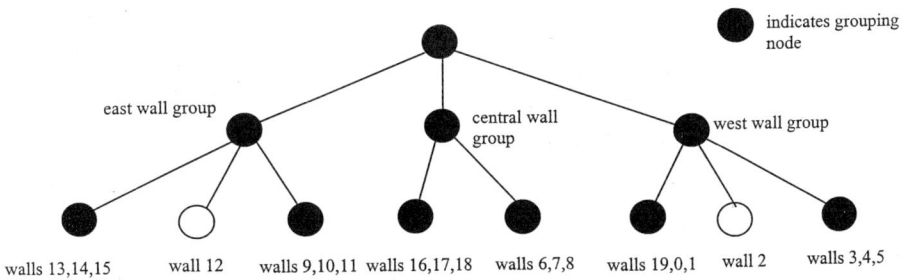

Figure 3.20 Scene graph design for effective culling.

sets as shown in the figure. When the walls are grouped in this fashion, the scene hierarchy can be culled at a high level (half the room can be culled with a single test). When you model the floor and ceiling for this room, you can group them with their associated walls.

In general, don't create large objects that span large scenes. If you have a large floor in your model, for example, break it up into several sections, putting each section under its own Transform or Group node to facilitate culling. If you do have large objects, avoid grouping other, smaller objects with them.

Example 4: Simple VR Factory

A virtual reality (VR)–aided factory simulator is a powerful tool that can enable a designer to understand the operations on a factory floor from the following perspectives:

- Overview and general acquaintance with factory operations (e.g., education and interactive training)
- Study of various detailed factory layout configurations (e.g., simulating various chaotic operations on a factory floor)
- Modeling events in space on a factory floor (e.g., modeling production-line stoppages)

Objects:

```
floor.hrc
cabinet_1.hrc
cabinet_2.hrc
coarse_cutter_0.hrc
coarse_cutter_1.hrc
coarse_cutter_2.hrc
coarse_cutter_3.hrc
coarse_cutter_4.hrc
coarse_cutter_5.hrc
coarse_cutter_6.hrc
coarse_cutter_7.hrc
coarse_cutter_8.hrc
coarse_cutter_9.hrc
conveyor_0.hrc
conveyor_1.hrc
conveyor_2.hrc
conveyor_3.hrc
conveyor_4.hrc
conveyor_5.hrc
conveyor_6.hrc
conveyor_7.hrc
fine_cutter_0.hrc
fine_cutter_1.hrc
fine_cutter_2.hrc
fine_cutter_3.hrc
fine_cutter_4.hrc
fine_cutter_5.hrc
forklift_1.hrc
forklift_2.hrc
forklift_3.hrc
```

```
furnace_1.hrc
furnace_2.hrc
green_1.hrc
green_2.hrc
housing_1.hrc
housing_2.hrc
milling_1.hrc
rack_1.hrc
rack_2.hrc
rack_3.hrc
rack_4.hrc
robot_0.hrc
robot_1.hrc
robot_2.hrc
robot_3.hrc
robot_4.hrc
shaft_1.hrc
shaft_2.hrc
shaft_3.hrc
shaft_4.hrc
skid_1.hrc
skid_2.hrc
truck_1.hrc
truck_2.hrc
truck_3.hrc
truck_4.hrc
truck_5.hrc
truck_6.hrc
```

Figure 3.21 shows a run-time database of the factory objects listed above. The following interactive controls of level of detail for individual objects are possible:

- A user can interactively specify the number of levels.
- Objects can be switched on and off.

The following task-based motion modeling primitives are also possible:

- Acceleration, deceleration
- Pick-and-place tasks

Following are the representative steps involved in developing the immersive display protocols:

1. Identify each composite object and the objects that make up each composite object. The objects are selected in such a way that they form one

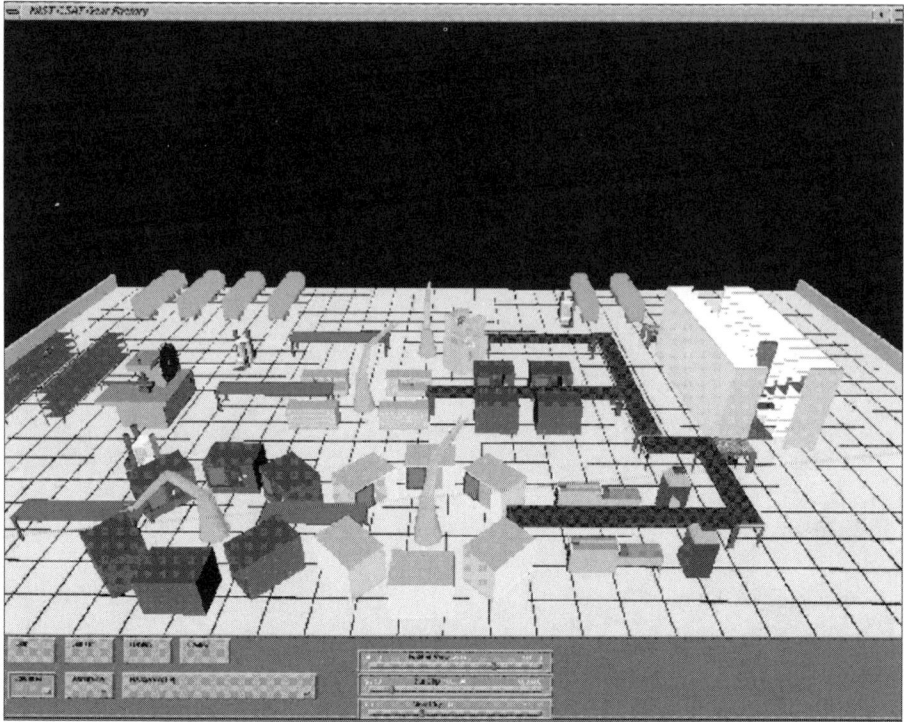

Figure 3.21 Run-time database of factory objects for interaction in immersive virtual reality. Data courtesy of NIST.

logical unit when certain properties (e.g., material, texture, motion) are applied to them. The dimensions of the objects are either obtained or computed as per the requirements. A hierarchy is established between these objects. This is of particular importance while defining motion properties of each object.

2. Once the details of the composite objects and the objects are obtained, a modeler is used to model these objects. In this case, SoftImage™ has been chosen to perform the modeling. Each object is modeled separately for convenience and ease of handling. An object is made up of basic primitive structures: circle, arc, spiral, square, cube, sphere, cylinder, cone, torus, tetrahedron, octahedron, icosahedron, dodecahedron, grid, and null. An object can be a simple primitive or a complex shape arrived at by performing a (constructive solid geometric) Boolean operation of union, intersection, or difference, or a combination of the three operations, on two primitives or on a primitive and a resulting object of an earlier Boolean operation.

3. After performing all the required Boolean operations, the final object is a single entity. This entity is attributed its own coordinate system, which is referred to as its *center*. It is normally located at the centroid of the

entity. The center is then shifted to the desired location within the entity. One of the reasons for shifting is for ease of defining certain types of motion, such as a rotational or revolving motion (e.g., a door can have its center located on the face that is hinged to the door frame).

4. Each object is assigned material properties from the Matter module of SoftImage (there are other means for doing this, the way described here is a convenient one). This includes assigning the following properties: Shading Model, Specular, Ambient and Diffuse Colors, Specular Decay, Reflectivity, Transparency, Refractive Index, and Static Blur. The objects are assigned texture properties. Texture includes the following properties: Texture Type (Marble, Wood, or Cloud with associated Spacing, Angle, Strength, Iteration, and Power factors); Transformation (Scaling, Rotation, and Translation for X, Y, and Z axes); Effect Value [Alpha Channel or RGB (i.e., red, green, blue) Intensity with Reflectivity, Transparency, and Roughness factors]; Blending (Without Mask, Alpha Channel Mask, or RGB Intensity Mask with Overall Blending, Ambient, Diffuse, and Specular factors).

5. A hierarchy structure is created to organize the various objects that have been created to compose the composite object. First the parent node is selected from the available objects, and then each child is attached to the parent. This process is repeated to attach all the objects in the hierarchy. A conceptual object is illustrated in Figure 3.22, whereas details from an actually implemented object are illustrated in Figure 3.23.

6. A composite object (or objects in certain cases) is stored as a Scene. This contains all the information regarding the object. The hierarchy is stored in file with the root node as filename and extension .hrc. (In the example cited in Figure 3.22, the hierarchy file will be prefix-Door.1-0.hrc. The middle extension is for the version number. Every time a scene is saved after some modification, a new version number is assigned. "prefix" is used for proper identification and some suitable value is supplied.)

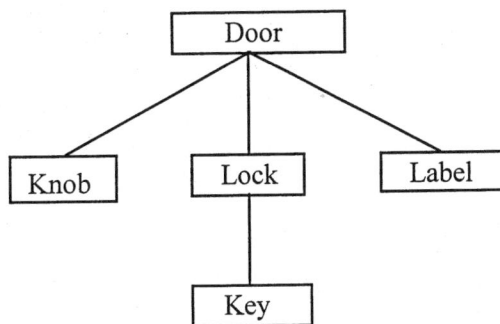

Figure 3.22 A simple object with its subobject hierarchy.

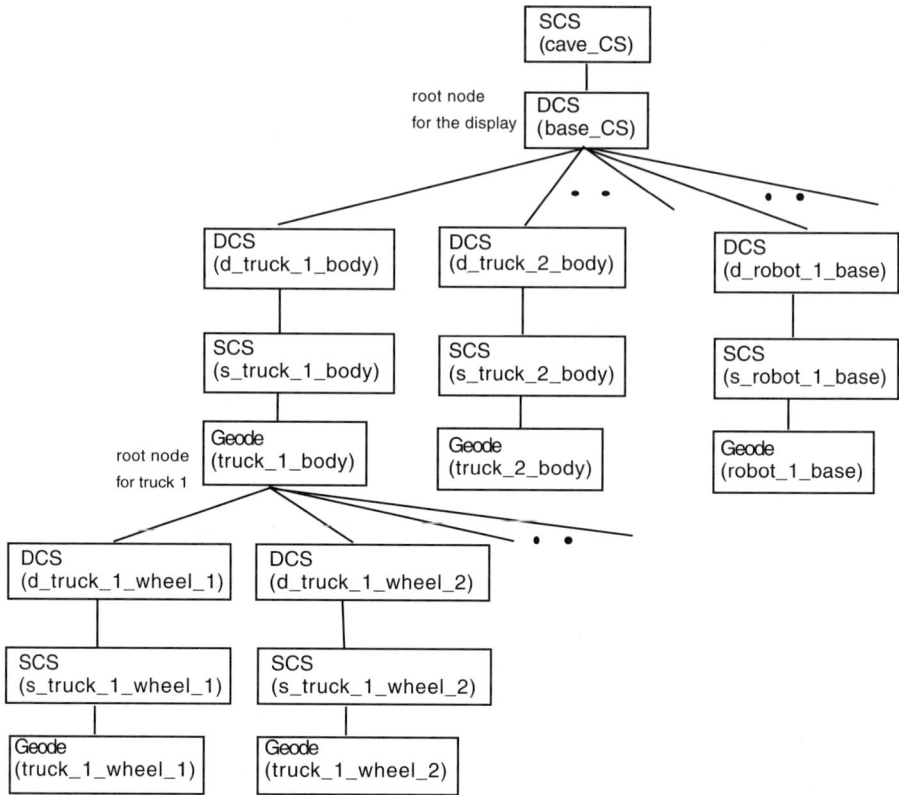

Figure 3.23 Portion of scene graph generated at run-time using IRIS Performer library.

7. The individual objects are stored in WaveFront® (.obj) format. The objects are selected and exported to .obj format in the Tools module. The associated material properties are stored in a .mtl file and the texture is stored in either a .rgb, .rgba, .int, .inta, or .bw file based on the type of texture. The .obj files and the associated .mtl and texture files are used to load the objects in Performer™. The .obj files contain information about all the vertex coordinate and vertex normal lists. All the faces are expressed using the vertex coordinates and the associated vertex normal vector. The file contains reference to the .mtl file to be used to obtain the material properties for the object.

8. IRIS Performer is an application development environment that combines a programming interface for creating visual simulation applications and a high-performance rendering library in a three-dimensional software tool kit. It provides an ANSI C application interface that incorporates the IRIS Graphics Library™ (GL) and the IRIX™ operating system as the foundation for a powerful suite of tools and features for creating real-time visuals on Silicon Graphics systems.

TABLE 3.4 NODE TYPES USED TO BUILD A RUN-TIME DATABASE USING THE PERFORMER LIBRARY

Node Type	Node Class	Description
Node	Abstract	Basic node type
Group	Branch	Groups zero or more children
Scene	Root	Contains the visual database
SCS	Branch	Static coordinate system
DCS	Branch	Dynamic coordinate system
Switch	Branch	Selects active children
Sequence	Branch	Sequences through a list of its children
LOD	Branch	Level-of-detail node
Layer	Branch	Renders coplanar geometry
LightPoint	Leaf	Contains light point(s) data and specification
Geode	Leaf	Contains geometry data and specification
Billboard	Leaf	Geometry that rotates to face the eye point

There is no archival database or file format for Performer. It defines a run-time-only database through its programming interface. The Performer run-time structures are loaded by the application from several popular database formats. A general database hierarchy is prepared in the form of a directed tree graph of nodes. There are specialized node types that are used for a visual simulation application. Table 3.4 lists the node types.

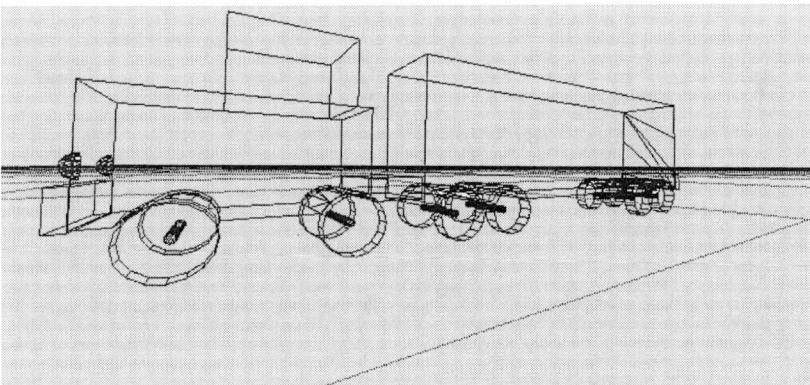

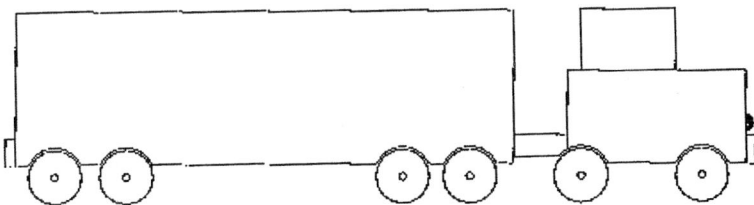

Figure 3.24 Semi truck in virtual factory.

Refer to the screen prints of a semi truck (Figure 3.24) used to simulate material handling in the loading/unloading area of a gear factory. Briefly the sequence is as follows. First, the .hrc file created from SoftImage is read and the hierarchy of all the objects are determined. Each object is added to the visual database as a Geode node. The Geode node is loaded by the application from the vertex coordinate list, vertex normal list, and the face data from the .obj file that was created from SoftImage. A set of SCS and DCS nodes are also attached to every Geode. See Figure 3.23 for details. The DCS nodes are required to perform the motion at run time by doing matrix manipulation for translational and rotational motion, and the SCS nodes are used to store this motion information for every frame.

```
/* THIS IS PART OF truck_1_body.obj FILE */
o truck_1_body
mtllib truck_1_body.mtl   /* reference to the material
file name */
v -21.352276 6.375434 -1.952915
v 6.490990 6.375434 -1.741475
v 6.445883 6.375434 4.198357
v -14.005706 -1.661633 -1.897126
v -17.893089 -1.872378 4.013528
v -13.909999 -1.872378 4.043776
# 324 vertices   /* Sample of six vertices out of 324 */

vn  0.000000 -1.000000 0.000000
vn  1.000000 0.000000 0.000000
vn  -1.000000 0.000000 0.000000
vn  0.000000 1.000000 0.000000
vn  0.000000 0.000000 1.000000
vn  0.000000 0.000000 -1.000000
# 885 normals  /* Sample of 6 out of 885 normals */

# 0 texture vertices   /* texture information will be
added later on */
usemtl truck_body1
f \
 321//885 274//885 269//885   /* A/B/C - A refers to the
vertex index
                    - B refers to the texture
                    index, currently empty
                    - C refers to the normal index
                    Three sets of A/B/C make up a face. */
```

```
f \
321//885 269//885 262//885
f \
 321//885 262//885 260//885
f \
 321//885 260//885 253//885
f \
 321//885 253//885 252//885
f \
 321//885 252//885 308//885
# 596 elements /* 6 faces out of 596 / are shown above */


/* THIS IS truck_1_body.mtl FILE */
newmtl truck_body1
Ka 1.0000 0.4431 0.4431  /*ambient color R G B */
Kd 1.0000 0.0000 0.0000  /*Diffusion color R G B */
Ks 1.0000 0.1098 0.1098  /* Specular color R G B */
Ns 50.0000               /* Shininess index */
Ni 1.0000                /* Refractive index */
illum 2
```

Movement Modeling in Performer.
This section tries to explain how translation and rotational motion is achieved
for an object loaded in Performer.

1. *First the object (pfGeode) should have a pfDCS node attached to it in order to achieve the motion. A pfDCS node is a 4 × 4 matrix* **M** *(having rows 0 to 3 and columns 0 to 3). Elements* **M[3][0]**, **M[3][1]**, *and* **M[3][2]** *contain the displacement in X, Y, and Z directions. The 3 × 3 matrix* **M[0][0]** *to* **M[2][2]** *contains information of the head, pitch, and roll angles.*
2. *Translational motion* It is relatively simple to make the pfGeode move. A call to the function,

```
    void pfGetDCSMat (pfDCS *dcs, pfMatrix m);
```

loads the dcs into the matrix m. If the object is to be moved by an amount
dx, dy, and *dz* in the *X, Y,* and *Z* directions, respectively, in the next frame,
the elements of the matrix are modified as follows:

```
    m[3][0] = m[3][0] + dx;
    m[3][1] = m[3][1] + dy;
    m[3][2] = m[3][2] + dz;
```

This is followed by a call to the function:

```
    void pfDCSMat (pfDCS *dcs, pfMatrix m);
```

which sets the dcs back to the new value of the matrix m. Or the following function sets the new values to the dcs:

```
void pfDCSTrans (pfDCS *dcs, float x, float y, float z);
```

where x = m[3][0], y = m[3][1] and z = m[3][2].

3. *Rotational motion.* The problem of rotating an object is comparatively complex. There are several approaches to achieving this.

Call to the function

```
void pfGetDCSMat (pfDCS *dcs, pfMatrix m);
```

loads the dcs into the matrix m. If m is orthonormal, a call to the function:

```
void pfGetOrthoMatCoord (pfMatrix m, pfCoord* c);
typedef struct
{
pfVec3 xyz;
pfVec3 hpr;
} pfCoord;
```

returns the x, y, and z values as well as the head, pitch, and roll (h, p, and r) angles. h is the rotation along the Z axis, p is the rotation along the X axis, and r is the rotation along the Y axis. pfGetOrthoMatCoord returns the translation and rotation of the orthonormal matrix, m. The pitch returned ranges from -90 to +90°. Roll and heading range from -180 to +180°.

If the object has to be rotated by an angle dh, dp, and dr, these values are added as follows:

```
c->hpr[0] = c->hpr[0] + dh;
c->hpr[1] = c->hpr[1] + dp;
c->hpr[2] = c->hpr[2] + dr;
```

If some translational motion is also required at the same time, it can be achieved by

```
c->xyz[0] = c->xyz[0] + dx;
c->xyz[1] = c->xyz[1] + dy;
c->xyz[2] = c->xyz[2] + dz;
```

The functions

```
void pfMakeCoordMat (pfMatrix m, pfCoord *c);
void pfDCSMat (pfDCS *dcs, pfMatrix m);
```

create the updated matrix m from the amended pfCoord *c, which is then used to update the dcs. All this is possible if the matrix m is orthonormal.

General case when the state of matrix m *is not known.* In such cases, the program needs to keep track of the h, p, and r angles of the object orientation. Every time the object is to be rotated,

```
h = h + dh;
p = p + dp;
r = r + dr;
```

followed by the call to the function:

```
void pfDCSRot(pfDCS *dcs, float h, float p, float r);
```

where h, p, and r contains the updated values. This function sets the dcs to the new values of h, p, and r and rotates the object by that amount in the next frame.

Example 5: Movement Modeling in VR Factory

The VR-compatible CAD information for the static model is stored in a file (see Figure 3.25). This output is in an object-oriented format with each object representing a logical entity in a factory (e.g., a machine, a robot, an ASRS). Every object has an attribute for indicating whether it has a moving component. This attribute is utilized to represent objects using a static coordinate system (scs) or a dynamic coordinate system (dcs). To efficiently manipulate the information about objects, it is organized hierarchically in the following format:

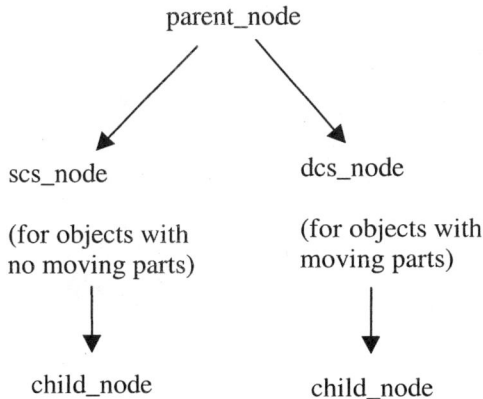

parent_node

scs_node

(for objects with
no moving parts)

dcs_node

(for objects with
moving parts)

child_node

child_node

We use the node format above as follows:

```
node_name attach_to_node  [optional data - based on
requirement]
node_name x y z h p r t
. . . . .
. . . . .
stop read_time
```

```
Sample simulation output file for immersive display interface
# name               x       y       z       h       p       r      Seconds
forklift_2_base    -175.0  -95.0   -4.4    90.0    0.0     0.0      2
stop                 2
forklift_2_base    -175.0  -95.0   -4.4   180.0    0.0     0.0      1
stop                 1
forklift_2_base    -175.0   60.0   -4.4   180.0    0.0     0.0      4
stop                 4
```

Sample Gear Factory

Forklift movement illustration

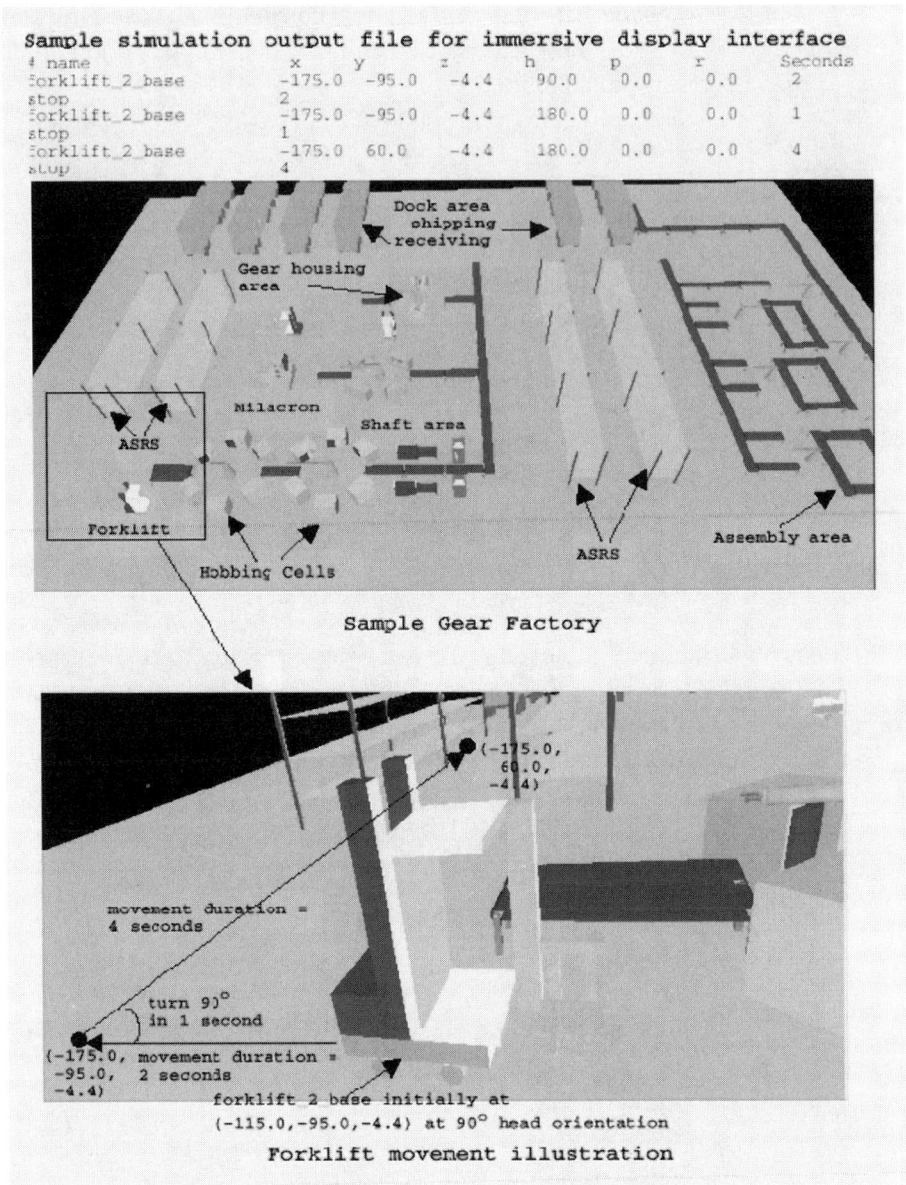

Figure 3.25 Example of movement modeling in virtual reality factory

where node_name is the name of the node to be moved or attached to another node as a child; attach_to_node is the name of the node which becomes the current parent of node_name; x, y and z are the coordinates of the destination to which the node is to move along the *x, y,* and *z* axes respectively; h, p, and r are the final head, pitch, and roll angles of the node, respectively; and t is the time in seconds to move to destination. The stop line indicates the end of input to the movement module, and the next set of simulation data will arrive after `read_time` seconds.

Figure 3.25 of a sample gear manufacturing factory illustrates the file format. The `node_name` in the example is the base of forklift truck 2. forklift_2_base is initially at $<x,y,z,h,p,r> = <-115,-95,-4.4,90,0,0>$. While at this location, the simulation output file for immersive display is read. The contents of the file instructs the truck to move to $<-175,-95,-4.4,90,0,0>$ in 2 s, then turn by a head angle of 90° at this location in the next 1 s, and then move to $<-175,60,-4.4,180,0,0>$ in the next 4 s.

For initiating the process, the data from this file are read into a C language structure (for illustrative purposes) termed `movement_path`, defined as follows:

```
typedef struct {
        char *node_name;
        float x;
        float y;
        float z;
        float h;
        float p;
        float r;
        int frame;
        int processing;
        } movement_struct;

movement_struct *movement_path;
```

The variable `processing` takes a value of either 0 or 1, indicating "not processing" or "processing done" or "being processed," respectively, for each element. The movement module is controlled by a flag, which has been termed a *simulation flag*. This flag governs the start and pause states. The flag variable is defined as

```
static int simulation;
```

The flag is initially set to 0, signifying a pause state. It can be set to 1 signifying the start state through a keyboard input of "S." It can be set back to 0 again through a keyboard input of "P."

There are two main modules for modeling motion in this example:

1. `movement_path` *loading module.* This task is performed periodically when the simulation output is read in from a file. The `read_time` variable in the `stop` line of the simulation output gives the time in seconds after which data should be read in again. This is converted to number of frames from the current frame, (`current_frame_number` + (`read_time` * `application_frame_rate`)); the movement mod-

ule should wait before reading the next set of simulation data. The `application_frame_rate` can be set externally in the application and is set to a default value of 30 frames/s. The current frame is maintained using a variable `current_frame_number` defined as

```
static long current_frame_number;
```

   This variable is incremented at the end of every frame as long as the simulation flag is set to "start." The `movement_path` loading is also responsible for detaching nodes from their current parents and attaching them to new parents. This operation is commonly used when a portion of an object is to be moved without having to move the entire object. For example, a certain number of boxes of material may be placed in a bin. Such a situation can be modeled by designating all the boxes of material as child nodes of the bin. When a forklift approaches the bin and picks up a box of raw material to transport it to another location, the box of raw material is detached from the bin and made a child of the forklift. So when the forklift moves on the floor, the box automatically moves along with it.

2. *Movement execution module.* This task is performed in every frame. This module contains the mechanics of moving the objects in an immersive environment.

   Having introduced the two modules above, more details on their operation is provided next. Both these modules become operational when the simulation flag is set to "start." The following processes are performed:

1. *Operational details of* `movement_path` *loading module.* Based on the type of simulation input, the module decides on the actions to be performed.

   a. If the input data are of type `attach_struct`, the parent of `attach_details.node_name` is removed and made a child of node `attach_details.parent`. It remains as the child of `attach_details.parent` until there is another simulation input which provides a new parent for `attach_details.node_name`.

   b. If the input is of type `movement_struct`, the following actions are performed: The array `movement_path` is made to grow dynamically while reading the simulation output data from a file according to the following logic: The module searches through the `movement_path` array and loads the first available index that has `processing` variable set to 0 with the simulation data item that has been read. It sets the `processing` flag to 1, indicating that this index is no longer free. The variable `frame` is computed as (`current_frame_number` + (`application_ frame_ rate` * `t`)), which gives the number of frames from the current frame that the movement is to continue.

The steps above are repeated for all the simulation data items until the stop line is encountered, signifying the end of input for the present cycle.

2. *Operational details of movement execution module.* This step is performed in every frame of the application. The moving objects are moved by a small amount delta in every frame. The module scans through the entire movement_path array and for each element that has processing variable set to 1, it computes the difference between the destination values stored in <x,y,z,h,p,r> variables and the current location and divides by the value of frames remaining to get the delta value. For example, to compute delta_X:

$$\text{delta\_X} = \frac{(\text{movement\_path[i].x} - (\text{x-coordinate of current\_location\_of\_object}))}{(\text{movement\_path[i].frame} - \text{current\_frame\_number})} \quad (3.3)$$

These delta values are then applied to the dcs matrix and the transformation of the object is achieved for every frame. As soon as current_frame_count is equal to movement_path[i].frame, it means that the transformation of the particular object is complete and the process sets the movement_path[i].processing flag to 0, thus making it available for future simulation data loading.

3.5 USER–OBJECT INTERACTIONS

The values of the VO-to-VE transform are made available in real time by head or hand tracking hardware that permits a VR user to explore the VE. In this scenario we have three frames of reference: the virtual world space storing the VE, the virtual observer, and the VO's hand.

Transform Notation

If a single subscript is used to remind us of the local frame of reference, the notation P_O represents a point P in its object space and P_W represents a point in world space. The transform $\mathbf{T}_{WO}$ relates the two frames of reference sharing the two subscripts and is employed as follows:

$$P_W = \mathbf{T}_{WO} P_O \quad (3.4)$$

Notice the juxtaposition of the subscripts — the W in P_W mirrors the W in $\mathbf{T}_{WO}$, while the O in $\mathbf{T}_{WO}$ mirrors the O in P_O. If the transform $\mathbf{T}_{WO}$ converts points from the frame O to the frame W, the inverse transform $\mathbf{T}_{OW}$ performs the reverse:

$$P_O = \mathbf{T}_{OW} P_W \quad (3.5)$$

3.5.1 Two-Dimensional Shape Picking

Figure 3.26 includes another frame of reference, H, in addition to W and O, which is related to W through the transform $\mathbf{T}_{WH}$, such that

$$P_W = \mathbf{T}_{WH} P_H \tag{3.6}$$

and

$$P_H = \mathbf{T}_{HW} P_W \tag{3.7}$$

We can also see that frames O and H are related as follows:

$$P_H = \mathbf{T}_{HO} P_O \tag{3.8}$$
$$P_O = \mathbf{T}_{OH} P_H \tag{3.9}$$

If we now wish to associate shape S with the movements of frame H, we can consider the action of this manipulation at two successive scene updates. Let the transform describing the current situation employ the foregoing notation and the transform describing the proposed movement be tagged with a prime ('). These two scenarios are shown in Figure 3.26.

If shape S is to share the behavior of frame H, the transform must be identical to the previous transform, $\mathbf{T}_{HO}$. Therefore, $\mathbf{T}'_{HO} = \mathbf{T}_{HO}$. In order to move shape S, the new transform $\mathbf{T}'_{WO}$ must be evaluated. Using the notation above, we can write

$$\mathbf{T}'_{HO} = \mathbf{T}'_{HW} \mathbf{T}'_{WO} \text{ and } \mathbf{T}_{HO} = \mathbf{T}_{HW} \mathbf{T}_{WO}$$

Using $\mathbf{T}'_{HO} = \mathbf{T}_{HO}$, we can write

$$\mathbf{T}'_{HW} \mathbf{T}'_{WO} = \mathbf{T}_{HW} \mathbf{T}_{WO} \tag{3.10}$$

Premultiplying both sides by $\mathbf{T}'_{WH}$, we get

$$\mathbf{T}'_{WO} = \mathbf{T}'_{WH} \mathbf{T}_{HW} \mathbf{T}_{WO} \tag{3.11}$$

and any point P_W can be evaluated as [note that $P'_O = P_O$ because any point P on shape S (or S') with respect to (w.r.t.) OCS will remain the same before and after movement]:

$$P'_W = \mathbf{T}'_{WO} P_O$$

or in its expanded form

$$P'_W = \mathbf{T}'_{WH} \mathbf{T}_{HW} \mathbf{T}_{WO} P_O \tag{3.12}$$

3.5.2 Three-Dimensional Object Picking

The two-dimensional concept of picking is now expanded to three dimensions. It is now convenient to isolate the virtual domain from the physical domain. The virtual domain is represented by the world frame and the physical domain is the room frame (attached to the CAVE room), as illustrated in

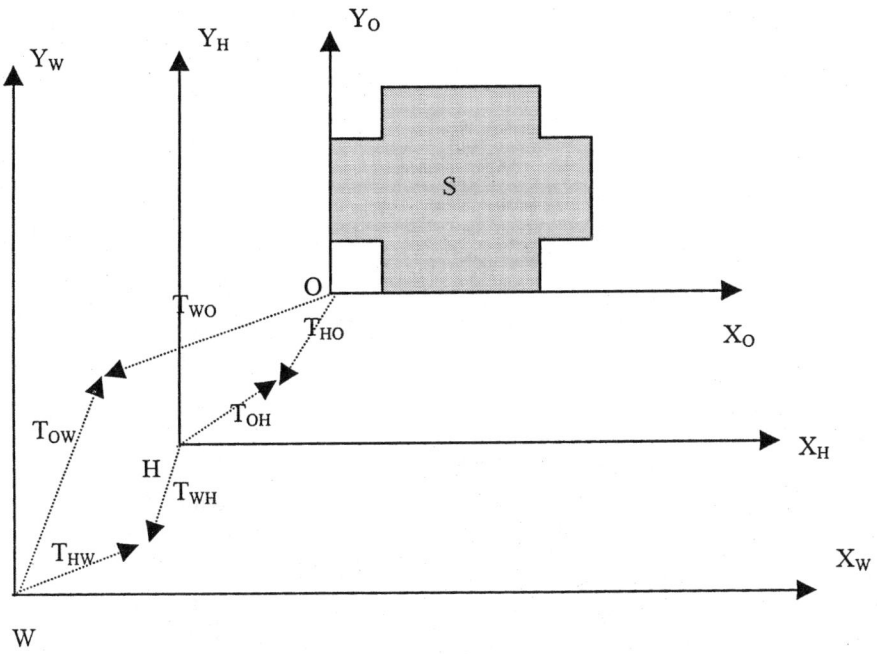

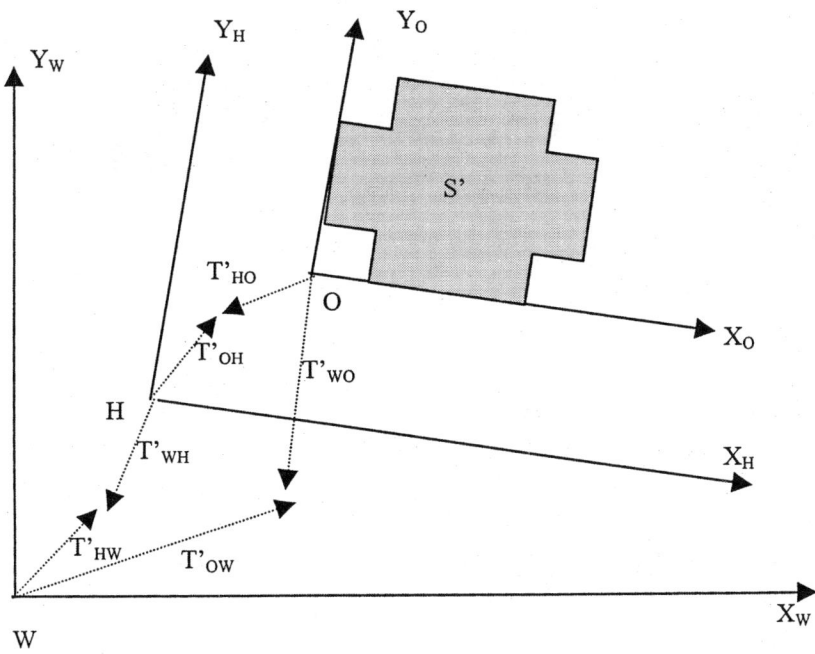

Figure 3.26 Hand *H* of VO picking up shape *S*.

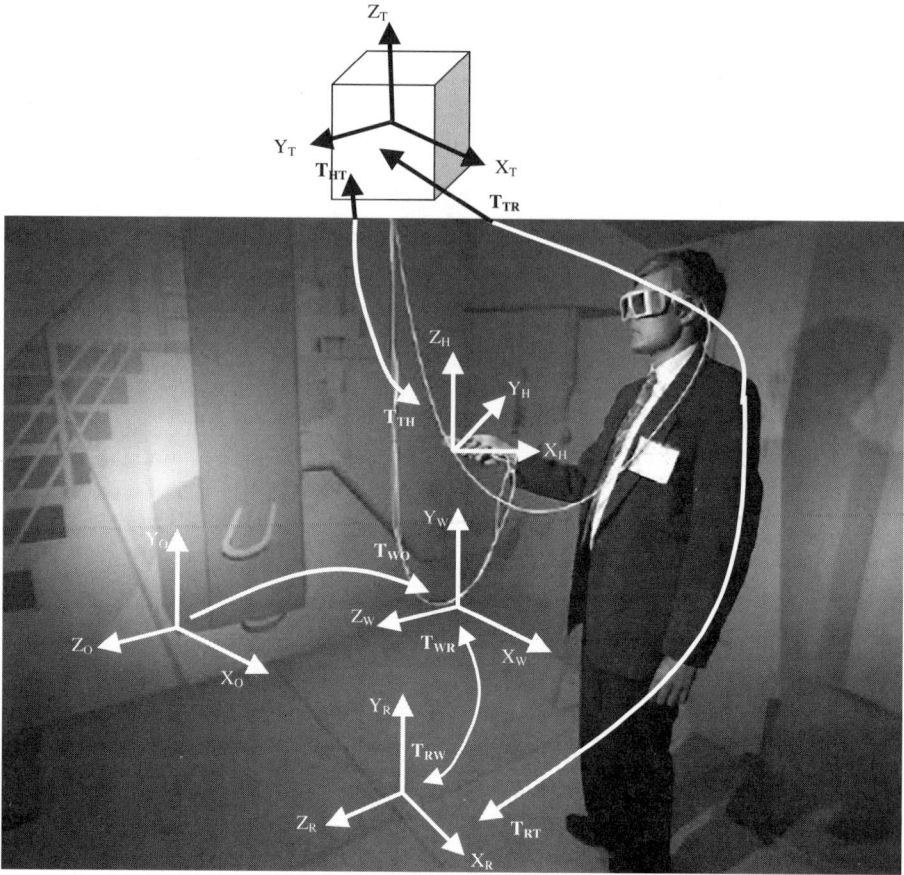

Figure 3.27 Three-dimensional object picking inside CAVE™ virtual environment.

Figure 3.27. A transform $\mathbf{T}_{RW}$ relates the two frames but plays no vital role in picking. Nevertheless, it will be used in our calculations for consistency.

The tracker emitter frame (attached to the room) is located inside the room and has a default position with respect to the room's origin. The VO also has a transform relating the head sensor (receiver) with the tracker. There is a similar transform relating the three-dimensional hand sensor with the tracker, and another transform relating an object with the origin of the VE. Note that for picking, it is mainly the VO's hand sensor that plays the most important role. Hence the head sensor is ignored, since its role is secondary.. These frames of reference are listed in Table 3.5, which shows the transform used to move from one frame of reference. $\mathbf{T}_{RW}$ is assumed invariant because the room's position is assumed static with respect to VE ($\mathbf{T}_{TR}$ and $\mathbf{T}_{RW}$ are invariant, which implies that $\mathbf{T}_{TW}$ is invariant).

Note that virtual domain is VE (or W) in Figure 3.27. While object picking is in progress, the transform relating the hand and object ($\mathbf{T}_{HO}$) is invariant. Recall from two-dimensional shape picking that $\mathbf{T}'_{HO} = \mathbf{T}'_{HW}\mathbf{T}'_{WO}$ and $\mathbf{T}_{HO} = \mathbf{T}_{HW}\mathbf{T}_{WO}$.

TABLE 3.5 VARIOUS COORDINATE TRANSFORMATIONS

		To				
		World	**Room**	**Tracker**	**Hand**	**Object**
From	World	—	$\mathbf{T}_{RW}$	$\mathbf{T}_{TW}$	$\mathbf{T}_{HW}$	$\mathbf{T}_{OW}$
	Room	$\mathbf{T}_{WR}$	—	$\mathbf{T}_{TR}$	$\mathbf{T}_{HR}$	$\mathbf{T}_{OR}$
	Tracker	$\mathbf{T}_{WT}$	$\mathbf{T}_{RT}$	—	$\mathbf{T}_{HT}$	$\mathbf{T}_{OT}$
	Hand	$\mathbf{T}_{WH}$	$\mathbf{T}_{RH}$	$\mathbf{T}_{TH}$	—	$\mathbf{T}_{OH}$
	Object	$\mathbf{T}_{WO}$	$\mathbf{T}_{RO}$	$\mathbf{T}_{TO}$	$\mathbf{T}_{HO}$	—

We can now rewrite these equations for three dimensions as follows:

$$\mathbf{T}'_{HO} = \mathbf{T}'_{HT}\mathbf{T}'_{TR}\mathbf{T}'_{RW}\mathbf{T}'_{WO} \tag{3.13}$$

$$\mathbf{T}_{HO} = \mathbf{T}_{HT}\mathbf{T}_{TR}\mathbf{T}_{RW}\mathbf{T}_{WO} \tag{3.14}$$

While object picking is in progress, $\mathbf{T}'_{HO} = \mathbf{T}_{HO}$; therefore,

$$\mathbf{T}'_{HT}\mathbf{T}'_{TR}\mathbf{T}'_{RW}\mathbf{T}'_{WO} = \mathbf{T}_{HT}\mathbf{T}_{TR}\mathbf{T}_{RW}\mathbf{T}_{WO} \tag{3.15}$$

Equation (3.15) is arrived at by noting that the hand position is normally recorded w.r.t. the tracker, and the tracker position is normally located w.r.t. the room. In our convention here, the virtual domain (or world) is located w.r.t. the physical world (or room) and the object location is normally specified w.r.t. virtual domain.

Premultiplying both sides by $\mathbf{T}'_{TH}$ produces

$$\mathbf{T}'_{TR}\mathbf{T}'_{RW}\mathbf{T}'_{WO} = \mathbf{T}'_{TH}\mathbf{T}_{HT}\mathbf{T}_{TR}\mathbf{T}_{RW}\mathbf{T}_{WO}$$

Further pre-multiplying both sides by $\mathbf{T}'_{RT}$ and finally by $\mathbf{T}'_{WR}$ produces:

$$\mathbf{T}'_{WO} = \mathbf{T}'_{WR}\mathbf{T}'_{RT}\mathbf{T}'_{TH}\mathbf{T}_{HT}\mathbf{T}_{TR}\mathbf{T}_{RW}\mathbf{T}_{WO}$$

Any point P_O associated with an object is transformed into world coordinates as follows:

$$P'_W = \mathbf{T}'_{WO}P_O$$

If the room is coincident with the world frame, $\mathbf{T}_{TR}\mathbf{T}_{RW}$ becomes $\mathbf{T}_{TW}$ and

$$\mathbf{T}'_{WO} = \mathbf{T}'_{WT}\mathbf{T}'_{TH}\mathbf{T}_{HT}\mathbf{T}_{TW}\mathbf{T}_{WO} \tag{3.16}$$

3.5.3 Flying

In flying through the VE, the VO is only required to indicate the direction of flight and instruct the system to fly in this direction until the command is terminated. Effectively, what happens is that the VE is moved past the VO in the direction opposite to that indicated. The flying direction is specified using a wand or a data glove, and the distance moved at each frame update can be a system-defined parameter or a value determined interactively by the VO.

Flying is achieved by updating the transform $\mathbf{T}_{WR}$ relating the room to the VE or world frame. Therefore, we need to find how far the room should be translated for each frame update and then move the world in the opposite direction. As the translation is defined in hand coordinates, it can be transformed into room coordinates using the following:

$$\mathbf{T}'_{\text{RtranslateR}} = \mathbf{T}'_{RT}\mathbf{T}'_{TH}\mathbf{T}'_{\text{H translate H}}\mathbf{T}'_{HT}\mathbf{T}'_{TR} \tag{3.17}$$

and the transform that can be applied repeatedly to effect flying is

$$\mathbf{T}'_{WR} = \mathbf{T}_{WR}\mathbf{T}'_{\text{R translate R}} \tag{3.18}$$

From the equations above we obtain

$$\mathbf{T}'_{WR} = \mathbf{T}_{WR}\mathbf{T}'_{RT}\mathbf{T}'_{TH}\mathbf{T}'_{\text{H translate H}}\mathbf{T}'_{HT}\mathbf{T}'_{TR} \tag{3.19}$$

FURTHER READING

For further details on VR principles and platforms, good references are Durlach and Mavor (1995) and Vince (1995). The reader interested in VRML should consult Hartman and Wernecke (1996).

REFERENCES

Durlach, N. I., and A. S. Mavor, *Virtual Reality: Scientific and Technological Challenges,* National Academy Press, Washington, DC, 1995.
Hartman, J., and J. Wernecke, *The VRML 2.0 Handbook,* Addison-Wesley, Reading, MA, 1996.
Vince, J., *Virtual Reality Systems,* Addison-Wesley, Wokingham, UK, 1995.

EXERCISES

3.1. A 21-inch computer terminal is approximated by a cube with dimensions of 0.5 m. It is being viewed by a user located 0.25 m from the perspective plane. One side of the cube is parallel to the perspective plane. The near clipping plane is 0.1 m from the user, and the far clipping plane is 3 m from the user. Find the dimensions of the side of the cube parallel to the perspective plane when it is located at a distance of **(a)** 0.2 m, **(b)** 0.25 m, **(c)** 1 m, and **(d)** 5 m from the user.

3.2. Consider a door 2 m high and 1 m wide with zero thickness (for simplicity) (Figure 3.28). The doorknob is centered at a height of 1 m and at a distance of 0.1 m from the nearest edge. The door is hinged as shown. Define a hinge-centered coordinate system, a door-centered coordinate system, and a

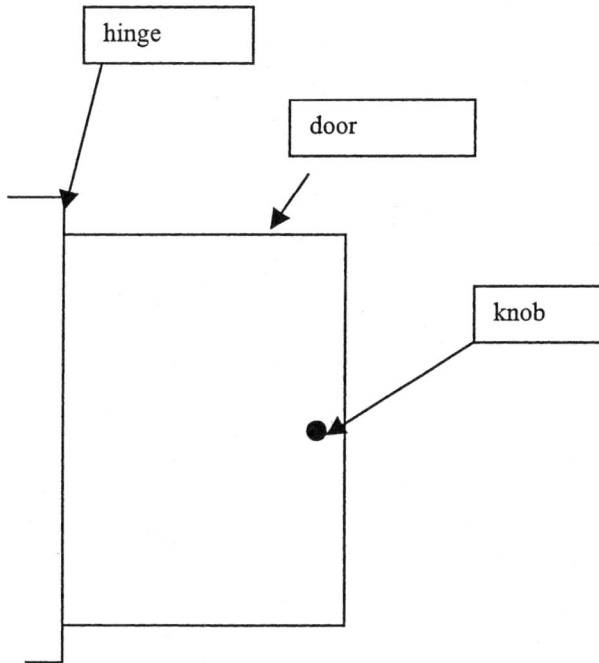

Figure 3.28 Door.

knob-centered coordinate system. Find the location of the door center and the knob center w.r.t. the hinge center when the door is closed, is half open (45°), and is completely open (90°). Consider a DCS and a SCS based on the Euler coordinate system, and use a door modeling hierarchy to show the interaction of the DCS and SCS of the three coordinate systems defined above. Indicate how one would traverse the hierarchical tree to indicate the various positions and orientations for the three door positions considered above.

3.3. How would you build a scene graph for the VR factory example given in Figure 3.21? Assume one instance of each type of equipment. How would you incorporate LOD? Suppose that you look at the scene from the front, how would you cull the database?

3.4. (a) In Figure 3.29, two coordinate systems are illustrated. Write the coordinate frame transformation (rotation plus translation) that relates the two frames. The origin of the second coordinate frame (O_2) has the coordinates (10,13) with respect to the first coordinate frame. Justify the result by considering an arbitrary point relative to the first coordinate frame and computing its coordinates relative to the second coordinate frame.

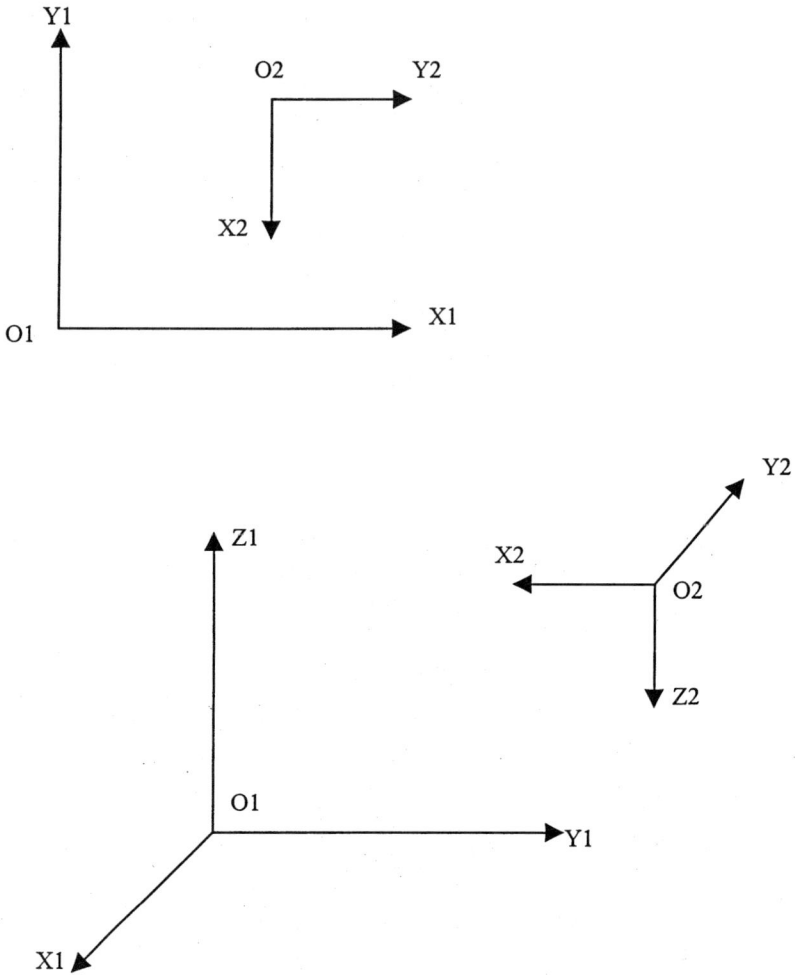

Figure 3.29 Coordinate systems used in Exercise 3.4.

(b) Two three-dimensional coordinate frames are depicted below. The relationship between the two coordinate systems is as follows:

- Axis X_2 parallel to Y_1
- Axis Z_2 parallel to Z_1
- Axis Y_2 parallel to X_1
- O_2 has coordinates (-50,100,50) relative to coordinate frame 1

Write the coordinate transformation relating the two coordinate frames and verify the result. Employ the methodology discussed in this chapter.

3.5. The following notation convention is applied below to denote the relationship between two axes of two different coordinate frames:

- $X_1 = Y_2$: axis X of the coordinate frame 1 is parallel to and has the same orientation as axis Y of the coordinate frame 2.
- $X_1 = -Y_2$: axis X of the coordinate frame 1 is parallel to and has opposite orientation relative to axis Y of the coordinate frame 2.

An object is placed in the virtual environment (VE). The location of the object is defined by means of an object-centered coordinate system, denoted OCS. The reference frame, associated with VE, is termed a world coordinate system (WCS). The following relationship exists between OCS and WCS.

$$X_O = -Y_W, \; Y_O = Z_W, \; Z_O = -X_W$$

The origin of OCS has coordinates (100,100,100) w.r.t. WCS. A user is immersed in the VE with the purpose of picking the object. Assume that the user's head remains fixed during picking. Attached to the room is a fixed coordinate system, called a tracker coordinate system (TCS). The relationship between TCS and WCS is as follows:

$$X_T = X_W, \; Y_T = Y_W, \; Z_T = Z_W$$

The origin of TCS has coordinates (50,150,50) w.r.t. WCS. The hand of the user is also tracked. The coordinate system associated with the hand is called a hand coordinate system (HCS). The object is to be picked and moved from its initial location to a different location. Accordingly, the hand coordinate frame is moved also (from HCS to HCS'). The relationships between HCS and WCS in both situations are given below.

- *Initial location:* $X'_H = X_W, \; Y_H = -Z_W, \; Z_H = Y_W$; the origin of HCS' has coordinates (150,150,150) w.r.t. WCS.

- *Second location:*
  $X'_H = -Y_W, \; Y'_H = -X_W, \; Z'_H = -Z_W$; the origin of HCS' has coordinates (150, -150,150) w.r.t. WCS.

   (a) Write the invariant transformation between HCS and OCS during the picking process.
   (b) What is the transformation between OCS and WCS in the second location of the object (OCS')?
   (c) What is the location of the head w.r.t. the object for both positions of the object?

3.6. This problem is aimed at introducing you to some aspects of virtual reality programming. Consider the following example already introduced in this chapter (the reference frame shown in Figure 3.30 is for the robot shown in Figure 3.17, whose scene graph is given in Figure 3.18). Please accomplish the

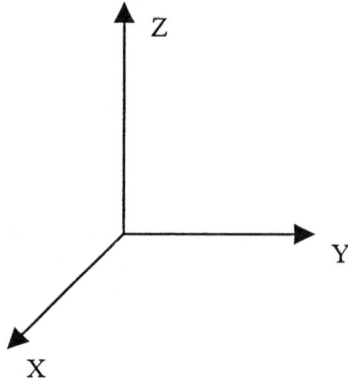

Figure 3.30 Reference frame for the robot shown in Figure 3.17.

following tasks using the scenegraph in Figure 3.18. First show the state of relevant portion of the scenegraph at the start of simulation and at the end of simulation. Then show the intermediate steps assuming a frame rate of 30 fps and a simulation duration of 1 min to accomplish each of the following tasks:

(a) Rotate head about Z (i.e., roll) by 120°
(b) Rotate left thigh and calf together by 30° w.r.t. body to take a step forward.
(c) Rotate left foot by 30° w.r.t. left thigh and calf to land flat on the ground

3.7. One of the very important applications of virtual reality is telecollaboration, that is, using two distant virtual environments to collaborate. How would you design coordinate systems and model three-dimensional picking and flying if you want to telecollaborate?

4

TELEMETRY-BASED DEPTH RECOVERY

4.1 INTRODUCTION

There are instances when virtual environments (VEs) have to be created in correlation to existing real environments. Instead of building the VE from scratch, using a CAD package or three-dimensional modeler by measuring the real objects physically, one can employ a technique that enables the fast construction of three-dimensional models from two-dimensional images. So far there has not been wide availability of such techniques, due primarily to the obstacles that hinder their development. In this chapter we present some of the solutions to this problem that lead to an integrated system aimed at building three-dimensional object models from camera images by employing stereo vision techniques. Before presenting the methodology details, we treat some of the computer vision and image processing topics that contribute to such a system.

4.2 RECOVERING THE THIRD DIMENSION FROM STEREO

An image is a two-dimensional projection of the three-dimensional world. Without dealing too much with the internal physics of a videocamera, we can approximate mathematically the mechanism of image formation as a perspective projection having as the center of projection the focal center of the camera lens. Such a camera model is termed *perspective camera* or *pinhole camera* (Figure 4.1). We can see from Figure 4.1 that by considering the inverse perspective projection (from *I* to *P*), point *I* can be the image of an infinite number of points located on the line *IV*. Therefore, a single image of a scene cannot reveal any three-dimensional information about that scene. If we imagine a second image of point *P* (say *I'*) with center of projection *V'* (see Figure 4.2), by intersecting *IV* and *I'V'* in space we can obtain the point *P* precisely. Therefore, *for recovering the third dimension* (also termed *depth*) *from two-dimensional images, at least two images of the same object are required.*

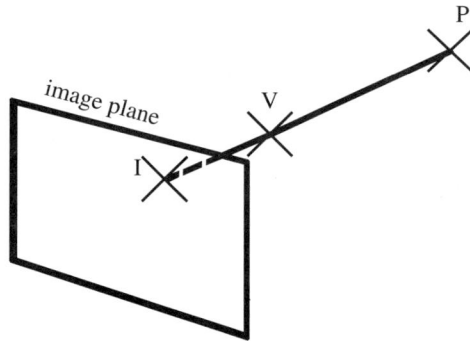

Figure 4.1 Perspective camera model: point *P* is projected onto the image plane at point *I*, through center of projection *V*.

To better illustrate the mathematics of depth computation, we can assume that the two image planes are perfectly aligned. A planar view of parallel camera configuration is shown in Figure 4.3. The purpose of depth computation is to retrieve the coordinates of point *P* relative to an object-centered coordinate system, termed world coordinate system (WCS). Also, the image plane has its

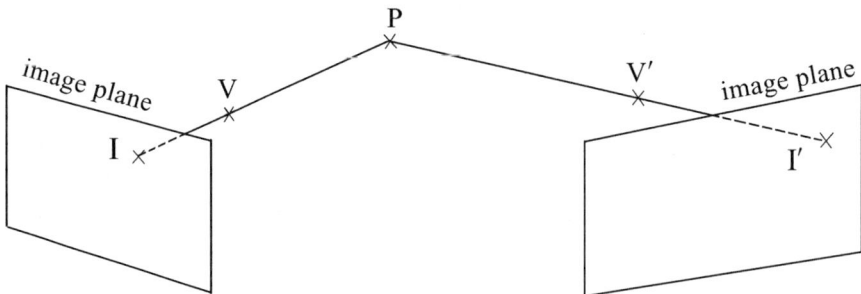

Figure 4.2 Recovery of the third dimension from stereo.

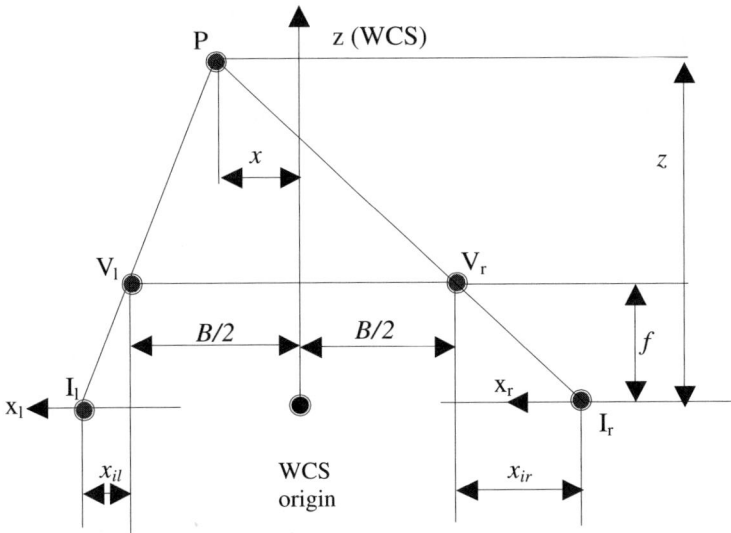

Figure 4.3 Parallel stereo configuration.

own coordinate system, called image coordinate system (ICS; more details about the coordinate systems involved in stereoscopy are given in Section 4.4.1). In this simplified stereo configuration, without any loss of generality, we consider WCS to be aligned with both left and right ICS. Thus the Z axis of WCS is parallel to the optical axes of the cameras, and by *depth* we understand the distance between point P and the image planes. Also, in this configuration, the line joining the two viewpoints is parallel to the image planes and its length is called *baseline*. Without any loss of generality, we can assume that the origin of WCS is located at the midpoint of the baseline. From Figure 4.3 we can write, based on the similarity of triangles, the following equations:

$$\frac{x_{il}}{f} = \frac{B/2 - x}{z - f}$$

$$\frac{x_{ir}}{f} = -\frac{B/2 + x}{z - f}$$

(4.1)

In equation (4.1), f is the focal length (assumed to be identical for both cameras), B the baseline, and x and z the corresponding coordinates of P with respect to WCS. From equation (4.1), the depth z is derived as follows:

$$z = f - \frac{fB}{x_{ir} - x_{il}}$$

(4.2)

In equation (4.2), the quantity $x_{ir} - x_{il}$ is called the *disparity* between two images and is, along with baseline and focal length, a major factor that influences the depth. We discuss depth recovery in the general stereo configuration (the image planes are not aligned and the optical axes are not parallel) in Section 4.4.4.

4.3 FEATURE EXTRACTION AND MATCHING

To build a three-dimensional model from two-dimensional images, certain features belonging to the visualized objects have to be extracted from the images. This is usually achieved by applying image processing algorithms. The extracted features can be edges, corners (vertices), and so on. After features are extracted in, say, two images of a stereo pair, they have to be matched automatically so that the algorithm knows which feature in one image corresponds to which feature in the other image. This is by far the most difficult problem associated with stereo vision. Unfortunately, perfect stereo matching is still impossible to achieve even by the state-of-the-art algorithms, so to create perfect models from stereo, some user intervention is still required. Therefore, the process of depth recovery from stereo stops short from being fully automatic. In this work, the features we extract from objects are corners and we will concentrate on matching corners. Some preliminary background material that is useful to the process of feature extraction and matching is presented first.

4.3.1 Digital Image Quantization

An image is a two-dimensional projection of a three-dimensional world. The aspect of an image is influenced by the light reflectance properties (brightness) of the objects in the image. In the ideal case, an image represents a continuous light intensity function $f(x,y)$, where x and y are spatial coordinates and the value of f at each point (x,y) is a measure of the brightness (gray level in the case of noncolor images, which we address mostly in this book) of the image at that point.

An image representation as a continuous light-intensity function is not suitable for computer implementation and processing of the image. Therefore, in digital image-processing applications, a discrete representation of function $f(x,y)$ is used. An image is discretized both in spatial coordinates and in brightness as a digital array. The basic element of such an array is called a *pixel,* whose spatial coordinates are associated with a point in the image, and its value represents the light intensity at that point (an integer value between 0 and 255 for gray-level images). This representation of an image enables the implementation of typically continuous operations on images (such as noise

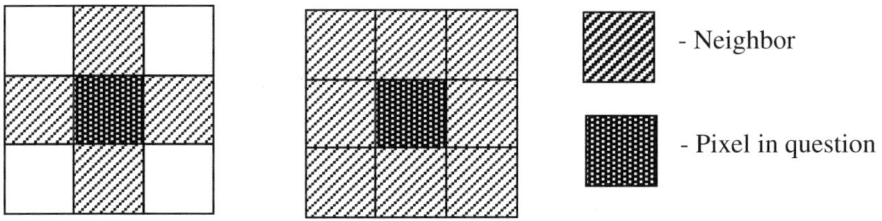

Figure 4.4 4- and 8-neighbors of a pixel.

suppression through filtering, edge detection, etc.) on a computer. Often, image operations are performed on one pixel in conjunction with the pixels neighboring the pixel in question. There are two types of pixel neighborhoods defined in digital image processing applications: 4- and 8-neighborhoods, illustrated in Figure 4.4.

In general, the output of an image operation is another image, which is different from the original image. Image operations (also known as *operators*) are applied for a wide range of purposes: image enhancement, noise reduction (filtering), edge detection, and so on. The image operations are applied on every pixel of the image (typically, starting from the upper left pixel) and, as a result of an operation, the value of the pixel on which the operator is applied is changed. There are two types of image operations:

1. *Point operations.* The output at each pixel depends only on the pixel value before applying the operator. The most popular point operation is thresholding, which is discussed later in this chapter.

2. *Neighborhood operations.* The output at each pixel depends on the initial value of the pixel and on the initial values of the pixels located in a neighborhood of the pixel on which the operator is applied. Customarily, the neighborhood has the shape of a square, with the pixel in question being the center of the square. Such a neighborhood is called a *mask,* and its size varies according to the purpose of the operator. The most widely used masks have the size of 3×3 pixels. Circular masks can also be used for certain purposes, especially edge and corner detection.

A widely used neighborhood image operation is *convolution.* Convolution operation for discrete domain is given by

$$f'(i, j) = fh = \sum_m \sum_n f(m,n)h(m,n) \qquad (4.3)$$

where f is the initial image intensity function and h is the convolution mask (operator). The summation is taken over the area where f and h overlap.

4.3.2 Image Filtering

Filtering is applied for the purpose of attenuating the effect of noise in the image, but this objective is achieved at the expense of degrading the image quality.

Mean Filtering

The noise in the image has high spatial frequency. The spatial frequency of the noise is related to the sampling frequency of the image acquisition system (the system that captures the image and converts it from analog to digital). It seems logical that to remove the noise, one has to apply a low-pass spatial filter that will attenuate the high spatial frequencies and allow only the low spatial frequency components to pass through to the output image. This implies that the high-frequency components of the image itself (such as sharp boundaries and corners) will be degraded after filtering.

There are filtering operations when convolution, discussed in Section 4.3.1, comes into action. The essence of convolution operation is to apply a convolution mask to each pixel in the image. The most typical average mask in shown in Figure 4.5. The value of each mask element has to be weighted equally to produce an averaging effect. The elements of an average mask should add up to 1. The output of the convolution operation superimposed on a 3 × 3 neighborhood of the image replaces the value of the center pixel of that neighborhood in the output image. Convolution is applied successively at each pixel in the input image by "sliding" the mask over the entire image. The pixels adjacent to the image boundary are either ignored or we can zero-pad the image to compensate for the boundary effect. This is valid for all operations involving the application of masks over images.

A numerical example of *average filtering* is shown in Figure 4.6. The center pixel of

1/9	1/9	1/9
1/9	1/9	1/9
1/9	1/9	1/9

Figure 4.5 Local average mask.

99 * 1/9	100 * 1/9	102 * 1/9	102	103	101
107 * 1/9	145 * 1/9	124 * 1/9	126	127	125
129 * 1/9	131 * 1/9	134 * 1/9	134	134	133
130	131	131	133	132	134
135	134	134	133	137	136
141	144	139	145	143	143

Figure 4.6 Average filter mask: numerical example.

the upper left 3 × 3 neighborhood of the image in Figure 4.6 is replaced, after convolution, with the following value:

$$(99 \times \tfrac{1}{9}) + (100 \times \tfrac{1}{9}) + (102 \times \tfrac{1}{9}) + (107 \times \tfrac{1}{9}) + (145 \times \tfrac{1}{9}) + (124 \times \tfrac{1}{9})$$
$$+ (129 \times \tfrac{1}{9}) + (131 \times \tfrac{1}{9}) + (134 \times \tfrac{1}{9}) = 119$$

As can be seen, the value 145 for the center pixel becomes 119 after filtering. This illustrates the smoothing effect that average filtering has on an image. An example of mean filtering on a real noisy image is shown in Figure 4.7. Figure 4.7*a* shows the original noisy image, and the output of the mean filter is shown in Figure 4.7*b*. We can see that the noise has been reduced (although not entirely removed), but the image becomes blurred after filtering.

(a)   (b)

Figure 4.7 Average filtering example: (*a*) original noisy image; (*b*) image after filtering.

Median Filtering

In *median filtering,* a center pixel of a neighborhood (of any size) is assigned the value of the median of pixel intensities forming the neighborhood. Median filtering introduces less blurring on the image and therefore is superior to average filtering, but it is computationally expensive and therefore is not widely used for image filtering. Figure 4.8 shows the result of applying a median filter on the image from Figure 4.7*a*. The outcome is a slightly less blurred image than the one in Figure 4.7*b*.

Figure 4.8 Median filtering example on the image from Figure 4.7*a*.

Gaussian Filtering

Gaussian filtering (also known as *Gaussian smoothing*) is perhaps the most widely used image filter, because of its useful properties, such as rotational symmetry and suitability for multiresolution implementations. A two-dimensional Gaussian function is represented by:

$$g(x,y) = \exp\left(-\frac{x^2 + y^2}{2\sigma^2}\right) \tag{4.4}$$

where σ defines the resolution (scale) of the Gaussian function. This function is approximated by a convolution mask (by the method described below) and applied at every pixel in the image. Gaussian functions have some important properties that make them useful in image processing applications. Some of these properties are as follows:

1. *Rotational symmetry.* In two dimensions, Gaussian functions are rotationally symmetric. This can be proven by transforming the expression (4.4) from Cartesian coordinates (x,y) to polar coordinates (r,θ). It is well known that $r^2 = x^2 + y^2$, so equation (4.4) becomes:

$$g(r,\theta) = \exp\left(\frac{-r^2}{2\sigma^2}\right) \tag{4.5}$$

which shows that $g(r,\theta)$ does not depend on θ and consequently is rotationally symmetric. This property tells us that the amount of smoothing performed by the Gaussian filter is the same in all directions and it will not bias subsequent edge detection (see Section 4.3.4) in any direction.

2. *Separability.* A two-dimensional Gaussian function can be separated as the product of two one-dimensional Gaussian functions. Let us consider the convolution of the image $f(x,y)$ with a gaussian $g(x,y)$:

$$g(x,y)f(x,y) = \sum_{k=1}^{m}\sum_{l=1}^{n} g(k,l)f(x,y) = \sum_{k=1}^{m}\sum_{l=1}^{n} e^{-\frac{k^2}{2\sigma^2}} e^{-\frac{l^2}{2\sigma^2}} f(x,y)$$

$$= \sum_{k=1}^{m} e^{-\frac{k^2}{2\sigma^2}} \left[\sum_{l=1}^{n} e^{-\frac{l^2}{2\sigma^2}} f(x,y)\right] \tag{4.6}$$

In equation (4.6), the summation in brackets is the convolution of the input image $f(x,y)$ with a one-dimensional Gaussian. The result of this convolution is a two-dimensional image, blurred in the vertical direction, which is then used as the input to a second convolution with a one-dimensional Gaussian, which blurs the image in the horizontal direction.

3. *Multiresolution.* In equation (4.4), the parameter σ is called the *space constant* of the Gaussian filter and controls the amount of smoothing performed by the filter. In literature, the value of σ is called the *resolution* of the

gaussian filter. The larger the σ, the higher is the amount of noise filtered, but at the same time the larger is the degradation of the image quality by the loss of fine details (especially sharp corners). This can affect the performance of subsequent operations, such as edge and corner detection. The exact size of a Gaussian filter cannot be determined unless we have some prior knowledge of the size and location of objects in an image. A possible approach to overcome the trade-off between the quality of the image after filtering and the amount of noise filtered is to apply Gaussians with different resolutions (an approach called *scale-space filtering*). At larger scales (large σ) we get robust but displaced edges and also a reduced amount of noise in images. The real locations of the edges can be determined accurately at smaller scales.

Discrete Approximation of a Gaussian Filter

A discrete approximation of a Gaussian is provided by the coefficients of the binomial expansion, as follows:

$$(1+x)^n = \binom{n}{0} + \binom{n}{1}x + \binom{n}{2}x^2 + \cdots + \binom{n}{n}x^n \tag{4.7}$$

For a mask of width n, we use the coefficients of the binomial expansion of exponent $n - 1$. For example, for $n = 3$, a one-dimensional mask will look as follows: $[1\ 2\ 1]$ ($[{}^2C_0\ {}^2C_1\ {}^2C_2]$) and for $n = 5$, the corresponding one-dimensional mask is: $[1\ 4\ 6\ 4\ 1]$. The two-dimensional mask is obtained by convolving the horizontal one-dimensional mask with the vertical one, as follows:

$$n = 3:$$

$$\begin{bmatrix} 1 & 2 & 1 \\ 1 & 2 & 1 \\ 1 & 2 & 1 \end{bmatrix} * \begin{bmatrix} 1 & 1 & 1 \\ 2 & 2 & 2 \\ 1 & 1 & 1 \end{bmatrix} = \begin{bmatrix} 1 & 2 & 1 \\ 2 & 4 & 2 \\ 1 & 2 & 1 \end{bmatrix}$$

$$n = 5:$$

$$\begin{bmatrix} 1 & 4 & 6 & 4 & 1 \\ 1 & 4 & 6 & 4 & 1 \\ 1 & 4 & 6 & 4 & 1 \\ 1 & 4 & 6 & 4 & 1 \\ 1 & 4 & 6 & 4 & 1 \end{bmatrix} * \begin{bmatrix} 1 & 1 & 1 & 1 & 1 \\ 4 & 4 & 4 & 4 & 4 \\ 6 & 6 & 6 & 6 & 6 \\ 4 & 4 & 4 & 4 & 4 \\ 1 & 1 & 1 & 1 & 1 \end{bmatrix} = \begin{bmatrix} 1 & 4 & 6 & 4 & 1 \\ 4 & 16 & 24 & 16 & 4 \\ 6 & 24 & 36 & 24 & 6 \\ 4 & 16 & 24 & 16 & 4 \\ 1 & 4 & 6 & 4 & 1 \end{bmatrix}$$

The multiplications above are point-by-point, not matrix multiplications. In Figure 4.9 an example of Gaussian filtering on the image from Figure 4.7a is shown.

Figure 4.9 Gaussian filtering example on the image from Figure 4.7a.

4.3.3 Image Segmentation

Image segmentation is the partition of an image into meaningful regions, as dictated by a particular application. Segmentation is a complex problem, which has not yet been fully solved. In general, there are two types of segmentation techniques: region-based segmentation and boundary-based segmentation. These approaches are complementary. *Region-based segmentation* techniques are concerned with grouping individual pixels sharing a common feature into connected regions. Typically, a region represents a distinct object in an image. Among the region-based segmentation techniques are thresholding and region growing. *Boundary-based segmentation* techniques aim at detecting the boundary pixels of objects in an image and subsequently isolating them from the rest of the image. The boundary (edge) detection process is usually just an intermediate step toward higher-level goals, such as retrieving the shape and location of objects in a three-dimensional scene.

Thresholding
Thresholding, the simplest and most popular region-based segmentation technique, converts a gray-level image into a binary image, with only two levels of gray (black – gray level 0 and white – gray level 255). The output image after thresholding is as follows:

$$f'(x, y) = \begin{cases} 0, f(x,y) < T \\ 255, \ f(x,y) \geq T \end{cases} \tag{4.8}$$

where T is the preset threshold and $f(x,y)$ is the intensity function of the input image. Typically, the image intensities are normalized so they vary between 0 and 1. In this case, after thresholding we would have a binary image with only 0 and 1 pixels. As can be seen from equation (4.8), thresholding works well when there is a good contrast between an object and the background and when the object exhibits a uniform gray level (Figure 4.10). As illustrated by equation (4.8) and Figure 4.11, thresholding does not work properly for images

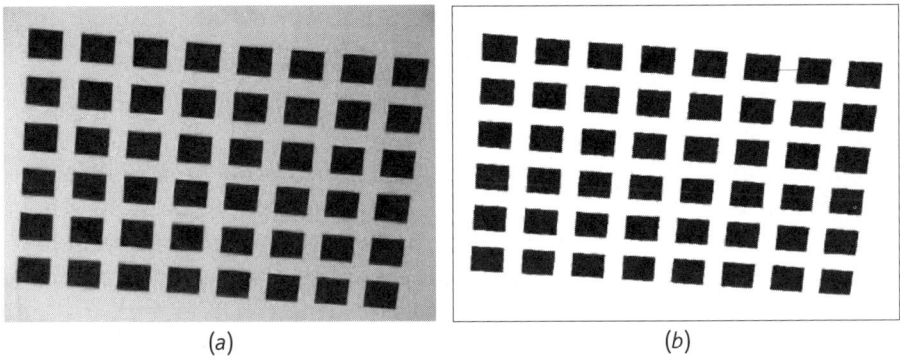

(a)   (b)

Figure 4.10 Thresholding example: (a) original image; (b) thresholded image.

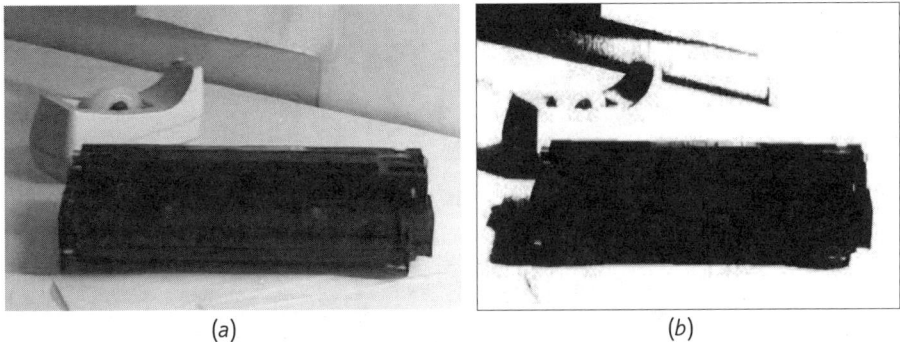

(a)   (b)

Figure 4.11 Thresholding does not work well for gradual variation in image intensity: (a) original image; (b) thresholded image.

portraying objects with a gradual change in image intensity. For this type of situation, more advanced techniques have been developed. We discuss adaptive thresholding next.

Adaptive Thresholding

Adaptive thresholding begins by partitioning the image into disjoint blocks that are typically rectangular (the blocks are specified by the user). The resolution of partitioning is dictated by the degree of variation in image intensity over some areas in the image. The resolution need not be uniform over the entire image. The algorithm for adaptive thresholding is as follows (for each image block in parallel):

1. Initialize the threshold as the average intensity in that specific block.
2. Threshold the image block using the threshold obtained in step 1.
3. Partition the pixels of the image block into two disjoint sets, resulting from the thresholding operation applied in step 2.
4. Threshold the newly formed sets (using the original intensity values) with the average intensity of the pixels forming the set.

5. Partition the pixels of each set into two disjoint subsets, resulting from the previous thresholding operation.

6. Repeat steps 4 and 5 until no change in threshold for two successive steps occurs.

An example of adaptive thresholding is provided in Figure 4.12 (the same original image as the one in Figure 4.11). As can be seen, adaptive thresholding tends to lose uniformity of regions with relatively uniform gray levels, but it captures features that a regular thresholding operation cannot capture in entirety (such as the toner's cavities and tape dispenser in its entirety in Figure 4.12). Usually, it is recommended to use different resolutions for partitioning, depending on the degree of intensity variation, but such an approach would be more difficult to implement.

Figure 4.12 Adaptive thresholding.

Connected Component Labeling

Connected component labeling is the operation of grouping together pixels belonging to the same region in an image after thresholding. In this way, each region is assigned its own label so it can be identified easily in subsequent operations. The labeling process takes place by scanning the image left to right, top to bottom. For each 1-pixel (since labeling is performed after thresholding, we can have only 1- and 0-pixels) the algorithm looks at the above and left 4-neighbors (which are the only ones already visited while scanning) and tries to assign the same label to the pixel in question. In case the above and left pixels are 0-pixels, a new label is assigned. Once a new label is assigned, it is introduced in an *equivalence table.* The purpose of this table is to keep track of equivalent labels (if the above and left pixels of a pixel in question have different labels, these labels are said to be *equivalent,* and their corresponding rows in the equivalence table are merged).

The algorithm described below is also called a *two-pass algorithm* and con-sists of the following steps:

1. Scan the image line by line, left to right, top to bottom.
2. If the pixel is 1, then:
   a. If only one of its upper and left 4-neighbors is labeled, assign this label to the pixel in question.
   b. If both above and left 4-neighbors have the same label, assign it to the pixel in question.
   c. If the above and left 4-neighbors have different labels, assign the label from above and merge the corresponding rows of the equivalence table.
   d. If none of the above occurs, assign a new label to the pixel in question and enter it in the equivalence table.
3. Repeat step 2 until no more pixels are left.
4. *Second pass.* For each row of the equivalence table (a row of the equivalence table stores equivalent labels, i.e., labels of pixels belonging to the same region), find the smallest label and assign it to all the pixels labeled with labels contained in this row. In this way, all the pixels of a region have the same label.

The example shown in Figure 4.13 represents a hypothetical binary image. The equivalence table after the first pass looks like Table 4.1. Note that

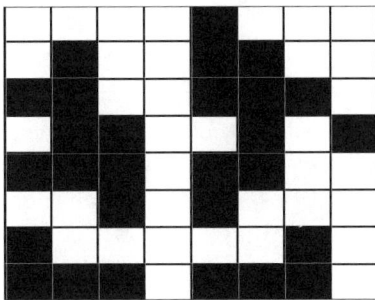

(a)

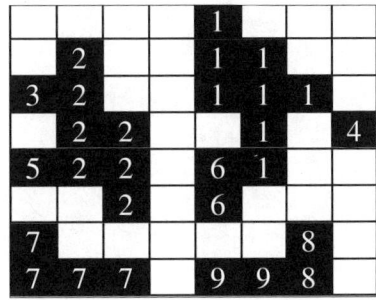

(b)

(c)

TABLE 4.1
Equivalence Table

1	6	
2	3	5
4		
7		
8	9	

Figure 4.13 Connected component labeling: (a) original image; (b) labeling after first pass; (c) labeling after second pass.

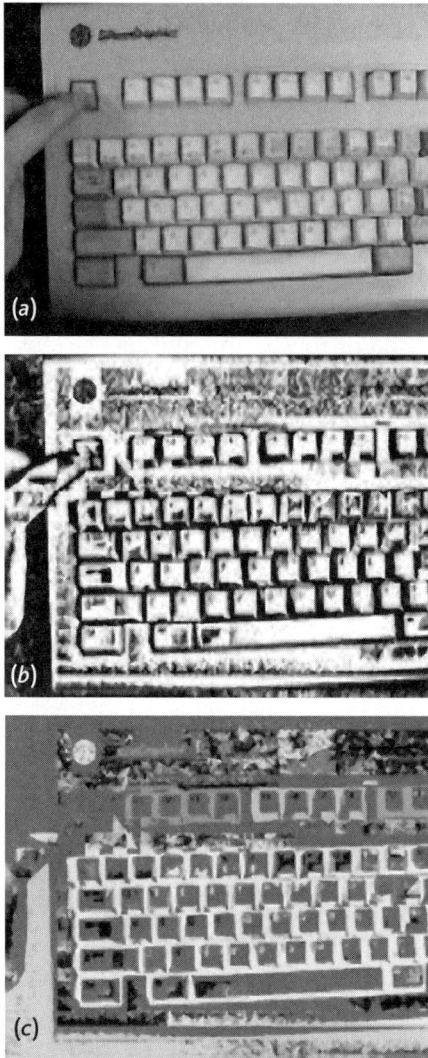

Figure 4.14 Connected component labeling; (a) original image; (b) image after adaptive thresholding; (c) regions after connected component labeling.

in Figure 4.13, the pixel labeled "4" is not assigned to the same region as the pixels labeled "1," due to the fact that the two-pass algorithm uses 4-connectivity. A more complex example is illustrated in Figure 4.14, where regions labeled by the algorithm are colored with different shades of gray. The labeling process has been performed after applying an adaptive thresholding algorithm on the image shown in Figure 4.14a.

4.3.4 Edge Detection

Edges in the image are located where sharp discontinuities in image intensity occur. There are different types of edges, according to the intensity profile in the neighborhood of the edge points, as shown in Figure 4.15. In this book we are concerned primarily with step edges.

Gradient Operators

If we consider a continuous image $f(x,y)$, we can conclude that the edge points will be located at points where the gradient magnitude of f is large (exceeds a preset threshold). In other words, the derivative of $f(x,y)$ assumes a local maximum in the direction of the gradient (which is perpendicular on the direction of the

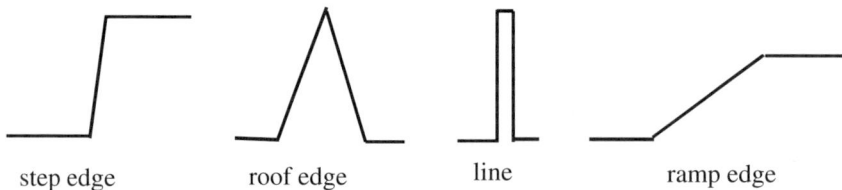

step edge   roof edge   line   ramp edge

Figure 4.15 Different types of edges.

edge). Therefore, a possible approach for edge detection would be to compute the gradient of the intensity function at each pixel and to classify pixels as edge pixels where the magnitude of the gradient exceeds a preset threshold. The gradient magnitude of $f(x,y)$ is given by

$$\sqrt{\left(\frac{\partial f}{\partial x}\right)^2 + \left(\frac{\partial f}{\partial y}\right)^2} \tag{4.9}$$

and the gradient direction θ (which is perpendicular on the edge direction) is given by

$$\theta = \arctan \frac{\partial f/\partial y}{\partial f/\partial x} \tag{4.10}$$

The gradient can be estimated if the directional derivatives along any two orthogonal directions are known. Of course, to determine the gradient value, we need a discrete approximation of the gradient, suitable for computer implementations. Often, the gradient is approximated by

$$\left|\frac{\partial f}{\partial x}\right| + \left|\frac{\partial f}{\partial y}\right| \tag{4.11}$$

Let us consider a 3 × 3 neighborhood, centered at the pixel with value $f(x,y)$. Consequently, the values of the 8-neighbors of the center pixels are represented as in Figure 4.16. The directional derivative in the positive x direction at the center pixel can be approximated as: $f(x + 1, y) - f(x,y)$, and the directional derivative in the positive y direction at the center pixel can be approximated as: $f(x, y + 1) - f(x,y)$. One of the first gradient operators is due to Roberts. The *Roberts operator* estimates the derivatives diagonally over a 2 × 2 neighborhood, with the magnitude of the gradient at point with intensity $f(x,y)$ being approximated by taking the root mean square (RMS) of the directional derivatives as follows [the following equation is actually an absolute value approximation of the RMS of the second derivatives, similar to the approximation of the gradient from equation (4.11)]:

$$R(x, y) = |f(x, y) - f(x + 1, y + 1)| + |f(x, y + 1) - f(x + 1, y)| \tag{4.12}$$

$f(x{-}1,y{-}1)$	$f(x{-}1,y)$	$f(x{-}1,y{+}1)$
$f(x,y{-}1)$	$f(x,y)$	$f(x,y{+}1)$
$f(x{+}1,y{-}1)$	$f(x{+}1,y)$	$f(x{+}1,y{+}1)$

Figure 4.16 Representation of the image intensity function in a 3 × 3 neighborhood.

(a)   (b)

Figure 4.17 Roberts operator: (a) original image; (b) edges found.

Figure 4.18
Roberts operators.

0	1
-1	0

1	0
0	-1

An example of output of the Roberts operator is illustrated in Figure 4.17. The masks representing the Roberts operator in two orthogonal directions are shown in Figure 4.18. As can be seen from Figure 4.18, the elements of the entire mask add up to zero, in order for the operator to produce no response in uniform areas of the image, where no edge is present. This is valid for all edge operators we encounter in this chapter. The problem with the Roberts operator is that it is a 2 × 2 operator and therefore cannot be applied in a unique fashion (i.e., cannot be "centered" at a unique pixel). This fact, combined with the fact that it is based solely on the first-order differences, makes the Roberts operator very susceptible to noise.

Another gradient operator has been proposed by Sobel, a 3 × 3 operator based on a combination between a difference operation and a local averaging. The *Sobel operators* in x and y directions are as follows:

$$S_x = [f(x+1, y-1) + 2f(x+1, y) + f(x+1, y+1)] - [f(x-1, y-1) + 2f(x-1, y) + f(x-1, y+1)]$$

$$S_y = [f(x-1, y+1) + 2f(x, y+1) + f(x+1, y+1)] - [f(x-1, y-1) + 2f(x, y-1) + f(x+1, y-1)]$$

(4.13)

The gradient may be estimated as $|S_x| + |S_y|$. The Sobel operators in two orthogonal directions are shown in Figure 4.19, and an example of applying Sobel operators is provided in Figure 4.20.

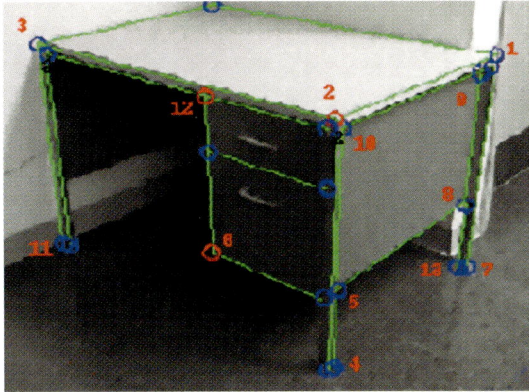

Figure 4.31 B-spline-based corner detection example.

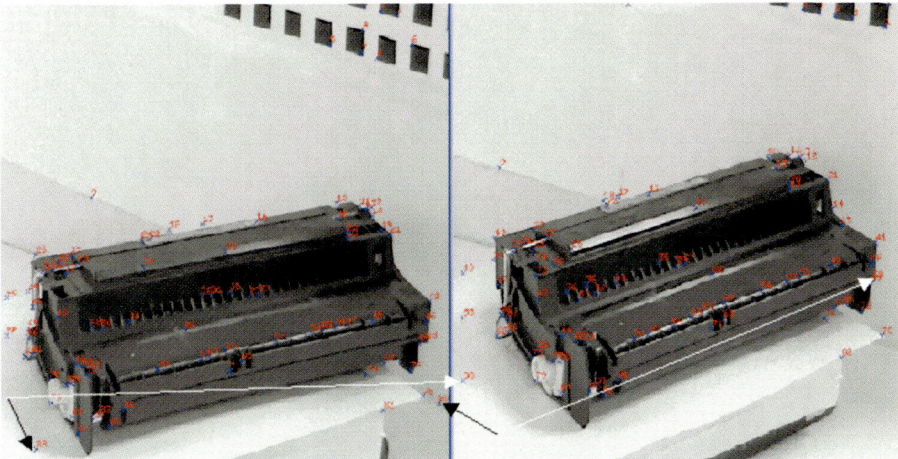

Figure 4.34 Matches through correlation.

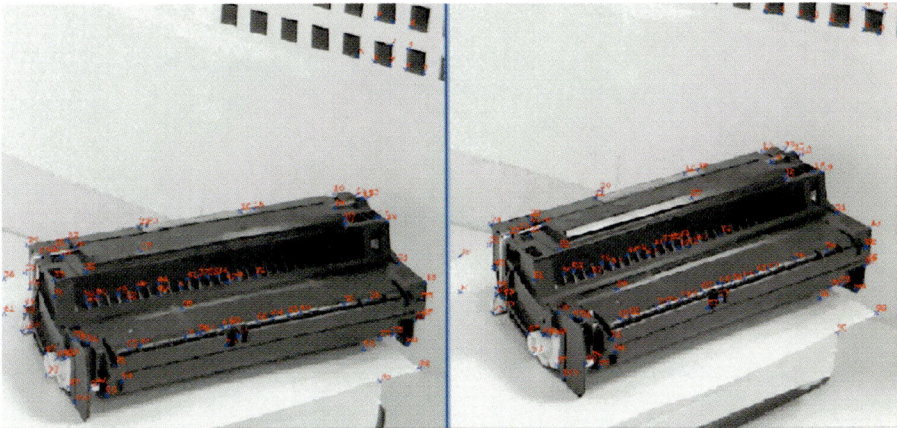

Figure 4.35 Disambiguating matches through relaxation.

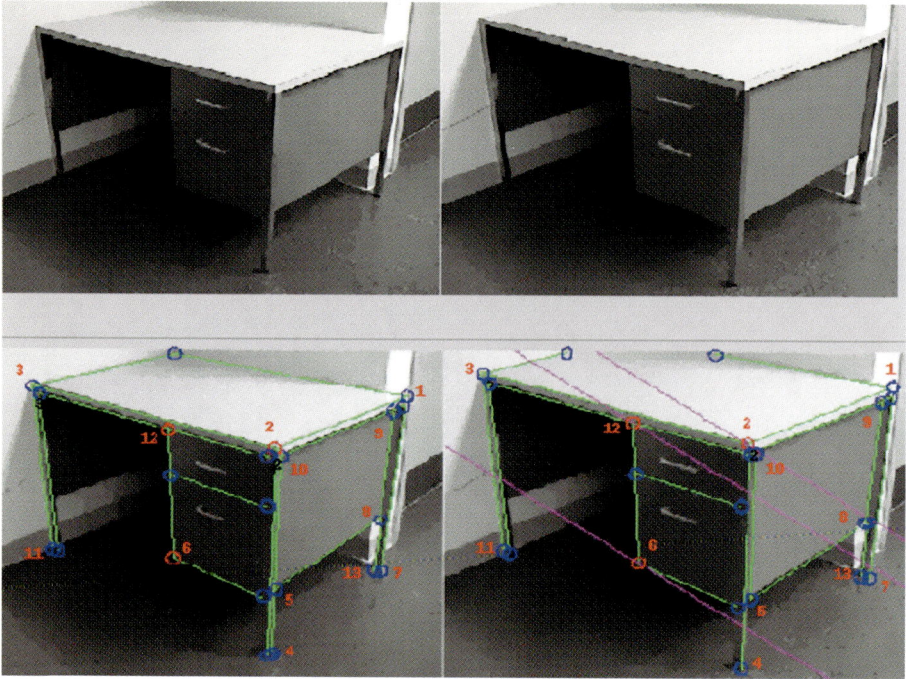

Figure 4.45 Images used in reconstruction and extracted points.

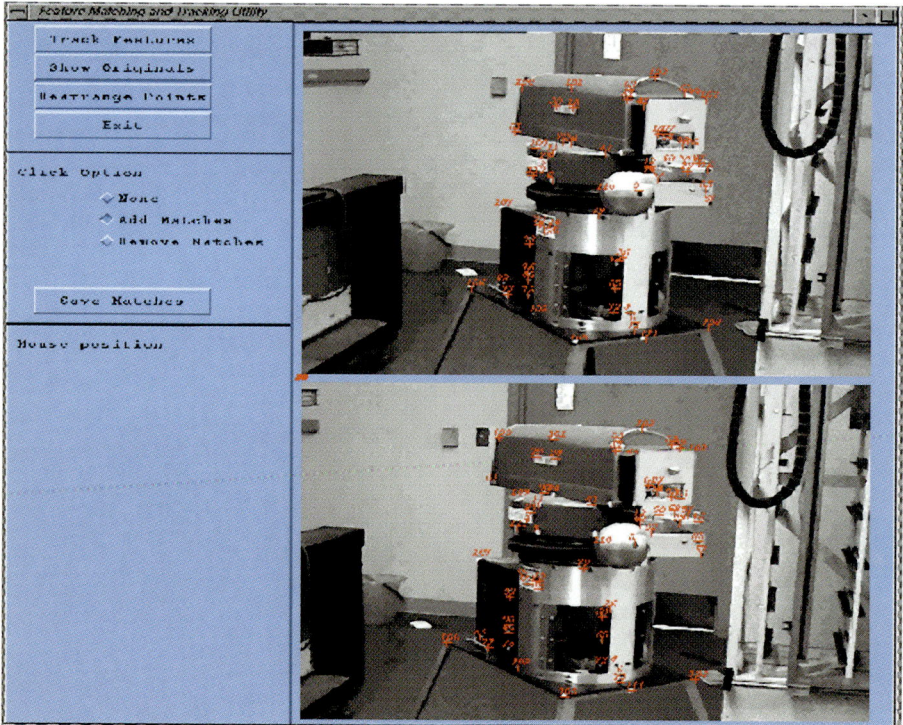

Figure 4.51 Points extracted from the object shown in Figure 4.50.

(a)

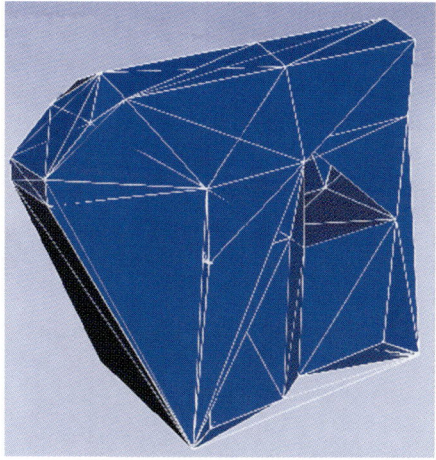

(b)

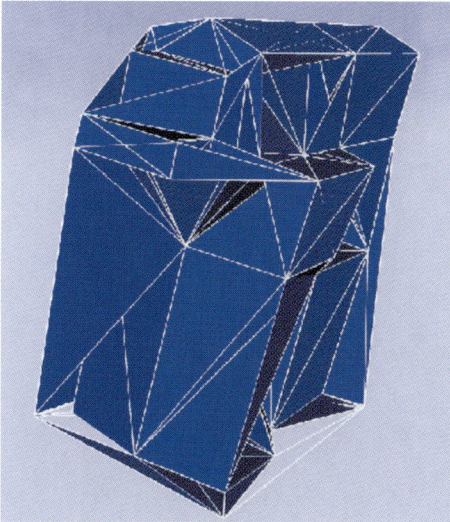

(c)

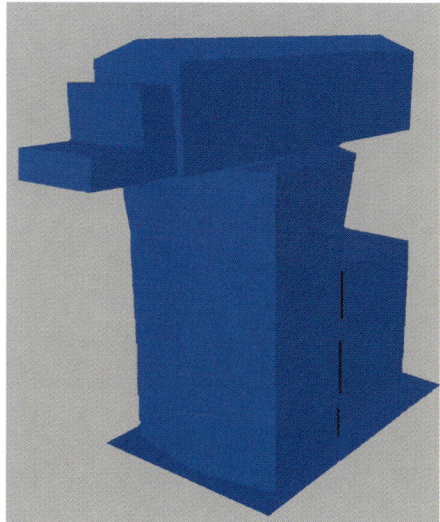

(d)

Figure 5.23 Object recovered with our methodology: *(a)* original object; *(b)* intermediate shape from one viewpoint; *(c)* intermediate shape from seven viewpoints; *(d)* final shape (17 viewpoints).

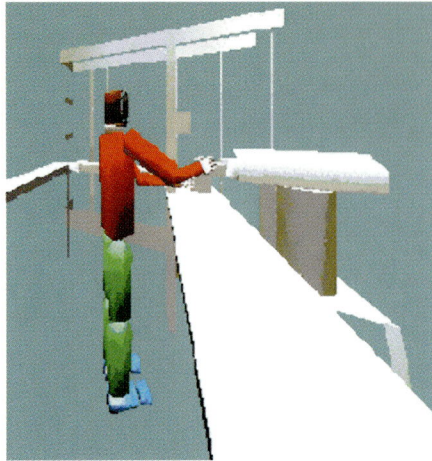

Figure 6.7 Human operating in a manufacturing system whose motion is replicated in a VE, a priori registered with the real system.

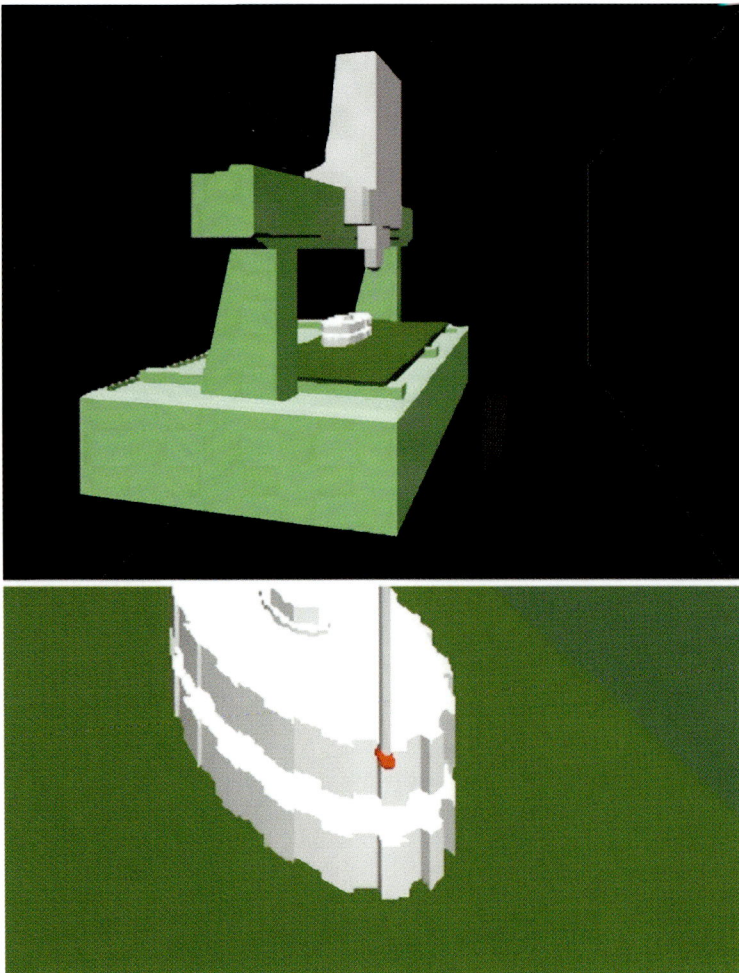

Figure 7.20 Virtual probe and collision detection.

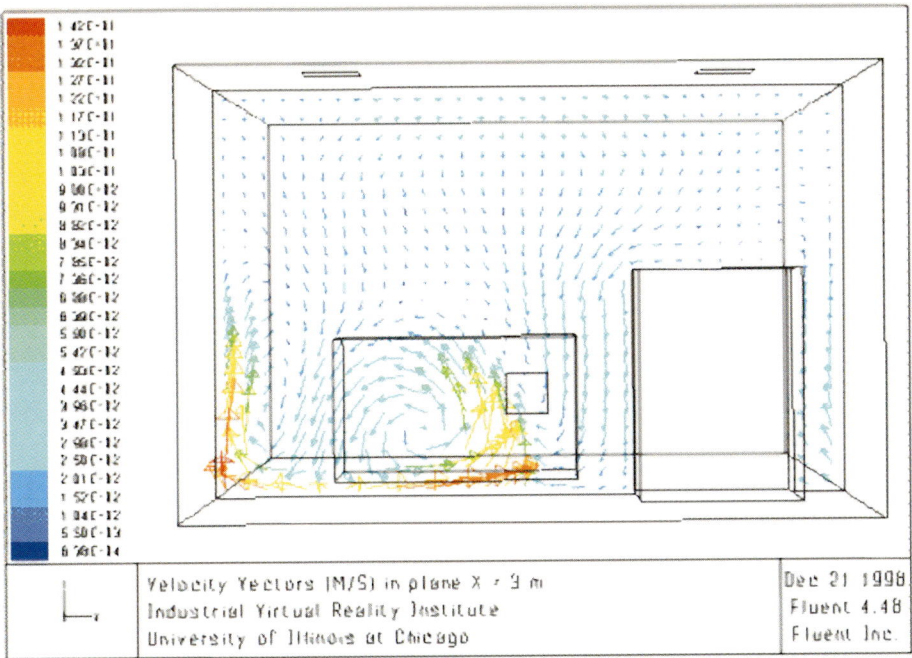

Figure 10.12 Air velocity fields in planes x = 3 m; the airflow pattern is equal for every air supply rate.

Figure 10.19 Preprocessing step. The user is immersed in the VR representation of the room (in this specific case, a pharmaceutical tablet press room) and can easily change parameters simply by "dragging and dropping" them (the long yellow line is the pointer the user employs for picking up and dragging objects).

Figure 10.20 Representation of the airflow in a recirculation region using fixed tetrahedral darts. *Source:* Lori Freitag, Argonne National Laboratory, and Bill Michels, Nalco Fuel Tech.

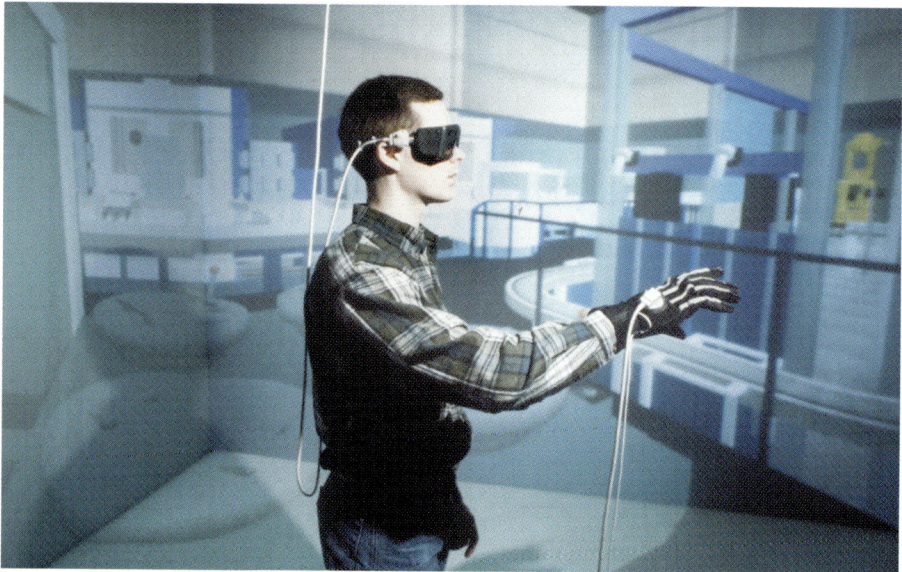

The results of a discrete event simulation program used as input to a virtual reality model of a factory work cell. Parts in the factory can be seen traveling through their manufacturing processes throughout the day. The time of day can be adjusted to examine bottlenecks in part flow. Different scenarios can be loaded and compared. *Source:* Jason Kelsick and Dr. Judy Vance, Virtual Reality Applications Center, Iowa State University.

Collaborative virtual design review work done at University of Illinois' National Computational Science Alliance (NCSA) for Caterpillar Inc. *Source:* Volodymyr Kindratenko, NCSA (http://www.ncsa.uiuc.edu/VEG/DVR).

GM Staff Research Scientist Randy Smith sits immersed within a virtual model of a Pontiac Sunfire, displayed with VisualEyes, an advanced 3-D visualization system created by GM Research and Development. A "virtual design review" is taking place, with an "avatar" representing a remote participant. Collaborative capabilities were provided by CAVERNsoft, developed by Jason Leigh at the Electronic Visualization Laboratory at the University of Illinois at Chicago.

ImmersaDesk R2™, which folds up into a compact unit, being used to simulate a building layout. *Source:* Electronic Visualization Laboratory, University of Illinois at Chicago. ImmersaDesk R2™ is a trademark of University of Illinois.

Army vehicle concept design using CAVE™. *Source:* picture courtesy of US Army TACOM National Automotive Center, Advanced Collaborative Environments Lab.

-1	0	1
-2	0	2
-1	0	1

-1	-2	-1
0	0	0
1	2	1

Figure 4.19 Sobel operators.

Figure 4.20 Sobel operator result for the image in Figure 4.13a.

An alternative gradient-based operator proposed by Prewitt approximates the partial derivatives as follows:

$$P_x = [f(x+1, y-1) + f(x+1, y) + f(x+1, y+1)] - [f(x-1, y-1) + f(x-1, y) + f(x-1, y+1)]$$

$$P_y = [f(x-1, y+1) + f(x, y+1) + f(x+1, y+1)] - [f(x-1, y-1) + f(x, y-1) + f(x+1, y-1)]$$

(4.14)

The gradient magnitude can be approximated as: $|P_x| + |P_y|$. The *Prewitt operators* in two orthogonal directions look as in Figure 4.21, and an example of Prewitt algorithm output is shown in Figure 4.22.

-1	0	1
-1	0	1
-1	0	1

-1	-1	-1
0	0	0
1	1	1

Figure 4.21 Prewitt operators in two orthogonal directions.

Figure 4.22 Prewitt operator result for the image in Figure 4.13a.

Second-Derivative Operators

The problem with first-derivative operators is that in practice step edges are not perfect in terms of steepness, but rather, they look more like ramp edges (see Figure 4.15), especially due to the blurring effect from previous filtering operations. So the local maxima of the first derivative cannot always be located accurately, or, in most cases, as can be seen from previous examples, the edges resulting from applying these operators are very thick. A possible solution to this problem would be to use second-derivative operators. The rationale behind this is that to the maximum of the first derivative corresponds a *zero crossing* of the second derivative (zero crossing is a point where the plot of a function intercepts the *x* axis). This is illustrated in Figure 4.23.

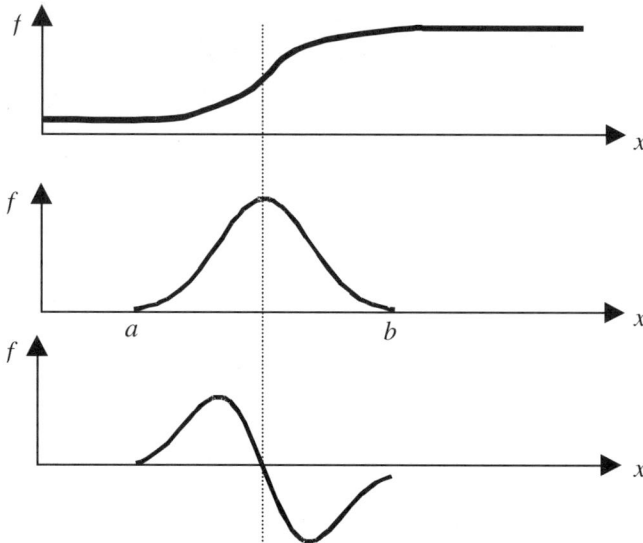

Figure 4.23 If a threshold is used for edge detection, all pixels between a and b will be marked as edge pixels. By using the zero crossings of the second derivative, points that are not local maxima are eliminated from consideration.

The second-derivative operator used for edge detection is an approximation of the *Laplacian,* which has the following form:

$$\nabla^2 = \frac{\partial^2}{\partial x^2} + \frac{\partial^2}{\partial y^2} \tag{4.15}$$

The Laplacian is approximated, for the purpose of computer implementation, as follows:

$$L(x,y) = f(x,y) - \tfrac{1}{4}[f(x, y + 1) + f(x, y - 1) + f(x + 1, y) + f(x - 1, y)] \tag{4.16}$$

Typical Laplacian masks are shown in Figure 4.24, and an example of Laplacian-based edge detection is shown in Figure 4.25.

Figure 4.24 Laplacian masks.

0	-1	0
-1	4	-1
0	-1	0

-1	-1	-1
-1	8	-1
-1	-1	-1

Figure 4.25 Laplace operator on the image from Figure 4.13a.

As can be seen from Figure 4.25, a major disadvantage of using the Laplace operator for edge detection is that by being a second-derivative operator, it is extremely sensitive to noise and it produces double edges (one for the negative direction of an edge and the other for the positive direction), since it contains no direction information in its mathematical expression. A major improvement to the Laplacian was introduced by Marr and Hildreth (1980), an approach based on first smoothing the image noise by convolving the image with a two-dimensional Gaussian filter and applying the Laplacian subsequently. The operator looks as follows:

$$\nabla^2[F(x,y)^x \ G(x,y)] \tag{4.17}$$

where *F(x,y)* represents the image intensity at a point *(x,y)* and *G(x,y)* is the two–dimensional Gaussian function, given by

$$G(x,y) = \exp{-\left(\frac{x^2 + y^2}{2\sigma^2}\right)} \tag{4.18}$$

Because of the properties of the Laplacian and convolution operation, we can derive a single filter, called in the literature the Laplacian of the Gaussian, as follows:

$$\nabla^2[F(x,y) \ G(x,y)] = \nabla^2 G(x,y)^x[F(x,y)] \tag{4.19}$$

Figure 4.26 Laplacian of Gaussian operator on the image of Figure 4.13a.

The Marr–Hildreth edge detector is used widely because it provides reliable results, as can be seen from Figure 4.26. As can be seen, it produces closed edges, an advantage for most applications and, because of the properties of the second derivative, the output edges do not need subsequent thinning as do the edges resulted from first-derivative operators.

4.3.5 Edge Linking

The output of edge detection algorithms, discussed in the preceding section, is not suitable for higher-level applications, such as object recognition and three-dimensional shape recovery from two-dimensional images. The reason is that by applying edge detection algorithms, we obtain only the edge pixels in the image, without any connectivity information and without any information about the features of the objects represented in a particular image. Certain applications require the grouping of the edge pixels into connected chains. From those chains we can extract useful information that can subsequently be used for different tasks, such as the ones mentioned above. The process of grouping edge pixels in connected chains is known as *edge linking*. There are multiple approaches for edge linking: edge relaxation, Hough transform, heuristic graph search, and so on. All the techniques use information resulting from edge detection, such as gradient magnitude and direction (recall that the edge direction is perpendicular to the gradient direction). In this section we are concerned with graph search techniques.

Edge Linking as Heuristic Graph Search

A graph (n_i, e_j) is a collection of nodes n_i, connected by edges (arcs) e_j. Often, the nodes or the edges have numerical weights associated with them. Suppose that the edge pixels in an image region look as in Figure 4.27. In this figure, the arrows denote that edge pixels have a direction (given by the gradient direction at a particular edge pixel). The direction of the arrow gives the edge direction. The boundary of an object in the image is associated with the collection of edge pixels of maximum "strength." The notion of *edge strength* is associated in our case with gradient magnitude. In Figure 4.27 the numerical values represent gradient magnitudes at edge pixels.

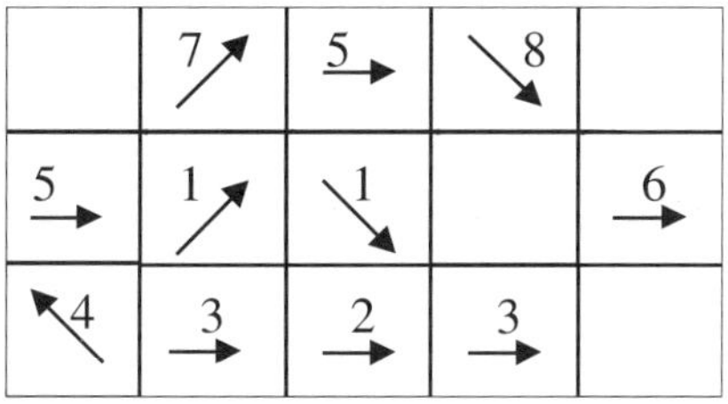

Figure 4.27 Gradient magnitudes and edge directions.

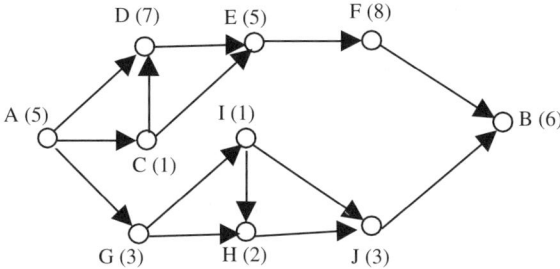

Figure 4.28 Graph version of the image from Figure 4.27.

According to Figure 4.27, a graph with nodes n_i can be formed, as in Figure 4.28. The nodes are denoted A to J, and in order to detect the boundary between A and B (start and end pixels of the boundary), we have to find a path from node A to node B in the graph from Figure 4.28, according to some criterion. The criterion for finding a path from A to B is as follows. Each possible path from A to B has associated an evaluation function $\phi(n_k)$, the path being constrained to pass through node n_k. The heuristic search algorithm works as follows:

1. Begin with the start node (in this case A).
2. Expand node A to its successors and select the node that maximizes ϕ (in this case, ϕ is the sum of weights of the two nodes forming an edge).
3. The node selected in step 2 becomes the start node and the process is repeated until the end node is reached (in this case node B).

The sequence of nodes selected constitutes the boundary path. One question remains unanswered: How are the successors of a node determined? The rule for classifying a node as a successor is shown in Figure 4.29 and states that node x is followed by a node y_i, $i = 1,2,3$, if the two nodes are connected

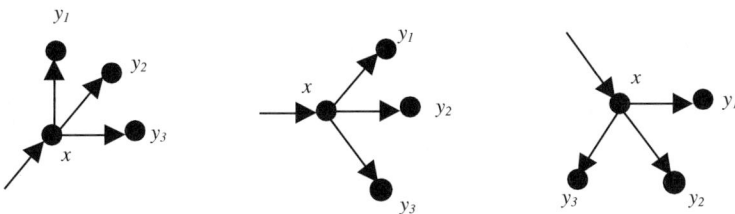

Figure 4.29 Connectivity rules for edge linking.

as in Figure 4.29 (the arrows represent edge directions), and the absolute value of the difference between the gradient directions at x and y_i is less than 90°. If we consider the graph in Figure 4.28, if we start with node A, its successors are D, C, and G, with $\phi(D) = 12$, $\phi(C) = 6$, and $\phi(G) = 8$. Therefore, node D is selected and C and G are discarded. Continuing with node D, we select subsequently the nodes E, F, and B, so the resultant path (boundary) is $ADEFB$.

4.3.6 Corner Detection

In computer vision applications, *corners* or *vertices* are called the points of interest belonging to the objects in a two-dimensional image. If we consider a planar contour of an image object (obtained by edge detection and edge linking), corners are the points of high curvature (points at which the curvature exceeds a preset threshold). A possible approach for corner detection is based on modeling the connected chains of edge pixels as B-spline curves. An overview of B-spline curve fitting of a set of planar points is described in Appendix A1.

The corners are sought among the knots of the B-splines approximating the given sets of connected pixels. (For a description of the terminology related to B-splines, see Appendix A1.) The problem here is to overcome the undesired effect of rounding (introduced by the mechanism of spline approximation) when we deal with sharp corners. The cornerness is tested at the knot corresponding to the parameter value $t = 0$. To overcome the rounding effects, the relative position (displacement) between the $t = 0$ knot and one of the control points is defined. If we consider the B-spline segment between the knots k_i and k_{i+1}, determined by the control points P_{i-1}, P_i, P_{i+1}, and P_{i+2}, we define the displacement between the $t = 0$ knot and the control point $P_i(x_i, y_i)$ as follows:

- Along the x axis: $d_x = k_{ix}(0) - x_i = d_1 - x_i$
- Along the y axis: $d_y = k_{iy}(0) - y_i = d_1 - y_i$

where d_1 is a factor of the polynomial describing the B-spline curve (see Appendix A1). If either d_x or d_y exceeds a preset threshold T_d, the knot is displaced by the value

$$d = \sqrt{d_x^2 + d_y^2} \tag{4.20}$$

in the direction of P_i (see Figure 4.30). In other words, the corner will be at point P_i. The conditions for a point to be classified as a corner are as follows:

1. The value of d_x or d_y is larger than a given threshold T_d.
2. The curvature C_v is larger than a given threshold T_c. (The expression of the curvature is given in Appendix A1.)

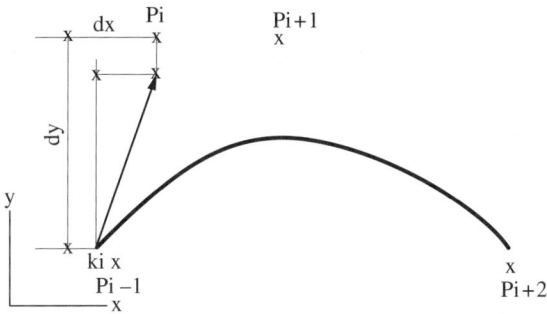

Figure 4.30 Overcoming the rounding effect.

Figure 4.31 B-spline-based corner detection example.

The thresholds are established depending on the level of detail in the object surface representation. In Figure 4.31, the results of applying the corner detection algorithm are shown.

Corners can also be detected directly from intensity images without prior edge detection and edge linking. Corners are not always interpreted as intersections of image lines or points of maximum curvature lying on detected edges. There are situations when patterns of intensities can be considered cornerlike features. This happens when the intensity function gradient can exceed a certain value along two distinctive directions at the same point. The point in question is then classified as a corner.

The intensity-based approach for corner detection is described next. We know from the discussion on edge detection that a pixel is classified as an edge pixel if the image gradient along a certain direction at that pixel exceeds a threshold value. In the case of corners, the image gradient should exceed the threshold along two distinct directions. This can be evaluated by means of the Hessian matrix of the image intensity function, as follows:

$$\mathbf{C} = \begin{bmatrix} \sum_{Q} I_x^2 & \sum_{Q} I_x I_y \\ \sum_{Q} I_x I_y & \sum_{Q} I_y^2 \end{bmatrix} \tag{4.21}$$

In equation (4.21), the subscripts indicate partial derivatives (i.e., $I_x = \partial I / \partial x$; I is the image intensity and Q is the neighborhood of an image pixel). The matrix $\mathbf{C}$ is symmetric, and therefore it can be diagonalized by a rotation of the coordinate axes. Therefore, matrix $\mathbf{C}$ can be approximated by

$$\mathbf{C} = \begin{bmatrix} \lambda_1 & 0 \\ 0 & \lambda_2 \end{bmatrix} \qquad (4.22)$$

where λ_1 and λ_2 are the eigenvalues of $\mathbf{C}$ (since $\mathbf{C}$ is symmetric, its eigenvalues are real and nonnegative).

If the neighborhood Q is perfectly uniform, the image gradient is zero throughout Q and therefore $\mathbf{C}$ becomes the null matrix and we have $\lambda_1 = \lambda_2 = 0$. If Q contains a corner, there are two principal directions in Q, and λ_1 and λ_2 each correspond to one of these principal directions. The larger the eigenvalues, the larger the strength of their corresponding edges and therefore the larger the strength of the corner (the intersection between the two principal directions). This corner detection methodology is faster than the one based on curve fitting but also produces corners that are not meaningful object vertices. A good example is illustrated in Figure 4.32, in which detected corners are marked with white dots.

Figure 4.32 Intensity-based corner detection.

4.3.7 Methods for Stereo Correspondence

As mentioned in the first paragraph of Section 4.3, the problem of stereo matching is to determine which parts of the left and right images of a stereo pair are projections of the same scene element. Two different classes of stereo-matching algorithms have been developed: correlation-based and feature-based. An exhaustive presentation of each stereo-matching algorithm is impossible to make in a single book and also goes beyond the scope of this book. We describe briefly next the basics of both classes of matching.

Correlation-based methods attempt to match all the pixels from both images. Image windows of certain sizes are defined, and matching is established based on a similarity criterion, which is a measure of the correlation between two windows belonging to different images. A good match is one that maximizes the similarity criterion (usually, a correlation score). The fundamental assumptions in this type of method are:

- Correspondent pixels in both images have the same intensity.
- One pixel in, say, the left image has one and only one correspondent in the right image, and vice versa.
- The disparity between two images is continuous (does not vary abruptly from pixel to pixel). This constraint is also called the *figural continuity constraint.*

Correlation-based methods produce dense sets of matched elements but are computationally expensive.

Feature-based methods are based, first, on the detection of features from images, and subsequently, matching those features, which can be edges, corners, or zero–crossings. They produce less dense sets of data and are therefore faster. Geometric constraints such as *epipolar constraints* (described in Section 4.4.4) are imposed to reduce the search space.

Note that the constraints mentioned above for both classes of stereo-matching algorithms are not restricted to one class of methods or another. In fact, there are many hybrid algorithms and these algorithms typically perform the best. One such algorithm (developed by Zhang et al. (1994)) attempts to match corners extracted from left and right images of a stereo pair. The basic idea behind this technique is to employ first some classical matching techniques (such as correlation) to establish a set of matches and subsequently to refine the matches through recovery of the epipolar geometry by means of the *fundamental matrix* (the fundamental matrix links the image coordinates of a corner from the left image to the image coordinates of its correspondent in the right image (described in more detail in Section 4.4.4). The first step of this method is to extract the points of interest from the images. The stereo-matching methodology consists of the following operations:

1. Apply a correlation technique to obtain an initial set of *candidate matches.*
2. Disambiguate matches through relaxation.
3. Recover the epipolar geometry by computing the fundamental matrix.
4. Refine the matches using the epipolar constraint.

The details of this methodology are beyond the scope of this work. In Figures 4.33 to 4.35, an example of applying this algorithm is shown. Figure 4.33 shows the original images, Figure 4.34 shows the matches after applying the correlation method, and in Figure 4.35 the matches after relaxation are shown. The results of applying correlation again after estimation of the fundamental matrix are omitted since no significant improvement has been obtained in this case. By comparison with correlation output (shown in Figure 4.34), we can see from Figure 4.35 that some matches (such as the ones labeled 80 and 88) have been disambiguated. But still some false matches remain, in particular for repetitive patterns, such as the array of squares located at the upper right corner of each image.

Figure 4.33 Original images for matching.

Figure 4.34 Matches through correlation.

Figure 4.35 Disambiguating matches through relaxation.

4.4 CAMERA MODEL AND CALIBRATION

4.4.1 Perspective Camera Model

As mentioned in Section 4.2, the type of projection that maps a three-dimensional space to a two-dimensional plane is called *perspective projection*. A camera maps a three-dimensional scene onto a plane, called an *image plane*. A more complete description of the geometry of the pinhole camera model is shown in Figure 4.36. In Figure 4.36, the significance of points *P*, *V*, and *I* is the same as in Figure 4.1. The following coordinate systems are associated with the mechanism of image formation:

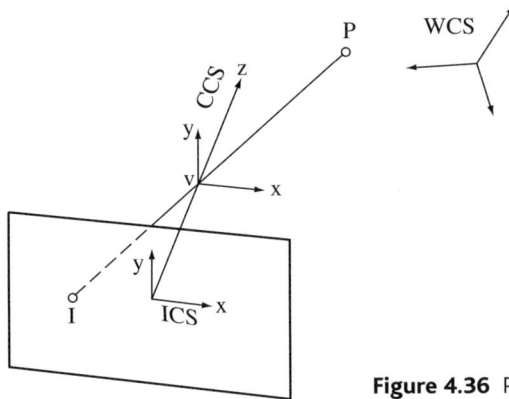

Figure 4.36 Pinhole camera model.

- The *image coordinate system* (ICS) is a planar coordinate system, with the origin at the point of intersection between the image plane and the optical axis of the camera (the perpendicular from *V* onto the image plane).
- The *camera coordinate system* (CCS) is a three-dimensional coordinate system with the origin at the viewpoint *V*. It has the *x* and *y* axes parallel with the *x* and *y* axes of ICS, respectively. The *z* axis coincides with the optical axis of the camera and points away from the image plane.
- In many applications (when the recovery of the third dimension out of two-dimensional images is required, for example), a scene-centered coordinate system is needed; this is called a *world coordinate system* (WCS) and its location in space is arbitrary (usually defined by the application designer).

Let us derive the pinhole camera equations, considering for the moment that the coordinates of the point *P* are expressed with respect to CCS and are denoted as (x', y', z'). In Figure 4.37, the perspective projection of *P* onto the

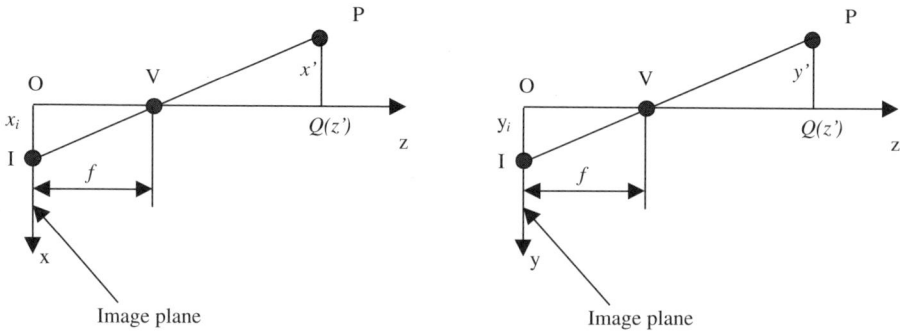

Figure 4.37 Geometry of image formation in the pinhole camera model.

image plane is viewed from above xz and yz planes of CCS. From the similarity between triangles IOV and PQV in Figure 4.37a and b, the following equations can be written:

$$\frac{x_i}{x'} = \frac{f}{z'}$$

$$\frac{y_i}{y'} = \frac{f}{z'}$$

(4.23)

In equation (4.23), f is called the focal length of the camera and it is the distance between the viewpoint and the image plane. In Figure 4.37, (x_i, y_i) are the coordinates of point I with respect to ICS. Equations (4.23) are called pinhole camera equations or perspective projection equations of the camera.

4.4.2 Camera Calibration

There are applications that require knowledge of the camera location relative to the scene. In such cases, the coordinate transformation between WCS and ICS has to be recovered. For this purpose, a regular pattern from which points with known WCS coordinates can easily be extracted is used. The technique of retrieving the coordinate transformation between WCS and ICS (or any other coordinate system associated with the image) is called *camera calibration*. Camera calibration algorithms take as input the WCS coordinates of the calibration points and their image coordinates obtained by image processing techniques and output the coordinate transformation between WCS and ICS. There are algorithms that alsso compute the location of the intersection between the optical axis of the camera and the image plane (called the *principal point*) with respect to the upper left corner of the image and others that account for lens distortion as well. An example of a calibration pattern is shown in Figure 4.38. The pattern from Figure 4.38 consists of two planar arrays of black dots on a white background. The angle formed by the two planar arrays is 120°. The cal-

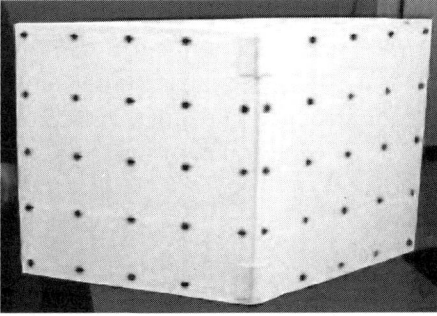

Figure 4.38 Calibration pattern.

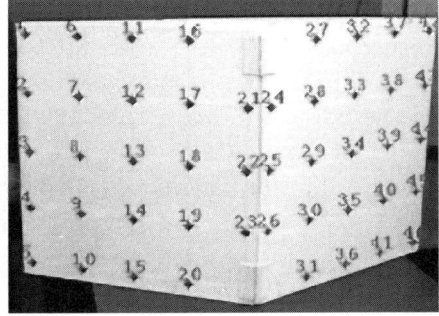

Figure 4.39 Calibration points are the centroids of the black dots.

ibration points, which have known coordinates relative to WCS, are the centers of the black dots (Figure 4.39). The mechanism of detecting the calibration points in an image is described in Section 4.4.3.

Before going into more detail regarding camera calibration procedure, a short digression regarding the coordinates systems associated with images has to be made. Recall that ICS related to perspective projection had the origin at the principal point. This point can be assumed for now to be located at the center of the pixel array as a result of image digitization by computer. In typical computer applications, the coordinate system associated with the pixel array has the origin at the upper left corner of the image frame. Let us call this coordinate system the *pixel coordinate system* (PCS). When applying image processing algorithms for edge and corner detection, for example, the coordinates of pixels that are the output of such algorithms are expressed relative to PCS. So the global transformation that results from calibration is from PCS to WCS, or vice versa. Therefore, the relationship between PCS and ICS that should convert pixels to meaningful units is needed and is treated next. The derivation of the coordinate transformation from PCS to ICS is illustrated in Figure 4.40.

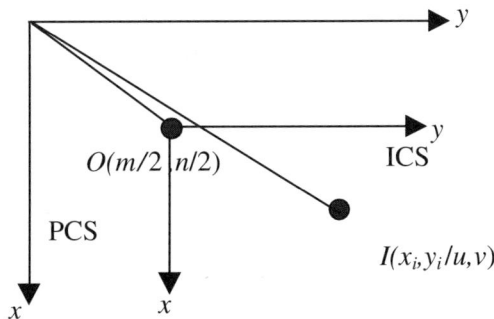

Figure 4.40 Conversion from PCS to ICS.

Let us assume that the image has $m \times n$ pixels in the x and y directions, respectively. Therefore, the principal point O has the coordinates $(m/2, n/2)$, according to the previous assumption that it is located at the center of the image frame. An image point I has the coordinates (u, v) relative to PCS and (x_i, y_i) relative to ICS. From Figure 4.40, the following equations, which represent the transformation from ICS to PCS, hold:

$$ud_x = \frac{m}{2}d_x + x_i$$

$$vd_y = \frac{n}{2}d_y + y_i$$

(4.24)

where d_x and d_y are pixel dimensions in the x and y directions, respectively. We can denote the coordinates of the principal point relative to PCS, in the general case, as (u_0, v_0). In matrix form, the transformation from ICS to PCS looks as follows:

$$\begin{bmatrix} u \\ v \end{bmatrix} = \begin{bmatrix} 1/d_x & 0 & u_0 \\ 0 & 1/d_y & v_0 \end{bmatrix} \cdot \begin{bmatrix} x_i \\ y_i \\ 1 \end{bmatrix}$$

(4.25)

The transformation from P to I is a perspective projection, which is modeled as a linear transformation in projective coordinates. If a point I has the coordinates $(u, v)^\mathrm{T}$ in PCS, its corresponding projective coordinates relative to PCS are denoted by $(U, V, S)^\mathrm{T}$, with

$$u = U/S$$

$$v = V/S$$

(4.26)

From equation (4.26), we can see that for the coordinates $(u, v)^\mathrm{T}$ to be defined, S has to be nonzero. Otherwise, if $S = 0$, P is in the *focal plane* of the camera (which is the plane that passes through V and is parallel to the image plane).

The global transformation from P to I (from WCS to PCS) is given by the following relationship:

$$\begin{bmatrix} U \\ V \\ S \end{bmatrix} = \mathbf{M} \begin{bmatrix} x \\ y \\ z \\ 1 \end{bmatrix}$$

(4.27)

where $\mathbf{M}$ is a 3 × 4 matrix, called the *perspective projection matrix,* and $(x, y, z)^\mathrm{T}$ are the coordinates of point P relative to WCS. The perspective matrix can be

decomposed into matrices corresponding to the following sequence of transformations:

- $WCS \rightarrow CCS$: $\mathbf{M}_1 = \begin{bmatrix} r_1 & r_2 & r_3 & t_x \\ r_4 & r_5 & r_6 & t_y \\ r_7 & r_8 & r_9 & t_z \end{bmatrix}$, the transformation matrix from WCS

  to CCS. This matrix embodies the rotational components
  r_i, $i = 1, 2, \ldots, 9$, and the translational components t_x, t_y, and t_z of the
  transformation.

- $CCS \rightarrow ICS$ transformation based on equations (4.23); its matrix is

  $\begin{bmatrix} f & 0 & 0 \\ 0 & f & 0 \\ 0 & 0 & 1 \end{bmatrix}$. Note that the transformation from CCS to ICS is expressed in

  projective space, since it is a mapping between a three-dimensional and
  a planar coordinate system. We can see that even though equations
  (4.23) are nonlinear, the transformation to projective space linearizes
  them.

- $ICS \rightarrow PCS$: given by equation (4.25).

We can combine the last two transformations into one, which gives

$$\mathbf{M} = \begin{bmatrix} f/d_x & 0 & u_0 \\ 0 & f/d_y & v_0 \\ 0 & 0 & 1 \end{bmatrix} \bullet \begin{bmatrix} r_1 & r_2 & r_3 & t_x \\ r_4 & r_5 & r_6 & t_y \\ r_7 & r_8 & r_9 & t_z \end{bmatrix} = \begin{bmatrix} m_{11} & m_{12} & m_{13} & m_{14} \\ m_{21} & m_{22} & m_{23} & m_{24} \\ m_{31} & m_{32} & m_{33} & m_{34} \end{bmatrix} \quad (4.28)$$

The left matrix elements f, d_x, d_y, u_0, and v_0 are called *intrinsic parameters* of
the camera, and the rotational and translational elements of the right matrix
are called *extrinsic parameters* of the camera. The purpose of camera calibration is to compute the extrinsic and intrinsic parameters of the camera, taking
as input the WCS coordinates of the calibration points and their image coordinates. As mentioned before, additional factors, such as lens distortion, have to
be compensated for.

The mathematics for solving the calibration problem is fairly intricate and
we do not address it in this book. Camera calibration source code is publicly
available on the Internet. The problem with traditional camera calibration
techniques is that cameras have to be calibrated each time an image is being
captured (more exactly each time the viewpoint changes). Even though intrinsic parameters can be considered as constant under vantage-point changes,
the relative position between WCS and CCS changes and therefore the camera
has to be recalibrated.

4.4.3 Finding the Calibration Points in an Image

This section is meant for the implementation-oriented reader and describes the mechanism of detecting the calibration points in an image of the calibration pattern (Figure 4.38). The following operations are performed:

1. Filter the image of the calibration pattern with a Gaussian filter to reduce noise.
2. Threshold the image of the calibration pattern to isolate the black dots from the background.
3. Label the regions obtained after thresholding by applying a connected component algorithm, described in Section 4.3.3.
4. Apply a size filter to eliminate the regions larger than a certain size (a size filter in this context would remove the regions with an area larger than a threshold; in this way we are left only with the regions representing the black dots).
5. Compute the centroids of the regions remaining after applying the size filter. This gives us the calibration points (as shown in Figure 4.39).

The calibration pattern image after thresholding is shown in Figure 4.41. It is obvious from Figure 4.41 that by applying a size filter, regions other than the ones representing the black dots (i.e., the background) will be removed.

Figure 4.41 Thresholded calibration pattern image.

The procedure for computing a region's centroid is described next. First, the area of the region has to be computed. For a binary image with only 1 and 0 pixels, the area of a region is simply the number of pixels contained in that region. The centroid of a region is computed using the following equations:

$$\bar{u} = \frac{\sum_R u}{A_R} \qquad \bar{v} = \frac{\sum_R v}{A_R} \qquad (4.29)$$

where the summation is made over the pixels of region R, A_R is the area of region R, and u and v are the coordinates of the pixels in region R with respect to the pixel array coordinate system (PCS).

4.4.4 Stereo Vision Revisited

We have shown in Section 4.2 the methodology of recovering the third dimension from stereo for parallel camera configuration. Usually, this kind of configuration is difficult to achieve in practice. In addition, WCS need not be aligned with both left and right ICS, especially when the reconstruction has to be achieved with respect to an object-centered coordinate system. In this case we need to detach WCS from CCS and to make WCS independent of the location and orientation of the camera. In this section we extend the presentation of a stereo vision system to a general configuration, in which cameras are facing the scene arbitrarily and their optical axes are not aligned. Such a configuration is illustrated in Figure 4.42. Considering Figure 4.42, a point P is projected onto an image plane (labeled in Figure 4.42 as the left image plane) at point I_l. If we consider another image of P, taken from V_r (Figure 4.42), P will project onto that image (right image) at point I_r. As can be seen from Figure 4.42, the optical rays I_lV_l and I_rV_r intersect in space at point P, and therefore by knowing the mathematical representation of those two rays we can actually recover the coordinates of P relative to WCS.

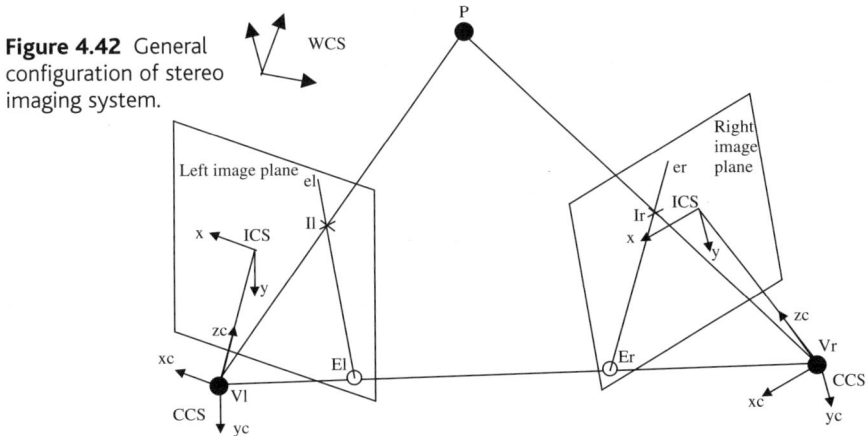

Figure 4.42 General configuration of stereo imaging system.

As was presented in Section 4.4.2, equation (4.27), a point P is projected through a linear transformation onto the image plane. This transformation gives us the coordinates of the projection point I in the projective space: $(U,V,S)^T$. Equation (4.27) can be reformulated as follows:

$$s \begin{bmatrix} u \\ v \\ 1 \end{bmatrix} = \mathbf{M} \begin{bmatrix} x \\ y \\ z \\ 1 \end{bmatrix} \tag{4.30}$$

In equation (4.30), s is called the *scale factor*. Scale can be eliminated by dividing each of the first two equations [obtained by developing equation (4.30)] by the third one, which contains s. Let us rewrite $\mathbf{M}$ as follows:

$$\mathbf{M} = \begin{bmatrix} \mathbf{m}_1^T & m_{14} \\ \mathbf{m}_2^T & m_{24} \\ \mathbf{m}_3^T & m_{34} \end{bmatrix} \tag{4.31}$$

where $\mathbf{m}_i$, $i = 1,2,3$ are the column vectors of rotational elements corresponding to the row i of $\mathbf{M}$ ($\mathbf{m}_i^T = [m_{i1}\ m_{i2}\ m_{i3}]^T$). It follows that

$$u = \frac{\mathbf{m}_1^T x + m_{14}}{\mathbf{m}_3^T + m_{34}}$$

$$v = \frac{\mathbf{m}_2^T x + m_{24}}{\mathbf{m}_3^T + m_{34}} \tag{4.32}$$

where $\mathbf{x}$ is the column vector of world coordinates of point P ($\mathbf{x} = [x\ y\ z]^T$). From equation (4.32), we have

$$(\mathbf{m}_3^T u - \mathbf{m}_1^T)x = m_{14} - u m_{34}$$

$$(\mathbf{m}_3^T v - \mathbf{m}_2^T)x = m_{24} - v m_{34} \tag{4.33}$$

This is a system of two equations and three unknowns: $(x,y,z)^T$. Now by adding another image, we get another two equations, and therefore the system becomes overdetermined. In general, one can use as many images as available (in which P is visible) and get WCS coordinates of P by using the *pseudoinverse* method for solving the system of the form given in equation (4.33). The pseudoinverse method is described in detail in Appendix A2. By solving equation (4.33) for at least two images, the depth computation is complete.

Epipolar Constraint.
We mentioned in Section 4.3 that epipolar constraint is an important geometric constraint of stereo vision and is used widely in stereo-matching algorithms. Now we have all the tools required to describe it in detail. We can see from Figure 4.42 that the intersections of the line joining the focal centers (viewpoints) V_l and V_r with the two image planes are the points E_l and E_r, respectively. These points are termed *epipoles*. The figure tells us that points P, I_l, I_r, V_l, and V_r are coplanar. If we consider, for example, the point I_l from the left image, its correspondent I_r from the right image lies on the line e_r passing through the epipole E_r. This line is termed an *epipolar line*. The same is valid the other way around, so to I_r corresponds the epipolar line e_l. If we had

another point in, say, the left image, its correspondent epipolar line in the right image would pass also through E_r. So all the epipolar lines in an image form a pencil of lines intersecting at the epipole. The property that the correspondent of a point in another image lies on the epipolar line is extremely important when looking for correspondent points in the sense that reduces the search space from two to one dimension. This property is called the *epipolar constraint*.

Fundamental Matrix.
We mentioned in Section 4.3.7 that the fundamental matrix relates the image coordinates of a point in the left image to the image coordinates of its correspondent in the right image. The fundamental matrix is a 3 × 3 singular matrix that describes the epipolar geometry relating two images of a stereo pair. Computation of the fundamental matrix relies on the availability of a set of corresponding points in the two images. Such a set can be obtained by using the method described briefly in Section 4.3.7 (through correlation and relaxation).

Let us recall equation (4.30), which describes the perspective projection of a point onto the image plane. If we denote the image coordinates vector as $\mathbf{I} = (u,v,1)^T$ and the world coordinates vector as $\mathbf{P} = (x,y,z,1)^T$, we have, from equation (4.30), $s\mathbf{I} = \mathbf{MP}$. This relationship can be written for the left and right images, respectively, with subscripts l and r for s, $\mathbf{I}$, and $\mathbf{M}$. Perspective matrix $\mathbf{M}$ can be decomposed as in equation (4.28). We can rewrite this equation, in short, as follows:

$$\mathbf{M} = \mathbf{A}[\mathbf{R}\ \mathbf{t}] \qquad (4.34)$$

where $\mathbf{A}$ is the intrinsic parameters matrix, $\quad \mathbf{A} = \begin{bmatrix} f/d_x & 0 & u_0 \\ 0 & f/d_y & v_0 \\ 0 & 0 & 1 \end{bmatrix}$

and $\mathbf{R}$ and $\mathbf{t}$ consist of the rotational components and the translational components of the transformation from WCS to CCS, respectively:

$$\mathbf{R} = \begin{bmatrix} r_1 & r_2 & r_3 & t_x \\ r_4 & r_5 & r_6 & t_y \\ r_7 & r_8 & r_9 & t_z \end{bmatrix}; \quad \mathbf{t} = \begin{bmatrix} t_x & t_y & t_z \end{bmatrix}^T$$

Let us consider, without any loss of generality, that WCS coincides with the CCS of the left camera. We then have the following equations:

$$s_l\mathbf{I}_l = \mathbf{A}_l[\mathbf{Id}\ 0]\mathbf{P}$$
$$s_r\mathbf{I}_r = \mathbf{A}_r[\mathbf{R}\ \mathbf{t}]\mathbf{P} \qquad (4.35)$$

where $(\mathbf{R}, \mathbf{t})$ is the rigid transformation from CCS associated with the left camera (WCS in this case) to right CCS and $\mathbf{Id}$ is the 3×3 identity matrix. By eliminating $\mathbf{P}$, s_l, and s_r from the equations above, we get

$$\mathbf{I}_r \mathbf{F} \mathbf{I}_l = 0 \tag{4.36}$$

where

$$\mathbf{F} = \mathbf{A}_r^{-T}[\mathbf{t}]_x \mathbf{R} \mathbf{A}_l^{-1} \tag{4.37}$$

In equation (4.37), $[\mathbf{t}]_x$ is the antisymmetric matrix defined by $\mathbf{t}$ such that $[\mathbf{t}]_x \mathbf{x}$ $= \mathbf{t} \times \mathbf{x}$ for all three-dimensional vectors $\mathbf{x}$. Geometrically, $\mathbf{FI}_r$ defines the epipolar line of point $\mathbf{I}_l$ in the right image. Equation (4.36) says no more than that, given $\mathbf{I}_l$, its correspondent in the right image ($\mathbf{I}_r$) lies on its epipolar line (recall the epipolar constraint). The 3×3 matrix $\mathbf{F}$ is called the *fundamental matrix*. Since $\det([\mathbf{t}]_x) = 0$, $\det(\mathbf{F}) = 0$, the rank of $\mathbf{F}$ is 2. $\mathbf{F}$ is defined up to a scale factor, because if $\mathbf{F}$ is multiplied by an arbitrary scalar, equation (4.36) still holds. Therefore, $\mathbf{F}$ has only seven independent parameters among its nine elements.

The definition of matrix $\mathbf{F}$ (equation (4.36)) assumes that $\mathbf{R}$ and $\mathbf{t}$ are known and they can be found through camera calibration. If we do not have the possibility of calibrating the camera, $\mathbf{F}$ can be computed based on a set of corresponding points in the two images. The minimum number of correspondent points for equation (4.36) to yield a solution is seven. Actually, in this case the solution is not unique (three distinct solutions are obtained). If we use eight points, we get a unique solution. But in practice more points are used to minimize the errors. There are multiple (linear and nonlinear) methods of computing the fundamental matrix. We will not go into any further detail here.

4.5 CALIBRATION-FREE DEPTH RECOVERY

Traditional calibration techniques rely on the presence, within the field of view of the camera, of a calibration pattern each time an image is being taken. This is virtually impossible to secure for most applications, especially the ones that require a high degree of autonomy. Therefore, a reliable technique for depth recovery which does not rely on the presence of a calibration pattern within the field of view of the camera is needed for this kind of application.

4.5.1 What Is State-of-the-Art?

Recovering the three-dimensional structure of a scene without pattern-based calibration has been an extensively studied area in recent years. Most researchers have started from the assumption that the only information available consists of the images themselves and that nothing is known about the camera pose (position plus orientation). This may or may not be very realistic, depending on the specifics of the application.

A brief, preliminary discussion about what kind of reconstruction can be achieved based on the information available is necessary. We have shown so far that when camera calibration is performed and therefore extrinsic and intrinsic parameters are known, the reconstruction is relatively straightforward. What happens when camera calibration based on a calibration pattern is not possible?

The image-only techniques for three-dimensional reconstruction rely on prior determination of a set of correspondent points in the two images of a stereo pair. These techniques center around the concept of *fundamental matrix,* described in Section 4.4.4. As mentioned in Section 4.4.4, the fundamental matrix can be computed uniquely from a set of minimum eight correspondent points. From equation (4.37), we can see that the fundamental matrix is a product between intrinsic parameters matrices for the two images, and the translation and rotation matrices between the camera coordinate systems in the two positions from where the images were taken. If we know the fundamental matrix and the intrinsic parameters of the camera(s), we can compute the relative transformation between the two cameras, but since the fundamental matrix is defined up to a scale factor (see Section 4.4.4), the reconstruction is possible *only up to a scale factor.* If the intrinsic parameters are not known, the reconstruction is still possible, but *only up to an unknown projective transformation.* Table 4.2 provides a summary of the reconstruction capabilities depending on the available information.

When no parameters are available, it is obvious that no meaningful reconstruction is possible since one cannot recover the unknown projective transformation (by meaningful reconstruction we mean that the reconstructed shape qualitatively resembles the original one). When only the intrinsic parameters are available, one can easily measure the distance between two arbitrary points in the scene and thus recover the scale factor. But what happens when a complex scene has to be reconstructed from a large sequence of images? The problem is that when a third image is captured and matched against one of the two previous images, the scale factor will be different. So the scale factor differs from pair of images to pair of images. It would be a tremendous task to

TABLE 4.2 SUMMARY OF RECONSTRUCTION CAPABILITIES

Information Available	*Three-Dimensional Reconstruction from Two Images*
Intrinsic and extrinsic parameters	Full-scale Euclidean reconstruction
Intrinsic parameters only	Euclidean reconstruction, up to an *unknown* scale factor
No information available	Reconstruction up to an *unknown* projective transformation

recover the scale factor for each pair of images by physically measuring a certain scene dimension.

The conclusion to this discussion is that for the creation of virtual manufacturing environments (which are expected to require fairly large numbers of images to be reconstructed completely), only the technique with both intrinsic and extrinsic parameters available is suitable. In the following sections we describe a technique for reconstructing a scene with all parameters available that, to the best of our knowledge, is the only one that does not use a calibration pattern throughout the image acquisition process and is still able to recover the scale factor.

4.5.2 Telemetry-Based Three-Dimensional Reconstruction

A possible approach for making both intrinsic and extrinsic parameters available is to use a commercially available tracking system attached rigidly to the camera. As new images are being captured without any calibration pattern within the field of view of the camera, telemetry is provided by the tracking system, which reports at any time the position and orientation of the camera with respect to WCS (subject to prior precalibration of the camera–tracker unit, which will be described shortly). CONAC™, the tracking system used in this application, is manufactured by MTI Research, Inc. The tracker is a laser-based active beacon positioning system. Typically, a laser tracker consists of two or more stationary beacons that broadcast navigation laser signals to one or more position transponders (PT), which record their position with respect to a coordinate system attached to one beacon [called a *tracker coordinate system* (TCS)]. A single transponder can provide only with information about its position, so to compute the orientation, one needs to use three PTs, attached to each other rigidly. By displacing the rigid body containing the three PTs, we can compute the rigid-body transformation (rotation plus translation) between two consecutive positions (typically, the positions when telemetry is recorded are the positions from where camera captures a scene image). The camera is attached rigidly to the transponders, so the coordinate transformation between the camera and the center of the circle circumscribing the triangle formed by the three PTs (hereafter referred to simply as the receiver) remains invariant under position and orientation changes. The coordinate system associated with the receiver is called the *receiver coordinate system* (RCS). The camera–tracker unit is depicted in Figure 4.43. (In this figure, for simplification, only one beacon is shown; in reality, CONAC consists of two stationary beacons.)

Using the CONAC tracking system, one can determine only the extrinsic parameters of the camera (i.e., the rotation and translation parameters between WCS and CCS). The principal point in the image plane is assumed to

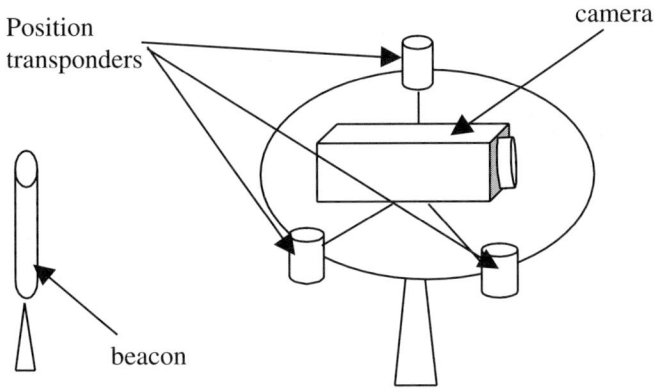

Figure 4.43 Camera–Tracker unit.

be located at the center of the image. An offset of up to 10 pixels does not produce a significant error in three-dimensional measurements. The focal length can be assumed constant (and determined at the initialization step of the module, described below) if the camera is believed not to undergo mechanical or thermal changes, or can be determined subsequently if the extrinsic parameters are known, as described later in this chapter.

Determining the Extrinsic Parameters

To retrieve the position and orientation of CCS automatically with respect to WCS, the initial calibration of the camera–tracker unit has to be performed, the purpose being to compute the coordinate transformation between CCS and RCS, which is invariant, since the camera and the receiver are rigidly attached. At this stage, a calibration pattern is used to initialize the transformation between WCS and CCS. Simultaneously, the position and orientation of the tracker are recorded within TCS. Subsequent images of the calibration pattern are shot and the same operations are performed. To compute the camera–tracker relative transformation, at least three images of the calibration pattern are needed, but we use more in order to minimize the effect of noise in measurements of camera and tracker position and orientation in a least-squares sense.

In Figure 4.44, the case of two consecutive positions of the camera–tracker ensemble is illustrated. The RCS for the two receiver positions are termed RCS_1 and RCS_2 and the corresponding CCS as CCS_1 and CCS_2. The following notations are used in Figure 4.44:

- **WT:** transformation matrix from WCS to TCS
- **$TR_{1,2}$:** transformation matrices from TCS to the corresponding location of RCS
- **$WC_{1,2}$:** transformation matrices from WCS to the corresponding location of CCS

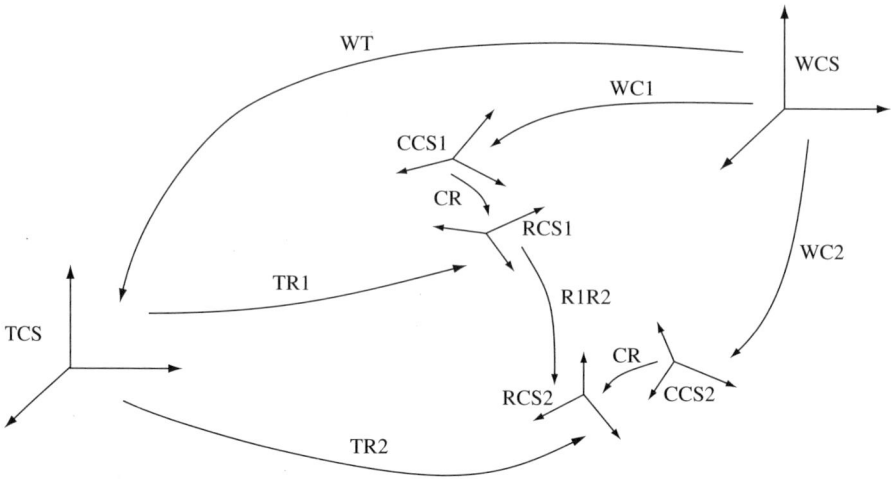

Figure 4.44 Geometry of Camera–Tracker unit.

- **CR:** transformation matrix from CCS to RCS (the parameter to be calibrated)
- R_1R_2: transformation matrix from RCS_1 to RCS_2.

For deriving R_1R_2, the following sequences of transformations are applied: WCS $\rightarrow$ TCS $\rightarrow$ RCS_1 and WCS $\rightarrow$ TCS $\rightarrow$ RCS_2, and the following equation holds:

$$R_1R_2 \cdot TR_1 \cdot WT = TR_2 \cdot WT \tag{4.38}$$

From (4.38), R_1R_2 is derived as follows:

$$R_1R_2 = TR_2 \cdot WT \cdot (TR_1 \cdot WT)^{-1} \tag{4.39}$$

From (4.39),

$$R_1R_2 = TR_2 \cdot TR_1^{-1} \tag{4.40}$$

With R_1R_2 known, the following equation can be written:

$$CR \cdot WC_2 = R_1R_2 \cdot CR \cdot WC_1 \tag{4.41}$$

which is equivalent to

$$CR \cdot WC = R_1R_2 \cdot CR \tag{4.42}$$

where

$$WC = WC_2 \cdot WC_1^{-1} \tag{4.43}$$

The relationship expressed by equation (4.41) is proved by the fact that all the transformations involved in this equation form a closed loop. From (4.42), **CR** can be computed. Thus the relative transformation between the camera and tracker's receiver is recovered, and this transformation is invariant under vantage-point changes, as can be seen from Figure 4.44. Once **CR** has been determined, the coordinate transformation between WCS and CCS can be computed for that particular viewpoint. We consider the previous viewpoint as the reference position in computation. Therefore, for any position k of the camera, the transformation $\mathbf{WC}_k$ can be computed if we rewrite the equation (4.42) as follows:

$$\mathbf{CR} \cdot \mathbf{WC}_k = \mathbf{R}_{k-1}\mathbf{R}_k \cdot \mathbf{CR} \cdot \mathbf{WC}_{k-1} \qquad (4.44)$$

Consequently,

$$\mathbf{WC}_k = \mathbf{CR}^{-1} \cdot \mathbf{R}_{k-1}\mathbf{R}_k \cdot \mathbf{CR} \cdot \mathbf{WC}_{k-1} \qquad (4.45)$$

Equation (4.42) is a matrix equation of the form $\mathbf{AX} = \mathbf{XB}$. There are different methods for solving this type of matrix equation (which is identical to the hand–eye calibration problem in robotics). Further details are omitted.

The initial calibration of the camera–tracker unit is done off-line, so the time required to perform it is not relevant for the time to complete the entire application. The advantage of using a tracking system for extrinsic parameters determination is that it requires neither a calibration pattern (except for the initialization phase) nor any information about the scene, such as the availability of a number of correspondent points in the images.

4.5.3 Simultaneous Depth and Focal Length Optimization in the Presence of Noise

We have seen in the preceding section that by using solely a tracking system, one can recover only the extrinsic parameters of the camera. The intrinsic parameters can be determined at the initialization phase of the camera–tracker unit, as the result of the standard camera calibration process. The principal point can be assumed to be located at the center of the computer frame that stores the captured image. A crucial factor in depth computation is the focal length. When camera undergoes small mechanical or thermal changes, which are beyond the control of the user, the assumption that it remains constant may affect the accuracy of depth computation. In this section we describe a technique, based on Kalman filtering, that deals simultaneously with the inaccuracies in feature extraction due to image noise and optimizes the focal length value automatically.

Addressing Noisy Measurements

Typically, when we extract specific features from images, the accuracy of the output is degraded by image noise. Even though we apply filtering operators prior to feature extraction, still some amount of noise remains, so that it has a negative influence on the precision with which the features are located. Since there is no way that we can completely eliminate the noise from our applications, we have to live with that and find a way to attenuate the effect of noise on the accuracy of measurements. Also, it would be useful to assess the level of confidence in one measurement by having an indication about the proportion of noise influence in the final result. There are some techniques proposed for dealing with noisy measurements. All these techniques originate from control and signal theory, but some of them can readily be adaptable to computer vision applications. Lately, a technique called *Kalman filtering* has become increasingly popular for such applications, due to its computational efficiency and to the fact that it performs well for both linear and nonlinear systems. Kalman filtering also provides a tool for numerically characterizing the uncertainty in measurements corrupted by noise. In this chapter we describe a Kalman filtering–based technique for depth recovery from noisy images and will apply Kalman filtering for extending the tracker-based depth recovery to the optimization of focal length. A description of Kalman filtering is provided in Appendix A3.

Depth Computation and Focal Length Determination

It is well known that to compute the WCS coordinates of a point extracted from an object of the scene visualized, at least two images in which the point is visible are needed. To improve the accuracy of depth computation, we can use all the images in which the point in question is visible and design a Kalman filter to incorporate this "redundant" information. We can see from Appendix A3 that use of a Kalman filter requires the definition of some state variables (that govern the state of a system) and some measurement parameters, based on which the state variables are determined. The need of measurement parameters is justified by the fact that typically the state variables are not directly observable; rather, they are determined indirectly from measurement parameters, which are directly observable. It is assumed that measurement parameters are not entirely accurate, due to noise. In Appendix A3 we see that a Kalman filter would attempt to attenuate the effect of noise by minimizing the error covariance.

The system that models the process of depth computation is time-invariant, because the three-dimensional coordinates of an object point relative to a fixed WCS are not expected to vary with time. Therefore, the state variables are considered to be the world coordinates of an object point (for each point, a separate Kalman filter is designed) and are grouped together in a vector of state variables called a *state vector:* $\mathbf{x} = (x,y,z)^T$. If the simultaneous optimization of the focal length (f) is desired, the state vector is augmented with a fourth variable, the focal length itself: $\mathbf{x} = (x,y,z,f)^T$.

Since the world coordinates of a point are obtained from the image coordinates of two projections of this point in two images, respectively, it seems logical to choose the measurement parameters as the image coordinates of these two projections. We denote the vector of image coordinates as $\mathbf{z} = (u,v)^T$, the measurement vector for the Kalman filter. The two vectors are related (for a single image) according to the following perspective projection relation:

$$\tilde{z} = \mathbf{M} \cdot \tilde{x} \qquad (4.46)$$

where the tilde indicates that the vectors are expressed in projective space $(U,V,S)^T$ and $\mathbf{M}$ is a 3×4 matrix, called the *perspective projection matrix,* whose elements depend on the extrinsic and intrinsic parameters of the camera, as described earlier in the chapter. Transforming equation (4.46) from projective space to Euclidean space, image coordinates u and v can be found as follows:

$$u = \frac{m_{11}x + m_{12}y + m_{13}z + m_{14}}{m_{31}x + m_{32}y + m_{33}z + m_{34}}$$

$$\qquad (4.47)$$

$$v = \frac{m_{21}x + m_{22}y + m_{23}z + m_{24}}{m_{31}x + m_{32}y + m_{33}z + m_{34}}$$

where m_{ij}, $i,j = 1,2,3,4$ are the elements of matrix $\mathbf{M}$. The elements m_{ij} are functions of the extrinsic and intrinsic parameters of the camera. At the stage of depth computation, we know only the extrinsic parameters. We can assume also that the principal point is at the center of the image frame. That leaves only one unknown in the expression of m_{ij}: the focal length f. The dependence between image coordinates and state variables is nonlinear, and therefore the Kalman filter is actually an *extended Kalman filter* (EKF), which is an extension of the classical Kalman filter (which was designed for systems characterized by linear state and measurement equations) to nonlinear systems. We could have chosen extrinsic parameters of the camera (given by the tracker) as measurement parameters as well, but this would have increased the degree of nonlinearity of the function that expresses the state-measurement dependence, thus increasing the possibility of the Kalman filter to diverge. The laser tracker employed in this application is very accurate, and we therefore do not see any reason not to trust the accuracy of extrinsic parameters. Much noisier are the image coordinates extracted from noisy images, so considering them as the measurement parameters seems more appropriate.

The state equations of the Kalman filter are

$$\mathbf{x}_{k+1} = \mathbf{I}\mathbf{x}_k + \mathbf{w}_k \qquad (4.48)$$

where the indices k and $k + 1$ show that the estimate is made at the discrete steps k and $k + 1$, respectively, $\mathbf{I}$ is the 3×3 identity matrix, and $\mathbf{w}_k$ is the process noise vector, zero-mean and with covariance matrix $\mathbf{Q}$. Since the state variables in equation (4.48) are modeled as constants, the matrix $\mathbf{Q}$ is constant throughout the process. Equation (4.47) can be rewritten as

$$\mathbf{z}_k = \boldsymbol{h}(\mathbf{x}_k) \qquad (4.49)$$

where h is a function of the relationship between the measurement and state vectors, with the following form:

$$h(\mathbf{x_k}) = \begin{bmatrix} m_{11}x_k + m_{12}y_k + m_{13}z_k + m_{14} \\ m_{31}x_k + m_{32}y_k + m_{33}z_k + m_{34} \\ m_{31}x_k + m_{32}y_k + m_{33}z_k + m_{34} \\ m_{21}x_k + m_{22}y_k + m_{23}z_k + m_{24} \end{bmatrix}$$

and

$$\mathbf{z_k} = \begin{bmatrix} u_k \\ v_k \end{bmatrix}$$

The function h can be linearized by a first-order approximation of its Taylor series expansion around the current estimate of the state vector (say, $\hat{\mathbf{x}}_k$). Based on this approximation, the measurement equation is

$$\mathbf{z}_k - h(\hat{\mathbf{x}}_k) = \mathbf{J}_k^h(\mathbf{x}_k - \hat{\mathbf{x}}_k) + \mathbf{v}_k \qquad (4.50)$$

In equation (4.50), $\mathbf{J}_k^h$ is the Jacobian of h around the current estimate and $\mathbf{v}_k$ is the 2 × 1 measurement noise vector, zero mean and with covariance matrix **R**. Since the uncertainty in image coordinates measurement is given by the stereo quantization error, whose upper bound is half a pixel, the diagonal elements of **R** are set to 0.5, and the off-diagonal elements are zero, based on the assumption that the errors in u and v are independent.

Filtering begins when for each point of interest extracted from images, the first depth estimate is available through the method described earlier in the chapter. This estimate serves as the initial value of the state variables representing the world coordinates of the point in question. Also, focal length is initialized with the value obtained when precalibrating the camera–tracker unit. The error covariance matrix is initialized to some arbitrary value. (For a thorough description of the elements involved in Kalman filtering, see Appendix A3.) The variances of the state variables representing world coordinates are computed as follows. The classical reconstruction method is applied (without Kalman filtering) for a certain number of points extracted from an object. The disparity between consecutive frames is of the same order of magnitude as the one expected when Kalman filtering is used. Ground truth data are also available. The depth computation results are compared with the ground truth, and for each coordinate, the mean-squared error is considered as the variance of depth computation along that particular coordinate. As for the focal length, multiple camera calibrations using a standard algorithm are performed (without zooming) and the mean-squared error of the focal length is considered its variance and supplied to the Kalman filter. As soon as the point in question becomes visible in another image, the depth is optimized based on standard Kalman filter equations as described in Appendix A3.

Figure 4.45 Images used in reconstruction and extracted points.

In Figure 4.45, an example of applying previous algorithms to an object is shown. The points of interest that have been extracted from the object are marked and the correspondent points in the left and right images are numbered accordingly. Also, three epipolar lines are drawn for the purpose of a better illustration. The best way to test the accuracy in depth computation is to compare the dimensions as obtained using the depth recovery algorithm with the physical dimensions of the object. The results are provided in Table 4.3 for three cases: without Kalman filtering, with a Kalman filter without focal-length autocalibration, and with a Kalman filter with focal-length optimization. As can be seen, the Kalman filter improves significantly the accuracy of three-dimensional measurements in both cases: with or without focal-length optimization.

In Figures 4.46 to 4.48, an object [automated storage and retrieval system (ASRS)] extracted from a manufacturing work cell is shown. Figure 4.46 shows one of the original images, Figure 4.47 shows the points extracted from two images of the sequence (the complete image sequence for this object contains 45 images), and the reconstructed model of the ASRS is shown in Figure 4.48. The numerical results are given in Table 4.4. In Table 4.4, the same kind of comparison between different depth extraction techniques as in Table 4.3 has been performed, for a few dimensions of the object.

TABLE 4.3 COMPARISON OF DEPTH COMPUTATION UNDER VARIOUS CONDITIONS

Dimension (mm)	Physical Dimension (mm)	Without Kalman Filtering			Kalman Filtering Without Focal-Length Optimization			Kalman Filtering With Focal-Length Optimization		
		Measured (mm)	Difference (mm)	Error (%)	Measured (mm)	Difference (mm)	Error (%)	Measured (mm)	Difference (mm)	Error (%)
1–2	762	773.44	11.44	1.5	767.243	5.243	0.68	766.97	4.97	0.65
2–3	1365	1378.3	13.3	0.9	1363.6	1.4	0.1	1363.7	1.3	0.095
4–2	790	780.43	9.57	1.2	789.3	0.7	0.08	789.42	0.58	0.073
5–6	600	604.55	4.55	0.75	600.335	0.335	0.05	600.32	0.32	0.05
3–12	765	758.47	6.53	0.85	763.223	1.78	0.23	763.47	1.53	0.2
3–11	790	779.43	10.57	1.34	785.18	4.82	0.61	786.22	3.78	0.47
4–5	250	243.32	6.68	2.67	250.794	0.794	0.31	250.67	0.67	0.27
5–8	730	741.48	11.48	1.57	738.35	8.35	1.14	738.12	8.12	1.11
8–9	535	522.28	12.72	2.37	530.38	4.62	0.86	530.83	4.17	0.78
7–13	20	18.65	1.35	6.75	19.92	0.08	0.4	19.92	0.08	0.4
6–12	530	524.36	5.64	1.06	528.65	1.35	0.25	528.83	1.17	0.22

TABLE 4.4 DEPTH COMPUTATION ACCURACY IN VARIOUS CASES

Dimension (cm)	Physical Dimension (cm)	Without Kalman Filtering			Kalman Filtering Without Focal-Length Optimization			Kalman Filtering With Focal-Length Optimization		
		Measured (cm)	Difference (cm)	Error (%)	Measured (cm)	Difference (cm)	Error (%)	Measured (mm)	Difference (cm)	Error (%)
1–2	76.2	77.344	1.144	1.5	76.7243	0.5243	0.68	76.697	0.497	0.65
2–3	536.5	537.83	1.33	0.7	536.96	0.46	0.1	536.37	0.13	0.085
1–5	250	243.32	6.68	2.67	250.794	0.794	0.31	250.67	0.67	0.27
3–4	20	18.65	1.35	6.75	19.92	0.08	0.4	19.92	0.08	0.4

Figure 4.46 ASRS used for reconstruction.

Figure 4.47 Points extracted from two images of the ASRS. Matches are numbered accordingly in the left and right image.

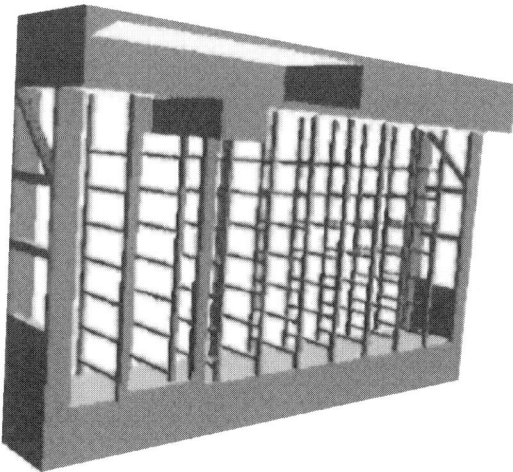

Figure 4.48 ASRS reconstructed from 45 images.

Figure 4.49 World-image projections of certain object points (and lines joining them) superimposed on the original image. Five points are numbered for numerical accuracy assessment purposes.

As can be seen from Figure 4.48, the reconstruction looks quite accurate qualitatively. Another way to assess the numerical accuracy of depth computation is to back-project some of the extracted three-dimensional object points onto one of the images. If the projection points are located close (less than 5 pixels) to the real points in the image, the reconstruction is accurate. In Figure 4.49, the projection results for an illustrative set of points are shown superimposed on the original image (for a better illustration, the projection points were joined with lines). The results shown in Figure 4.49 confirm the accuracy of our method.

Figures 4.50 to 4.53 show another example (a machine from the same manufacturing work cell as the ASRS shown in Figure 4.46). In Figure 4.50, one of the original images is shown (the complete sequence for this object contains 22 images). Figure 4.51 (also shown in the color section) shows the

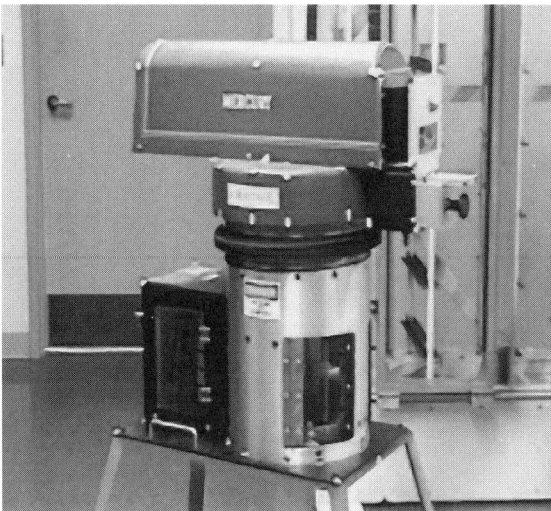

Figure 4.50 Original image of a machine-tool.

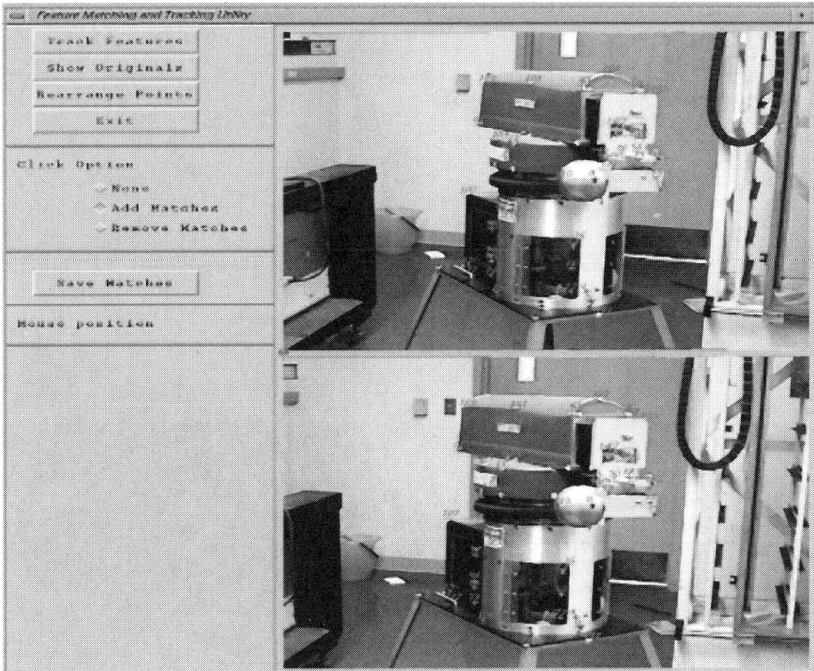

Figure 4.51 Points extracted from the object shown in Figure 4.50 (also shown in the color section).

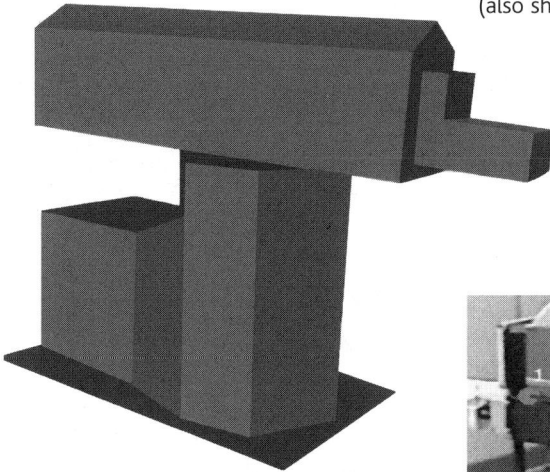

Figure 4.52 Machine reconstructed from 22 images.

Figure 4.53 World-image projections of certain object points (and lines joining them) superimposed on the original image. Seven points are numbered for numerical accuracy assessment purposes.

(a)

Figure 4.54 Manufacturing work cell reconstructed with our algorithms: (a) one of the original images; (b) final object. (Unwanted details have been removed by the user.)

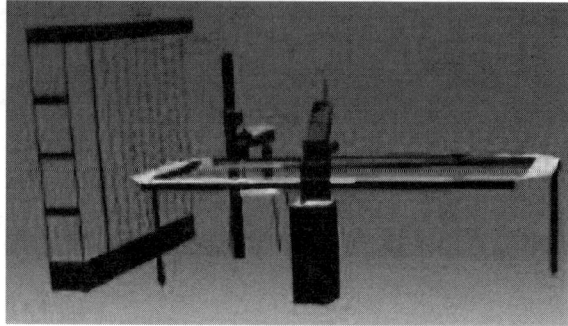

(b)

points extracted from two stereo images of the object. The final object is shown in Figure 4.52. In Figure 4.53, the back-projections of a few points are shown superimposed on the original image. The complete work cell (composed of the objects shown in Figures 4.46 and 4.50 and others) is shown in Figure 4.54.

FURTHER READING

For more details regarding the algorithms presented in this chapter, the interested reader can consult computer vision books, such as Ayache (1991), Jain et al. (1995) and Trucco and Verri (1998). The stereo-matching algorithm described briefly in this chapter is due to Zhang et al. (1994). Other stereo-matching algorithms can be found in Marr and Poggio (1976), Medioni and Nevatia (1985), Ohta and Kanade (1985), Pollard et al. (1985), Prazdny (1985), Ayache and Faverjon (1987), Horaud and Skordas (1989), and Robert and Faugeras (1991). Stereo reconstruction techniques based on the fundamental matrix have been described in Sparr (1991), Faugeras (1992), Luong and

Faugeras (1996), and Zhang (1996). Approaches for recovering the Euclidean structure from projective reconstruction have been developed in Devernay and Faugeras (1995), Hartley (1995), and Pollefeys et al. (1996). The telemetry-based reconstruction technique has been described in more detail in Zetu et al. (1998). Details about traditional camera calibration techniques can be found in Faugeras, Toscani (1986), and Tsai (1987). Methodologies for solving equations of the form $\mathbf{AX} = \mathbf{XB}$ have been described in Chiu and Ahmad (1989) and Tsai and Lenz (1989). Kalman filtering was first described in Kalman (1960), and good references on the topic are Maybeck (1979) and Brown and Hwang (1992).

REFERENCES

Ayache, N., *Artificial Vision for Mobile Robots*. MIT Press, Cambridge, MA, 1991.

Ayache, N., and B. Faverjon, "Efficient Registration of Stereo Images by Matching Graph Descriptions of Edge Segments," *International Journal of Computer Vision*, Vol. 1, No. 2, pp. 107–131, 1987.

Brown, R. G., and P. Hwang, *Introduction to Random Signals and Applied Kalman Filtering*, Wiley, New York, 1992.

Chiu, Y. C., and S. Ahmad, "Calibration of Wrist-Mounted Robotic Sensors by Solving Homogeneous Transform Equations of the form $AX = XB$," *IEEE Transactions on Robotics and Automation*, Vol. 5, No. 1, pp. 16–27, 1989.

Devernay, F., and O. D. Faugeras, *From Projective to Euclidean Reconstruction*, Technical Report 2725, The French National Institute for Research in Computer Science and Control (INRIA), Sophia-Antipolis, France, 1995.

Faugeras, O. D, "What Can Be Seen in Three Dimensions with an Uncalibrated Stereo Rig?" *Proceedings of the 2nd European Conference on Computer Vision*, Santa Margherita, Italy, pp. 563–578, 1992.

Faugeras, O. D., and Toscani, G., "The Calibration Problem for Stereo," *Proceedings of the IEEE Computer Vision and Pattern Recognition Conference*, Miami Beach, FL, pp. 15–20, 1986.

Hartley, R. I. "In Defense of the 8-Point Algorithm." *Proceedings of teh 5th International Conference on Computer Vision*, Cambridge, MA, pp. 1064–1070, 1995.

Horaud, R., and T. Skordas, "Stereo Correspondence through Feature Grouping and Maximal Cliques," *IEEE Transactions on Pattern Analysis and Machine Intelligence*, Vol. 11, pp. 1168–1180, 1989.

Jain, R., *Machine Vision*. Academic Press, San Diego, CA, 1995.

Kalman, R. E., "A New Approach to Linear Filtering and Prediction Problems," *Transactions of the ASME—Journal of Basic Engineering*, Vol. 82, pp. 35–45, 1960.

Luong, Q., and O. Faugeras, "The Fundamental Matrix: Theory, Algorithms and Stability Analysis," *International Journal of Computer Vision*, Vol. 17, pp. 43–75, 1996.

Marr, D., and E. Hildreth, "Theory of Edge Detection," Proc. Roy. Soc. Lond. B207, pp. 187–217, 1980.

Marr, D., and T. Poggio, "Cooperative Computation of Stereo Disparity," *Science*, Vol. 194, pp. 283–287, 1976.

Maybeck, P. S., *Stochastic Models: Estimation and Control,* Vol. 1, Academic Press, San Diego, CA, 1979.

Medioni, G., and R. Nevatia, "Segment-Based Stero Matching," *Comp. Vision and Image Proc.,* Vol. 31, pp. 2–18, 1985.

Ohta, Y., and T. Kanade, "Stereo by Intra- and Inter-scanline Search Using Dynamic Programming," *IEEE Transactions on Pattern Analysis and Machine Intelligence,* Vol. 7, pp. 139–154, 1985.

Pollard, S. B., J. E. Mayhew, and J. P. Frisby. "PMF: A Stereo Correspondence Algorithm Using a Disparity Gradient Limit," *Perception,* Vol. 14, pp. 449–470, 1985.

Pollefeys, M., L. Van Gool, and M. Proesmans, "Euclidean 3-D Reconstruction from Image Sequences with Variable Focal Lengths," *Proceedings of the European Conference on Computer Vision,* Cambridge, UK, pp. 31–42, 1996.

Prazdny, K., "Detection of Binocular Disparities." In M. A. Fischler and O. Firschein, eds., *Readings in Computer Vision: Issues, Problems, Principles, and Paradigms,* Morgan Kaufman Publishers, New York, pp. 756–764, 1987.

Robert, L., and O. D. Faugeras, "Curve-Based Stereo: Figural Continuity and Curvature," *Proceedings of the IEEE Computer Vision and Pattern Recognition Conference,* pp. 57–62, 1991.

Sparr, G., "An Algebraic-Analytic Method for Reconstruction from Image Correspondences," *Proceedings of the 7th Scandinavian Conference on Image Analysis,* pp. 274–281, 1991.

Trucco, E., and A. Verri, *Introductory Techniques for 3D Computer Vision,* Prentice Hall, Upper Saddle River, NJ, 1998.

Tsai, R., "A Versatile Camera Calibration Technique for High-Accuracy 3D Machine Vision Metrology Using Off-the-Shelf TV Cameras and Lenses," *IEEE Journal of Robotics and Automation,* Vol. 3, No. 4, pp. 323–344, 1987.

Tsai, R., and R. Lenz, "A New Technique for Fully Autonomous and Efficient 3D Robotics Hand/Eye Calibration," *IEEE Transactions on Robotics and Automation,* Vol. 5, No. 3, pp. 345–358, 1989.

Zetu, D., P. Banerjee, and P. Schneider, "Data Input Model for Virtual Reality-Aided Facility Layout Design," *IIE Transactions,* Vol. 30, No. 7, pp. 597–620, 1998.

Zhang, Z., *A New Multistage Approach to Motion and Structure Estimation: From Essential Parameters to Euclidean Motion via Fundamental Matrix,* Technical Report 2910, The French National Institute for Research in Computer Science and Control (INRIA), Sophia-Antipolis, France, 1996.

Zhang, Z., R. Deriche, O. Faugeras, and Q. Luong, *A Robust Technique for Matching Two Uncalibrated Images Through the Recovery of the Unknown Epipolar Geometry. Technical Report 2273,* The French National Institute for Research in Computer Science and Control (INRIA), Sophia-Antipolis, France, 1994.

EXERCISES

4.1. In what situations is adaptive thresholding preferred to classical thresholding?

4.2. Discuss the advantages and disadvantages of classical thresholding versus adaptive thresholding. Are there any regions that should not be thresholded when applying adaptive thresholding?

4.3. What are the assumptions used in image averaging?

0	0	0	0	0	0	0	0	0	0
0	2	9	9	0	0	0	0	0	0
0	1	1	1	1	1	1	1	1	0
0	0	1	1	1	0	0	0	1	0
0	0	1	40	40	40	40	40	2	0
0	0	1	40	40	40	40	40	2	0
0	5	6	40	40	40	40	40	3	0
0	7	7	10	10	10	10	10	5	0
0	0	0	0	0	0	0	0	0	0
0	0	0	0	0	0	0	0	0	0

4.4. Consider the image above. A 3×3 mean filter mask is applied on the image. Plot the result of filtering and stress which pixels have been smoothed the most. What is the result of mean filtering?

4.5. Apply 3×3 median filtering on the same image. What are the differences between mean and median filtering?

4.6. Apply 3×3 Gaussian filtering on the same image. State the effects of the Gaussian filter on the given image.

4.7. State three important properties of the Gaussian filter that make it desirable for image filtering.

4.8. Why is rotational symmetry an important property of a Gaussian function, as applied to image processing?

4.9. What is scale-space filtering and how does it work?

4.10. Why should the sum of the elements of an edge detection mask be zero? Exercises 4.11–4.13 are based on the following image.

2	2	2	6	6	2	2	2
2	2	2	6	6	2	2	2
2	2	2	6	6	2	2	2
6	6	6	6	6	6	6	6
6	6	6	6	6	6	6	6
2	2	2	6	6	2	2	2
2	2	2	6	6	2	2	2
2	2	2	6	6	2	2	2

4.11. Plot the response of the Roberts operators along x and y directions of the image above.

4.12. Perform the same operations as in 4.11 using Sobel operators.

4.13. Perform the same operations as in 4.11 using Prewitt operators.

4.14. How can the use of second-derivative operators improve the localization of edge pixels?

4.15. Why is the Laplacian not a good edge operator?

4.16. Describe the Laplacian of Gaussian edge detector. Why is this a good edge detector, as compared to the Laplacian?

4.17. The following control points of a two-dimensional B-spline are given: A(2,2), B(4,2), C(5,3), D(6,3), E(6,4), and F(7,5). The parameter t of the B-spline is assumed to vary between 0 and 1 for each curve segment.

(i) How many different segments does the B-spline have based on the given data?

(ii) Find the B-spline curve equations for each curve segment. Calculate the coordinates of the endpoints (knots) for all segments.

(iii) Calculate the slope of the curve at each knot (if a knot joins two curve segments, calculate the slope for both segments at that particular knot). Prove C^1 continuity.

(iv) Calculate the curvature of the curve at each knot (if a knot joins two curve segments, calculate the curvature for both segments at that particular knot). Prove C^2 continuity.

(v) Sketch by hand a plot of the B-spline.

5

VIEWPOINT-BASED SHAPE RECOVERY FROM MULTIPLE VIEWS

5.1 INTRODUCTION

We presented in Chapter 4 an approach for recovering three-dimensional structure from a sequence of images using stereo vision techniques. The three-dimensional structure recovered from images is represented as a set of unconnected three-dimensional points, extracted from the scene. This is not a suitable representation for VR applications, because a set of three-dimensional points cannot provide a meaningful qualitative representation of a scene. Therefore, a technique for recovering the true shape of the reconstructed objects from the given set of points is needed. A useful geometric structure in this kind of situations is *Delaunay triangulation,* which fills the space occupied by the three-dimensional points with disjoint tetrahedra such that each point of the given set is a vertex of at least one tetrahedron in the triangulation. Other properties of the Delaunay triangulation are discussed in Section 5.2.

There are general cases when no a priori connectivity information between the points extracted from an object is available. The assumption is that the sets of points are relatively dense. In these cases, some general-purpose techniques for building the topology of the object surfaces are available. Such techniques are described briefly in Section 5.3. As will be seen, these techniques do not work well in all cases, especially for objects with small cavities, and they can certainly fail when the sets of points are not dense. Such is the case for sets of points recovered as a result of applying stereo vision algorithms. But in this case, additional information is available and it can be exploited for accurately recovering the shape of the extracted objects. Such a technique is being described in Section 5.4.

5.2 DELAUNAY TRIANGULATION PRELIMINARIES

A triangulation T of a three-dimensional point set S is a collection of disjoint tetrahedra filling the space covered by S such that each point of S is a vertex of at least one tetrahedron in T. The triangulation T is called a *Delaunay triangulation*, denoted by D, if the circumsphere of each tetrahedron *abcd* contains no point of S other than its endpoints *a, b, c,* and *d*. This property is called *insphere criterion* and is illustrated in Figure 5.1, where the points *a, b, c,* and *d* form a tetrahedron that is a part of an already constructed triangulation. Point *e* is not included in the interior of the circumsphere of the tetrahedron *abcd,* so the newly formed local triangulation will be Delaunay, according to the definition. The point e_1 is inside the sphere, so the Delaunayhood is not preserved. Points *a, b, c, d* and e_1 have to be retriangulated to obtain a local Delaunayhood. The point e_2 is located exactly on the sphere. In this case the

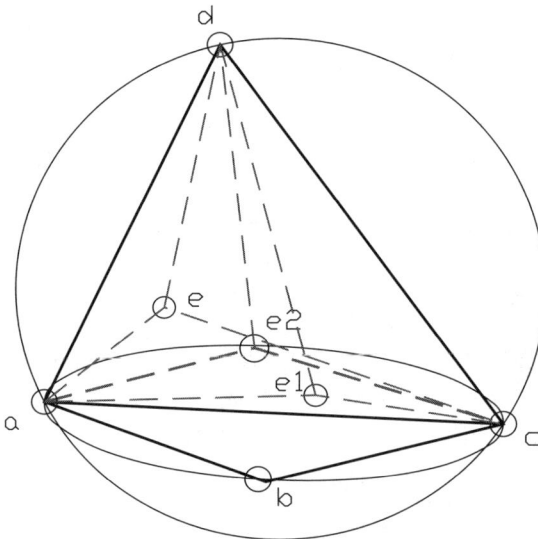

Figure 5.1 In-sphere criterion for preserving the Delaunay triangulation at the local level of five points.

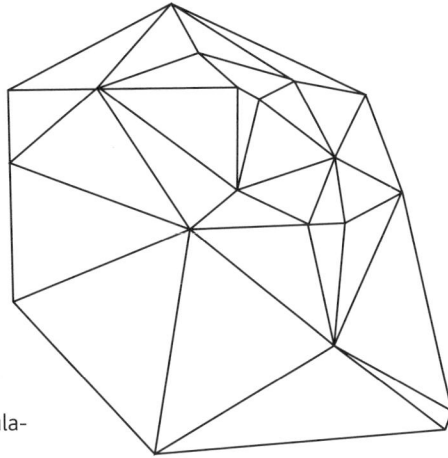

Figure 5.2 Delaunay triangulation of the points.

set of five points is degenerate, and the triangulation is not unique.

Delaunay triangulation is directly related to another geometric structure, the *convex hull* of the set of points. The *convex hull* of a set of points in space is the enclosing convex polyhedron with the smallest volume. In Figure 5.2, an example of Delaunay triangulation of a small set of two-dimensional points is shown. Some of the important properties of Delaunay triangulation are as follows:

1. $D(S)$ is *unique* if no five points of S are cospherical. If this case is encountered, we deal with a *degeneracy* and $D(S)$ is no longer unique (see Figure 5.1).
2. The boundary of $D(S)$ is the convex hull of S.
3. The interior of any tetrahedron of $D(S)$ contains no points of S.

In the remainder of this section, we describe in detail an algorithm that constructs a Delaunay triangulation of a set of three-dimensional points. This algorithm is called the *incremental flip algorithm,* for reasons that will become obvious as we go along. We will not go into the nuts and bolts of the algorithm, but rather, provide brief explanations of the basic operations involved. The incremental flip (IF) algorithm was first proposed for the two-dimensional case. The basic idea behind any Delaunay triangulation algorithm is to construct, at the local level of five points, an arbitrary triangulation and subsequently to restore the local Delaunay property (the observance of the in-sphere criterion). The IF algorithm does that through a sequence of "flips." In two dimensions, the flipping mechanism is illustrated in Figure 5.3.

In Figure 5.3, four points are triangulated first as shown in the left side of the figure. The circle circumscribing the triangle *abc* contains point *d,* and the circle circumscribing *acd* contains point *b.* This means that the triangulation is not locally Delaunay. To restore Delaunayhood, the algorithm flips edge *ac*

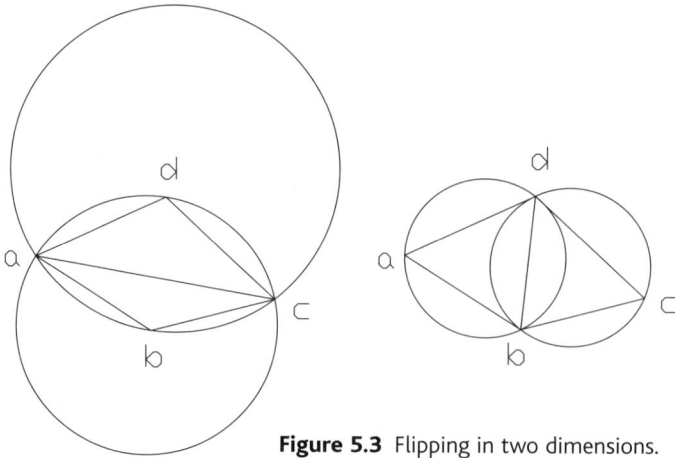

Figure 5.3 Flipping in two dimensions.

with edge *bd* (i.e., edge *ac* is removed and subsequently, edge *bd* is created). In this way, the new triangulation is Delaunay.

In three dimensions, there are two types of flips: edge flips and face flips. To explain flipping in three dimensions, we can use Figure 5.4 as a visual aid. A triangle *t* of a triangulation of the set *S* is locally Delaunay if one of the following conditions is fulfilled:

- *t* lies on the convex hull of the set *S*
- *t* is a common facet of two disjoint tetrahedra of the triangulation whose vertices opposite *t* fulfill the in-sphere criterion (i.e., no one lies inside the circumscribing sphere of the other tetrahedron).

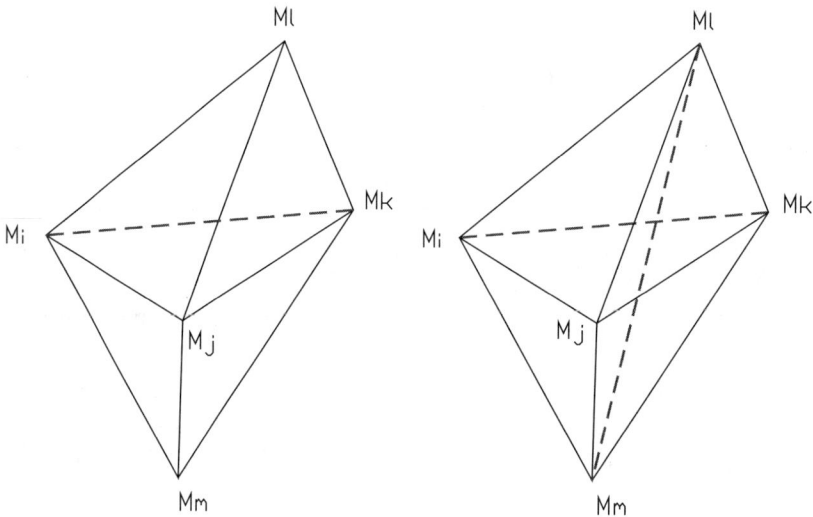

Figure 5.4 Flipping in three-dimensions.

As mentioned before, if a triangle t is not locally Delaunay, the Delaunayhood can be restored by edge or face flips as follows (see also Figure 5.4):

1. In Figure 5.4a, let us assume that triangle $t = \{M_i,M_j,M_k\}$ is not locally Delaunay. Let us denote tetrahedron $\{M_i,M_j,M_k,M_l\}$ by t_1 and tetrahedron $\{M_i,M_j,M_k,M_m\}$ by t_2. If the union between t_1 and t_2 is a convex polyhedron, t can be eliminated from triangulation and the edge $e_{lm} = \{M_l,M_m\}$ is formed, thus creating the tetrahedra $\{M_i,M_j,M_l,M_m\}$, $\{M_j,M_k,M_l,M_m\}$ and $\{M_i,M_k,M_l,M_m\}$ (as in Figure 5.4b). In this way, the local Delaunayhood is restored. This technique is called a *face flip*.

2. Suppose that edge $e_{lm} = \{M_l,M_m\}$ is a part of the triangulation (Figure 5.4b) and it is not locally Delaunay. The Delaunayhood is restored by removing e_{lm} from the triangulation and forming the triangle $t = \{M_i,M_j,M_k\}$ (as in Figure 5.4a). This technique is called an *edge flip* and is the reverse of case 1.

A straightforward generalization of two-dimensional flipping to three dimensions does not work (i.e., it is not possible to create an arbitrary triangulation first and then restore the local Delaunayhood through a sequence of edge and/or face flips). Some tetrahedra might not be transformable and still not be locally Delaunay. This happens when two tetrahedra with a face in common do not form a convex polyhedron. Fortunately, there is a case when, by inserting a point in an already built triangulation, a sequence of edge and/or face flips would always lead to a Delaunay triangulation of the new set of points (to prove this concept is beyond the scope of this book). In this case, the vertices are sorted along a specific direction (say, the x axis). Initially, a tetrahedron is built from the first four points, and subsequently, the remaining points are added to the triangulation incrementally. In this way, a newly added point is always outside the convex hull of the set of points previously inserted into the triangulation. This algorithm can be improved, timewise, by selecting the points of insertion randomly rather than sorting them in increasing order along a coordinate system axis. Randomized point insertion increases the average speed of the algorithm. By inserting a new point s_i randomly in a triangulation D_{i-1}, the point can be either inside D_{i-1} or outside. If s_i is inside D_{i-1}, the tetrahedron that contains s_i is found and s_i is connected with the four vertices of this tetrahedron, thus creating four additional tetrahedra. If s_i is outside D_{i-1}, first the triangular facets of D_{i-1} that are visible from s_i are determined, and subsequently, s_i is connected with the vertices of these triangular facets. After the formation of the new tetrahedra, edge and/or face flips are performed, if necessary, to restore Delaunayhood.

Before we conclude this section, we present the concept of *visibility:* A triangular facet is said to be *visible* from a point if the point is located in the positive half-space of that triangular facet. The positive half-space (a^+) of a trian-

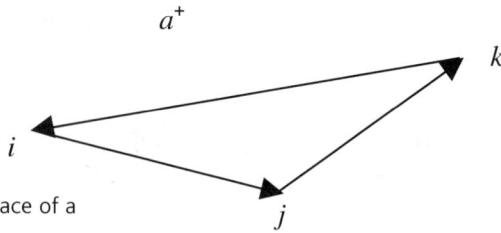

Figure 5.5 Positive half-space of a triangular facet.

gular facet $a = t_{ijk}$ is the space that contains all the points that "see" the points i, j, and k oriented in counterclockwise order (see Figure 5.5). Usually, the order in which a point sees three points of a triangular facet is the order in which the algorithm stores the three points.

5.3 CURRENT SHAPE-RECOVERY TECHNIQUES AND LIMITATIONS

Most techniques for surface reconstruction from a set of unconnected points start with the Delaunay triangulation of the set. The final product of the Delaunay triangulation is the convex hull of the set of points. The convex hull cannot capture any information about cavities an object might have. Therefore, we need to "carve" the convex hull until we are left with a set of triangles that best approximates the real shape of the object.

5.3.1 Boissonnat's Technique

Boissonnat (1984) suggested a Delaunay triangulation-based technique for shape recovery. In this technique, the convex hull of an object is sculptured by iteratively eliminating the tetrahedra that lie in the exterior of the object, according to a specific rule. Each tetrahedron belonging to the Delaunay triangulation of the object points has associated with it a value V that is given by the maximum distance between the faces of the tetrahedron and the corresponding sectors of its circumscribing sphere. *Boissonnat's rule* is to select for elimination, at each iteration, only tetrahedra with exactly one face or exactly two faces on the convex hull. At each iteration, the tetrahedron with the largest V value is eliminated from the triangulation, and the algorithm stops when no more tetrahedra can be eliminated.

When the sets of points are relatively dense and cavities are relatively large, the exterior tetrahedra always have a larger maximum distance from their faces to the corresponding sectors of the circumscribing sphere than the interior ones, and therefore Boissonnat's technique works well and does not fail. Not all sets of points enjoy the above-mentioned properties. There can be

Figure 5.6 Example of an object (desk) for which Boissonnat's technique fails.

special cases where, according to Boissonnat's rule, an interior tetrahedron could have a higher priority than an exterior one. For example, in Figure 5.6, which shows a desk with a few tetrahedra resulting from the Delaunay triangulation superimposed on it, the tetrahedron $ABEF$ (with base ABE drawn with dark lines) is an interior one, and the tetrahedron $ABCD$ (with base ACD drawn with white lines) is exterior to the object. The angles at F and B are right angles, so both tetrahedra are enclosed by a hemisphere, respectively. Since $BF > BD$ and $AF > AD,$ the radius of the sphere circumscribing $ABEF$ is greater than the radius of the sphere circumscribing $ABCD,$ so the value (V) associated with $ABEF$ (as defined above) is greater than the value associated with $ABCD$. Both tetrahedra have two faces on the convex hull formed by the set of points defining the object. So, according to Boissonnat's rule, the tetrahedron $ABEF$ will be eliminated first, and consequently, the shape of the desk will be destroyed.

5.3.2 Three-Dimensional α-Shapes

More recently, a new geometric structure proposed to define the shape of certain three-dimensional objects is the α-shape concept, developed at the University of Illinois at Urbana–Champaign between 1990 and 1995 (see "Further Reading" for more information). Conceptually, α-shapes are a family of shapes, parameterized by a real α, ranging from 0 to ∞. The α-shape of a point set is a polytope that is neither necessarily convex nor necessarily connected. When $\alpha = \infty$, the α-shape is identical to the convex hull of the point set. When $\alpha = 0$, the α-shape is the point set itself, with no connection between any two points. When α decreases, certain triangles or edges disappear from the triangulation. Intuitively, a geometrical entity (triangle or edge)

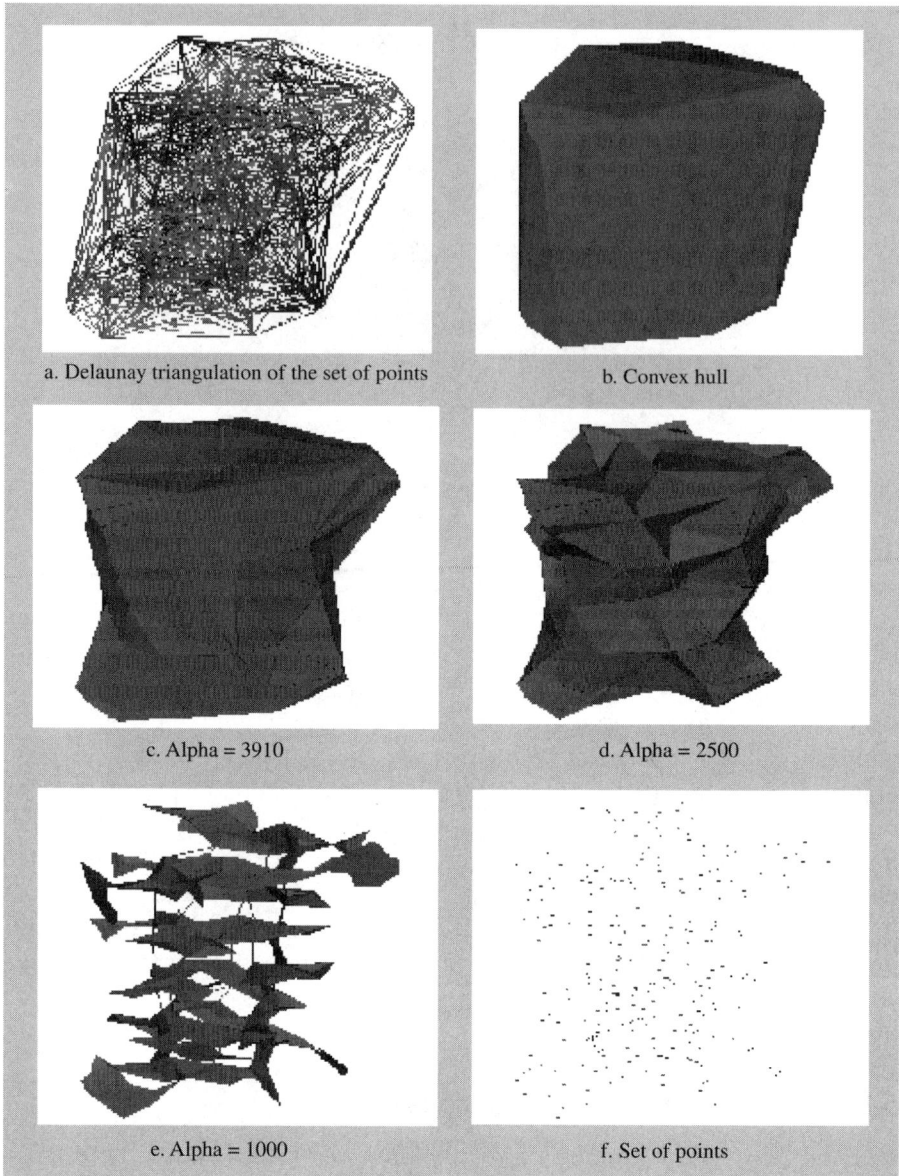

a. Delaunay triangulation of the set of points

b. Convex hull

c. Alpha = 3910

d. Alpha = 2500

e. Alpha = 1000

f. Set of points

Figure 5.7 Alpha-shapes of a molecule in various cases.

disappears from a triangulation when α becomes small enough so that a sphere with radius α can occupy the space of the geometrical entity, without enclosing any of its vertices.

Initially, α-shapes were designed to represent certain geometric objects (such as molecules) under different situations and meanings. In Figure 5.7, an example of a molecule is shown for different values of α. As can be seen, the

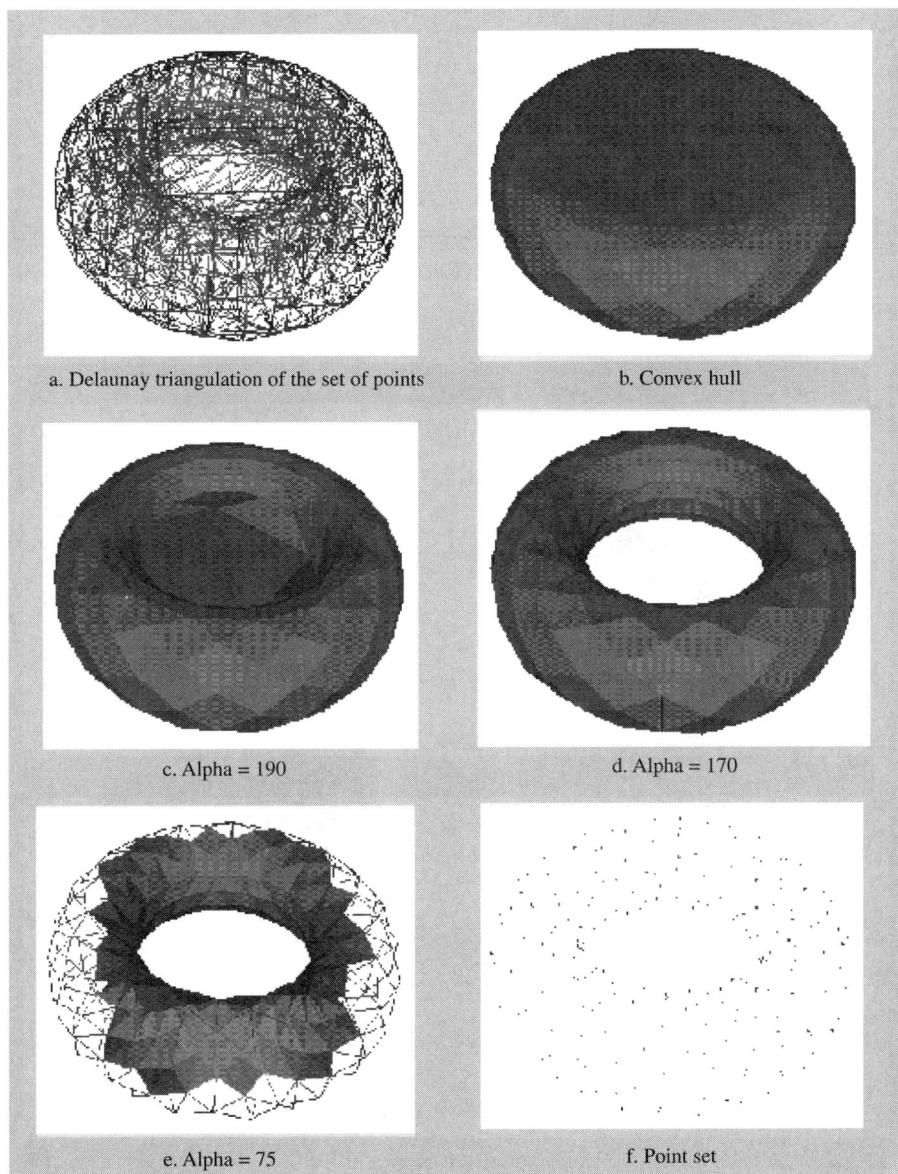

a. Delaunay triangulation of the set of points

b. Convex hull

c. Alpha = 190

d. Alpha = 170

e. Alpha = 75

f. Point set

Figure 5.8 Alpha-shapes of a torus in various cases.

α-shapes of the molecule have no immediate meaning to the inexperienced viewer, but can very well represent a time-varying molecule. In Figure 5.8, an example of a torus is shown. In this case, the α-shape concept works quite well, in the sense that there exist certain values for α that can accurately preserve the real shape of the object. Note that the torus is a good example of a dense set of points and large cavities.

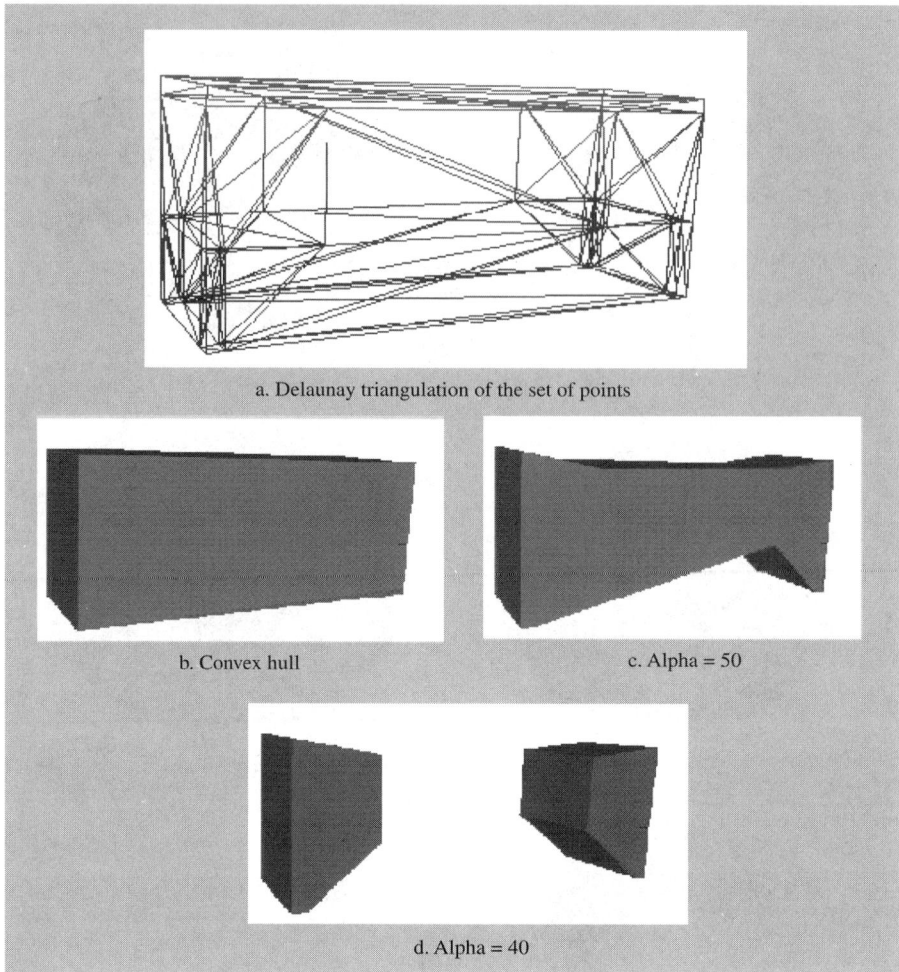

a. Delaunay triangulation of the set of points

b. Convex hull

c. Alpha = 50

d. Alpha = 40

Figure 5.9 Alpha-shapes of a desk in various cases. Note that the true shape of the desk cannot be recovered for any value of α.

Figures 5.9. and 5.10 illustrate two cases where the α-shapes technique cannot provide an accurate representation of the "real" object. The two objects shown in these figures are defined by nondense sets of points, and there are cavities that are "smaller" than the density of points. In Figure 5.9, the example of a desk is provided. No suitable α-value was found to accurately represent the desk. The same situation occurs with the cubicles shown in Figure 5.10. A few α-shapes are shown, but no value of α can recover the true shape of the object. It is obvious from previous examples that a different approach is needed in order to accurately recover the shape of objects in a wider variety of cases, possibly with additional information.

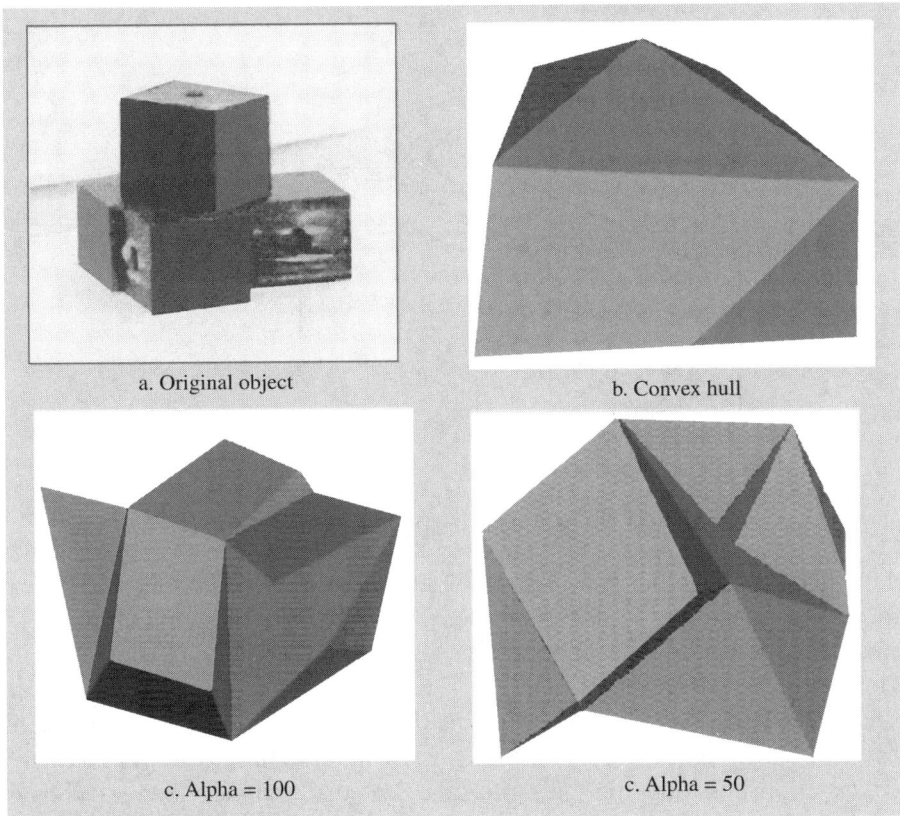

a. Original object

b. Convex hull

c. Alpha = 100

c. Alpha = 50

Figure 5.10 Example in which α-shapes do not work.

5.4 VIEWPOINT-BASED APPROACH FOR SHAPE RECOVERY

There are situations when more than the set of points is available. Such is the situation described in Chapter 4. In this case, the set of points has been obtained from two-dimensional images of an object. As described in Chapter 4, a characteristic of the depth recovery process is that camera location is known at the time of image capture. As we describe in this section, we can exploit this additional information to improve the process of shape recovery.

A two-dimensional image of an object portrays that object as seen from a certain viewpoint. It is not necessary to see all the points that define that particular object completely. In fact, in a particular image, only a subset of the entire set of points is visible from the viewpoint from where the image was taken. When applying a Delaunay triangulation on the set of points, the convex hull is obtained. The convex hull fills the cavities of the object with unwanted triangles. Our approach is intuitively as follows: Define imaginary

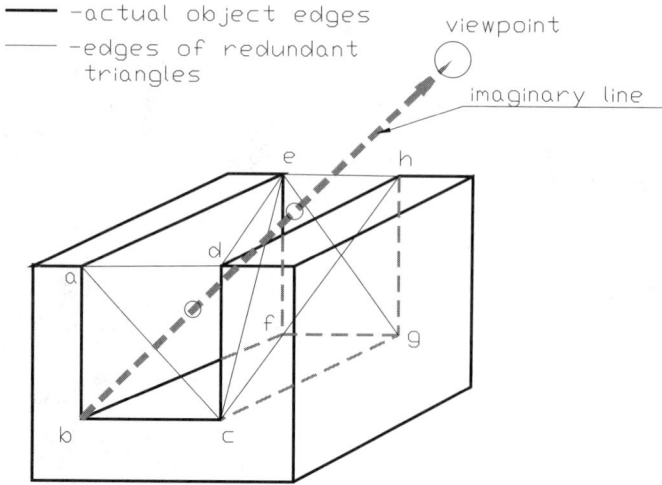

Figure 5.11 Imaginary line intersecting redundant triangles.

lines joining a viewpoint with the points visible from that viewpoint. Then those lines will intersect some exterior triangles. The triangles that are intersected by imaginary lines at least once are eliminated from the triangulation. This concept is illustrated in Figure 5.11. In this figure, the imaginary line intersects the triangles *ace, cde, ceg,* and *dhe* (for simplicity, only the intersections between the imaginary line and triangles *ace* and *ceg* have been marked with circles). Consequently, triangles intersected by the imaginary line are considered redundant (exterior) and are therefore eliminated from the triangulation.

5.4.1 Shape Recovery Algorithm

The shape recovery procedure starts by grouping the point set S into subsets $S_i, i = 1,2,\ldots,n$. Each subset contains points that were extracted from a particular image, corresponding to viewpoint V_i. The subsets can overlap. The grouping of the point set into subsets is illustrated in Figure 5.12. The flow-

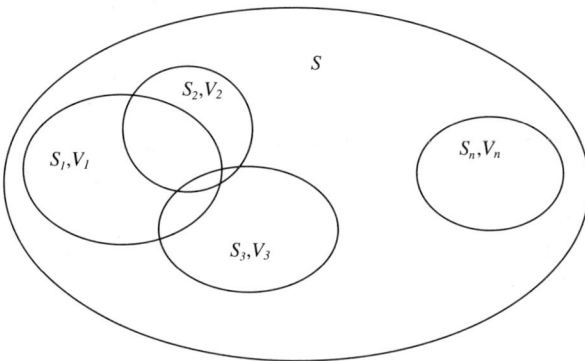

Figure 5.12 Grouping the set of points S into subsets S_i, associated with viewpoint V_i.

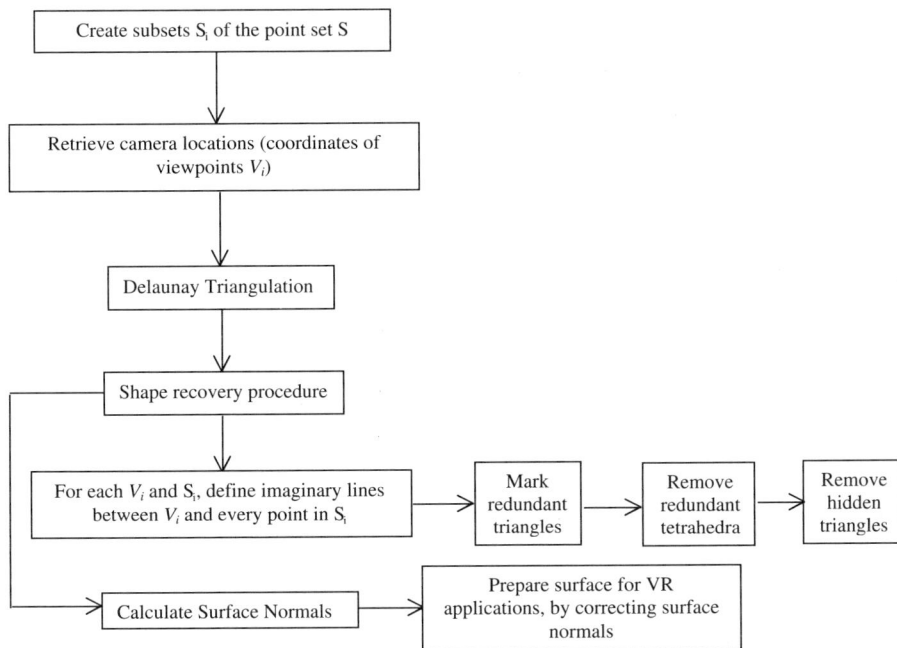

Figure 5.13 Shape recovery procedure.

chart of shape recovery algorithm is shown in Figure 5.13 and the pseudocode looks as follows:

```
- For(all viewpoints V_i and subsets S_i)
{
• Define imaginary lines between V_i and every point in S_i
• Mark all triangles intersected by imaginary lines for
deletion
}
- Remove all tetrahedra that contain at least one marked
triangle
- Eliminate all triangles that lie in the interior of the
object
```

The key point of the algorithm is to find triangles obstructing the visibility of a point from a certain viewpoint. If there is such a triangle, that triangle is classified as redundant and consequently, is marked for deletion. To determine whether an imaginary line intersects a triangle, we need to determine whether the point of intersection between the line and the plane formed by the vertices of the triangle is located in the interior of the triangle.

5.4.2 Point-in-Polygon Testing

The problem of determining whether a point is inside or outside a polygon is an important one in computational geometry applications. A very popular algorithm for solving this problem is known as the *Jordan curve theorem*. This algorithm works by shooting a ray in an arbitrary direction and counts the number of polygon edges crossed. If the number of crossings is odd, the point is inside the polygon; otherwise, it is outside. An example is given in Figure 5.14. In Figure 5.14, the ray R_1 crosses the polygon

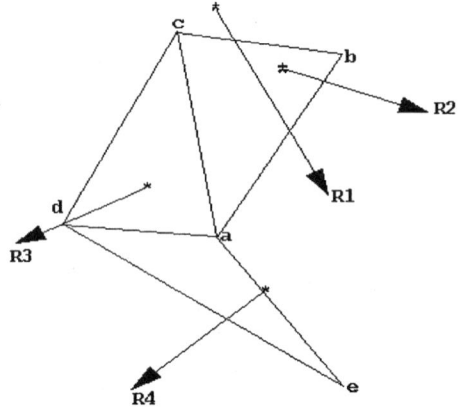

Figure 5.14 Jordan curve theorem.

twice, so its point of origin is outside the polygon. Ray R_2 crosses only once, so it originates from a point that is inside the polygon. This technique gets complicated when a ray intersects a vertex of the polygon (R_3 in Figure 5.14), or the point of origin is located on an edge of the polygon (R_4 in Figure 5.14).

When the polygons are triangles, a simpler technique exists. First project the triangle in question onto the planes formed by the axes of the Cartesian coordinate system. Thus three projections are obtained: onto the *xy, yz,* and *zx* planes, respectively (see Figure 5.15). Out of the three projections, the one

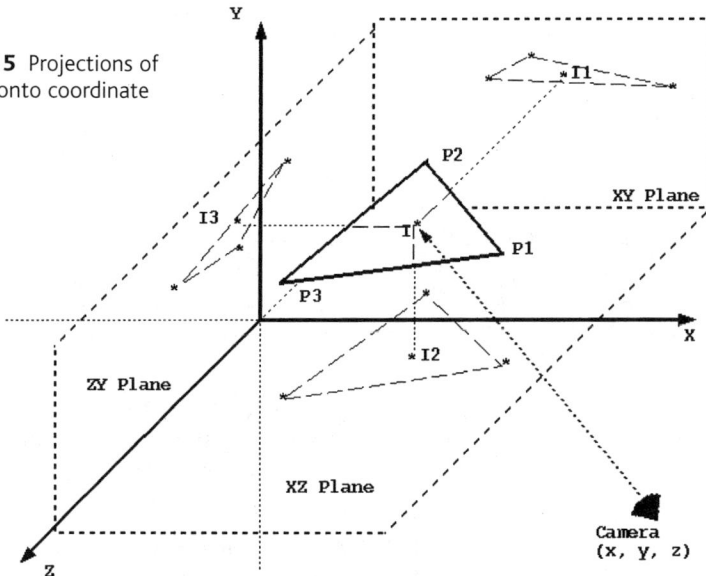

Figure 5.15 Projections of a triangle onto coordinate planes.

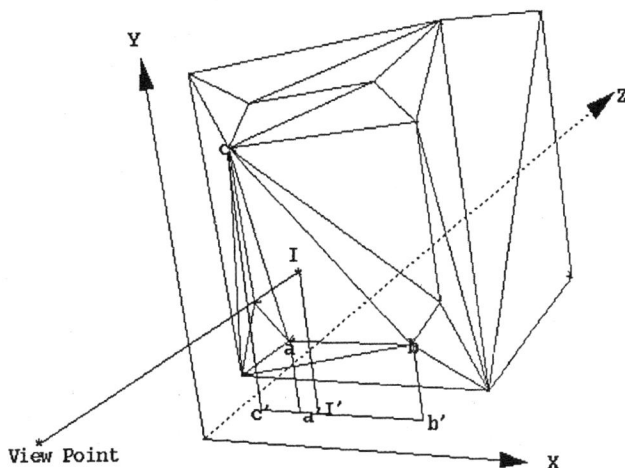

Figure 5.16 Triangle *abc* is projected onto the *zx* plane into a line.

with the largest area is selected. The rationale for selecting the projection with the largest area is to avoid degenerate cases, such as when two vertices of a triangle have the same coordinate along a coordinate axis. If two vertices have the same *x* coordinate, for example, the projection onto *xy* plane will have zero area and we cannot ascertain whether or not a point is inside or outside the triangle; see Figure 5.16, where triangle *abc* has the edge *ac* parallel to the *y* axis, so projection onto the *zx* plane is reduced to a line).

To speed up the computations, instead of computing the area of each projection triangle, we can compare all the *X, Y,* and *Z* coordinate values of the triangle's vertices, and select a plane which includes the largest min/max differences, as follows:

$$X_i = \max\{X_1,X_2,X_3\} - \min\{X_1,X_2,X_3\} \qquad X \text{ coordinates of three vertices of triangle}$$

$$Y_i = \max\{Y_1,Y_2,Y_3\} - \min\{Y_1,Y_2,Y_3\} \qquad Y \text{ coordinates of three vertices of triangle}$$

$$Z_i = \max\{Z_1,Z_2,Z_3\} - \min\{Z_1,Z_2,Z_3\} \qquad Z \text{ coordinates of three vertices of triangle}$$

$$i = \{0,1,2\}$$

The largest two values from the comparison above will be selected as a projection plane (if the *X* and *Z* values are the largest two values, then the *xz* plane will be used as projection plane for further in/out evaluation). The advantage of selecting the projection with the largest possible area for a particular triangle is that it eliminates many possible exceptions (if one of the coordinates is the same for two vertices, it is still possible to identify if the line–plane intersection would occur inside the triangle).

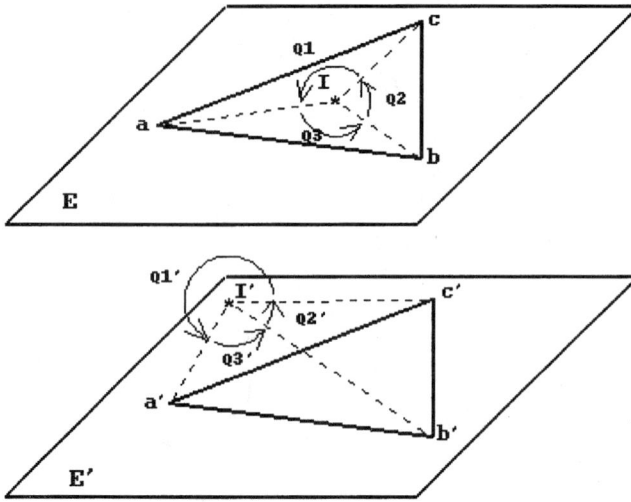

Figure 5.17 Intersection point inside or outside a triangle.

Once the largest possible two-dimensional projection of the triangle is selected, there can be two different relationships between the intersection point and triangle (Figure 5.17). As shown in Figure 5.17, if we connect the three vertices of the triangle with the intersection point, we obtain three angles, and if one of these angles is larger than or equal to 180°, we can conclude that the point is outside the triangle. If all three angles are less than 180°, we conclude that the intersection occurs inside the triangle, and this triangle shall be marked as exterior. In Figure 5.17, point I is inside the triangle abc, and the point I' is outside the triangle $a'b'c'$ (angle Q'_1 is greater than 180°).

To avoid ambiguities generated by rounding errors, we add a tolerance (ϵ). If an angle is greater than (180° + ϵ), the intersection point is exterior to the triangle (if one angle is exactly 180°, the intersection point is located on an edge of a triangle and this could generate an ambiguity due to rounding errors). In our experiments, $\epsilon = 0.01$.

5.4.3 Removing Redundant Tetrahedra

The goal of the shape recovery algorithm is to remove exterior triangles that obstruct the visibility of object points from a particular viewpoint. The algorithm demonstrated above is able to detect triangles that fall between camera and object points. It is not guaranteed that all the exterior triangles will be intersected by an imaginary line at least once. This depends on the number of images available. To ensure that all the exterior triangles will be intersected, the operator in charge with image acquisition should consider all possible situations and capture images from various viewpoints. One case is studied in

Figure 5.18, where the triangle formed by points 1, 4 and 5 is not intersected by any imaginary line defined by two cameras oriented toward the object. When the size of the triangle is small relative to the object, or the number of viewpoints available is not sufficient, it is possible to have exterior triangles left undetected in the first stage of our algorithm. Two camera images are sufficient to recover the depth of a particular point (recall Chapter 4), but if we are dealing with small triangles, this number may not be enough to recover the shape of the object. Even though it cannot guarantee the removal of all the exterior triangles, an alternative approach exists to improve the shape recovery process: working with tetrahedra instead of triangles. For example, in Figure 5.18, even though the triangle t, formed by points 1, 4, and 5, is not intersected by any imaginary line, those lines intersect other triangles, adjacent to t in the Delaunay triangulation, that are faces of the same tetrahedra as t. Such triangles can be the ones formed by points 4, 5, and 6 and 2, 4, and 5, respectively. (These triangles are not drawn in Figure 5.18, for reasons of simplicity.) If one tetrahedron contains at least one triangle that was previously marked as redundant, it is removed from the triangulation.

5.4.4 Removing Hidden Triangles

The first two steps of the shape recovery procedure are sufficient in providing the final shape of the object, but they are not able to remove unnecessary triangles, located in the interior of the object. These triangles could cause a computational overhead especially for VR applications, where rendering speed is of utmost importance. We do not really need the interior triangles for most applications. All that is needed is a suitable representation of the surface of an object. Therefore, the interior (hidden) triangles should be removed.

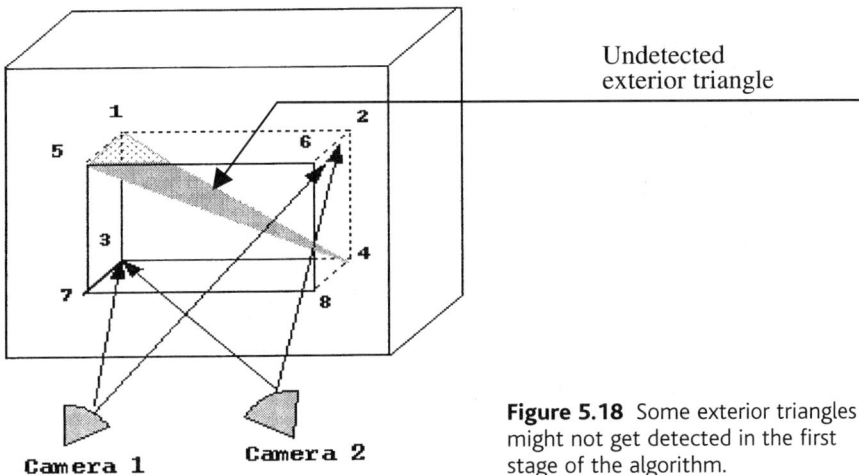

Figure 5.18 Some exterior triangles might not get detected in the first stage of the algorithm.

The removal of hidden triangles is conceptually simple. Up to this point, we have eliminated the triangles located in the exterior of the object, so we are left with a collection of triangles that best preserves the shape of the object. Intuitively, we can imagine that a triangle that belongs to the surface of the object can be a facet of only one tetrahedron of the triangulation. On the other hand, the hidden triangles are common faces of two tetrahedra. Therefore, after exterior triangle removal, the triangles that are common faces of two tetrahedra are considered interior and therefore eliminated from the triangulation. After removal of exterior and interior triangles, we are left with a collection of triangles that represent the object's surface.

5.4.5 Correcting Face Normals

After all the redundant triangles (exterior and interior) have been removed, the surface of the object has to be prepared properly for VR applications. This entails computation of the triangular facets normals for proper visualization. Typically, in VR applications, Inventor™ or Performer™ compilers are used to view three-dimensional models. For the VR simulation, Performer format for the CAVE™ is used. All these visualization techniques require calculation of surface normals. In our technique, surfaces are created from scattered point sets, and initially, directions of normals are not known.

After retrieving the surface of the object, the normals of the triangular facets are set to point toward the positive half-space of each triangle, as stored by the Delaunay triangulation algorithm. The problem with this approach is that Delaunay triangulation stores the vertices of each triangle to serve the visibility purposes as internally required by the algorithm. This order might not be suitable for visualization purposes. It is well known from computer graphics that a surface patch is visible from a particular viewpoint if the normal of the patch points toward the viewpoint. As a result of the shape recovery algorithm, this might not be the case with all facets and all available viewpoints (the viewpoints are the camera locations when capturing images of the scene).

To correct the visualization problem, the following algorithm is designed. Consider the same viewpoints as the ones used for shape recovery, and consider all triangular facets of the object surface (already known) visible from each viewpoint (the triangular facets that are visible from a particular viewpoint are defined by points visible from that viewpoint). For each viewpoint, we check whether the normal of each visible triangular facet points toward the viewpoint. This is done as follows (the illustration is provided in Figure 5.19). Consider the triangular facet defined by points a, b, and c and a viewpoint V (Figure 5.19). Suppose that the normal of the triangle $t = abc$ (denoted $\mathbf{N}_t$) points towards the positive half-space of t. To check whether $\mathbf{N}_t$ points toward V, we simply check whether the angle (θ) defined by the positive direction of $\mathbf{N}_t$ and the direction vector ($\mathbf{N}_a$) of the line that joins V with an arbitrary vertex

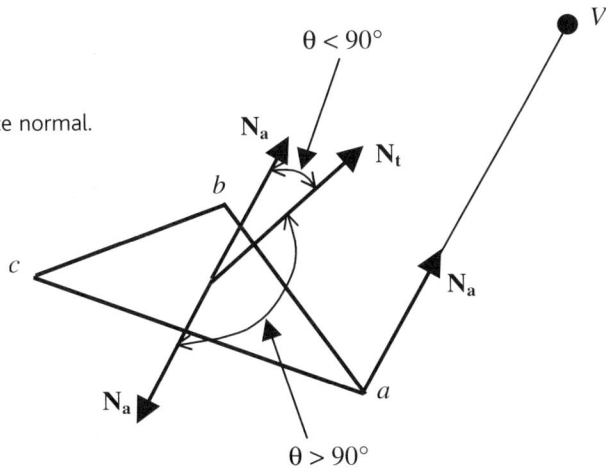

Figure 5.19 Correcting face normal.

of t (say, a) is less than 90°. The vector $\mathbf{N}_a$ points from a to V. If $\theta < 90°$, t is visible from V in any VR modeling software. Otherwise, the direction of $\mathbf{N}_t$ should be reversed. The pseudocode of the above-mentioned algorithm looks as follows:

```
For(all viewpoints V_i)
    For(all surface triangular facets t_i visible from V_i)
    {
    •  define triangle normal N_t
    •  define direction vector N_a from a vertex a of t_i to V
    •  if the angle θ between N_t and N_a is greater than 90°
          revert the direction of N_t
    }
```

5.4.6 Performance and Complexity Analysis

Our shape recovery algorithm has five major components that can have an impact on the overall CPU time required by the algorithm:

1. Delaunay triangulation
2. Checking for redundant triangles
3. Removing redundant tetrahedra
4. Removing hidden triangles
5. Correcting normals

Delaunay triangulation performance is dependent on the number of points in the set and independent of the number of viewpoints. The other operations (2 to 5 above) depend on both the number of vertices and the number of viewpoints. We analyze the CPU cycle distribution only as a function of the number of viewpoints (it is anticipated that as the number of points in the set increases, CPU time increases as well). The analyses below have been performed on a 200-MHz SGI O2 workstation, running IRIX 6.5, with 256 megabytes of main memory. The number of points in the set is 29. The times required by each operation when the number of viewpoints varies, and the CPU cycle distribution are given in Table 5.1. From Table 5.1 we can see that most of the CPU time is taken by Delaunay triangulation and the detection of redundant triangles. When the number of viewpoints increases, CPU time and percentage of CPU cycle used by redundant triangle check increase also.

TABLE 5.1 CPU CYCLE DISTRIBUTION FOR MODULES OF THE SHAPE RECOVERY ALGORITHM

	Time (s)				
Number of Viewpoints	*Delaunay Triangulation*	*Checking for Redundant Triangles*	*Removing Redundant Tetrahedra*	*Removing Hidden Triangles*	*Correcting Face Normals*
7	0.641	0.5	0.0719	0.05	0.1
10	0.641	0.635	0.072	0.057	0.11
15	0.641	0.79	0.083	0.055	0.12
23	0.641	1.1	0.08	0.062	0.12
Percentage of CPU time	39	45	4.8	3.4	7.8

The computational complexity of our algorithm depends on the number n of points in the set and the number p of viewpoints used for shape recovery. The complexity of the incremental flip algorithm for Delaunay triangulation is $O(n^2)$. It is well known that the expected number of triangles created by Delaunay triangulation is at most $9n + 1$. The worst-case scenario for our algorithm occurs when all triangles are visible from every viewpoint (although this never happens in practice). The point-in-triangle testing is being performed in $O(1)$ time. Therefore, the process of marking redundant triangles has a worst-case complexity of $O((9n + 1)p) = O(np + p)$. The other operations of the algorithm are performed in constant time. Consequently, the total computational complexity of our shape recovery algorithm is $O(n^2 + np + p)$. We can see that the complexity of the algorithm is dominated by the complexity of the Delaunay triangulation for small p. Since, typically, $p < n$, our algorithm can be performed in a reasonable amount of time.

5.4.7 Examples

In this section, a few examples of shapes recovered with the algorithm described here are provided. Figures 5.20 and 5.21 show two artificially gener-

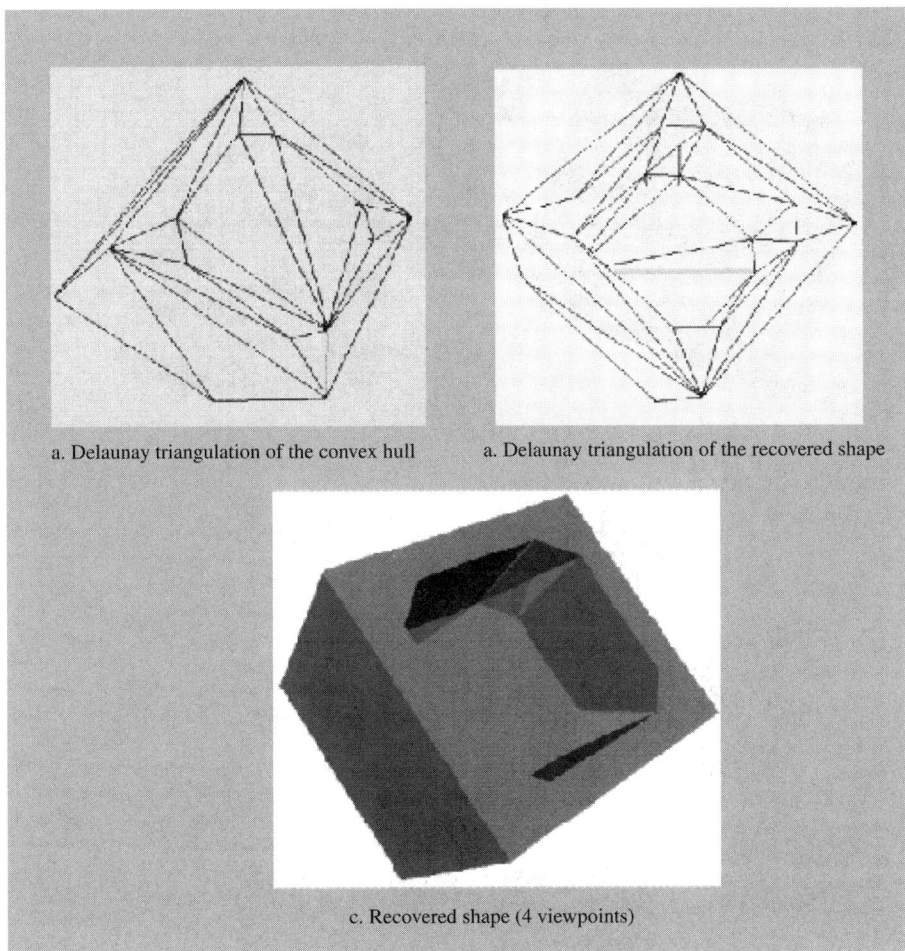

a. Delaunay triangulation of the convex hull   a. Delaunay triangulation of the recovered shape

c. Recovered shape (4 viewpoints)

Figure 5.20 Cube with one cavity recovered with our procedure.

ated objects whose shapes have been recovered by applying our shape recovery methodology. The same methodology is applied on two real objects, shown in Figures 5.22 and 5.23, respectively. The object in Figure 5.23 is the same as the one illustrated in Figure 5.10. It can be seen that unlike Boissonnat's technique or α-shapes, our method can accurately recover the shape of these objects, provided that a sufficient number of images (and, implicitly, viewpoints) is available. The examples provided in this section show that our methodology can handle a wider variety of shapes (in particular, shapes recovered from nondense sets of points) than can current techniques, by using specific information derived from the methodology the sets of points were obtained (i.e., extracted from two-dimensional images).

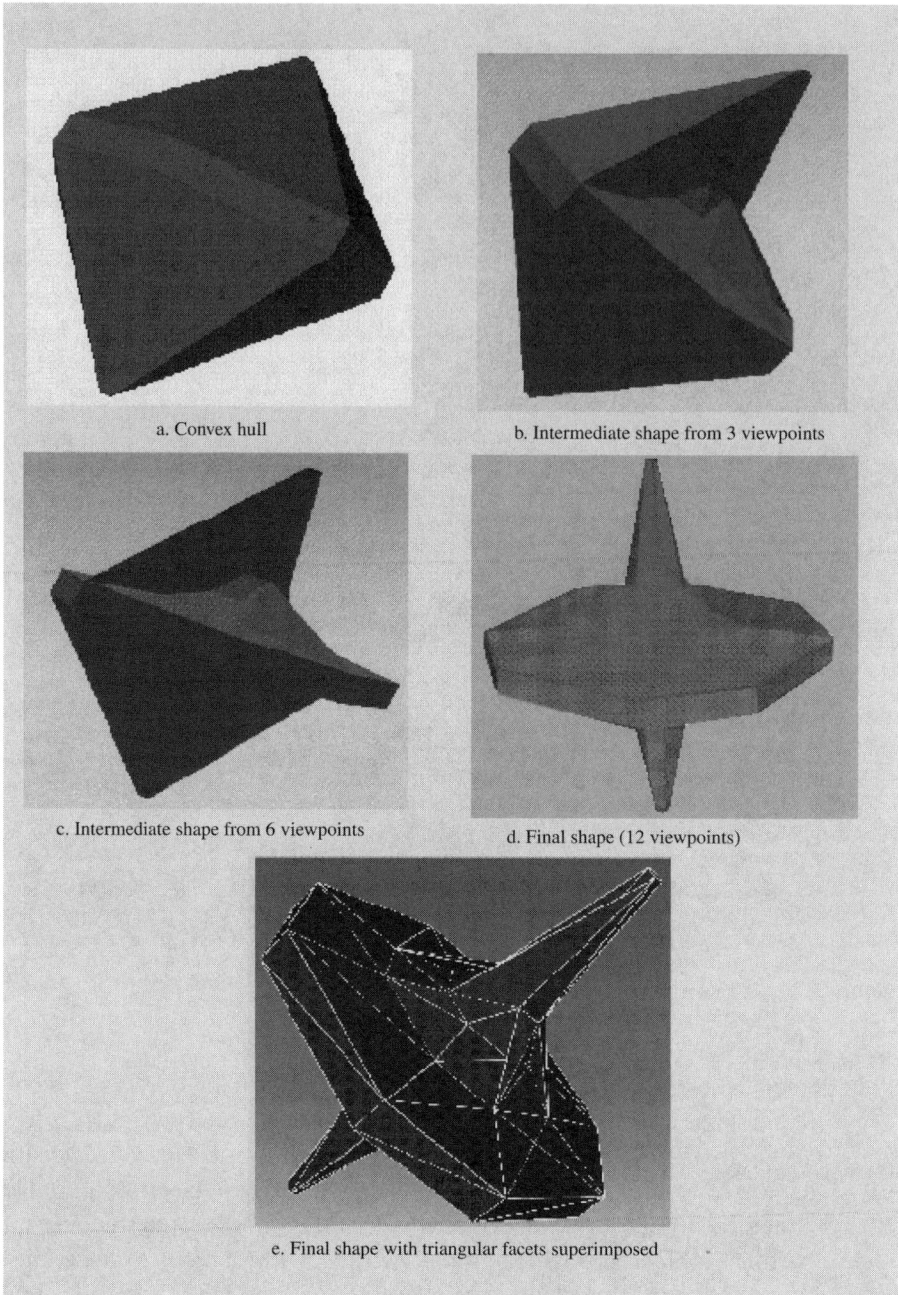

a. Convex hull

b. Intermediate shape from 3 viewpoints

c. Intermediate shape from 6 viewpoints

d. Final shape (12 viewpoints)

e. Final shape with triangular facets superimposed

Figure 5.21 Synthetic object recovered with our procedure.

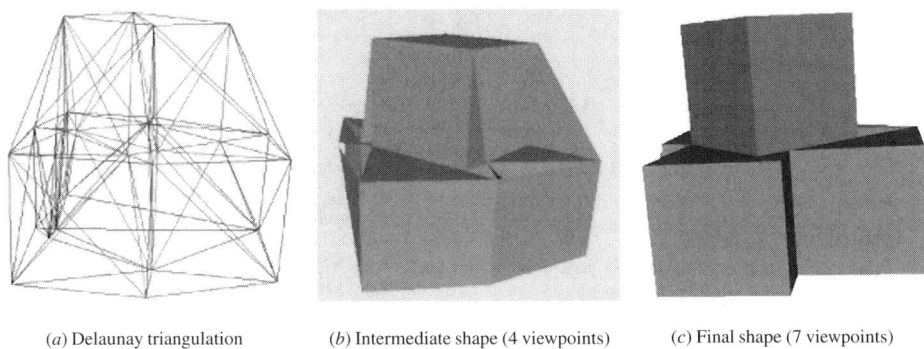

(*a*) Delaunay triangulation   (*b*) Intermediate shape (4 viewpoints)   (*c*) Final shape (7 viewpoints)

Figure 5.22 Object from Figure 5.10 recovered with our methodology.

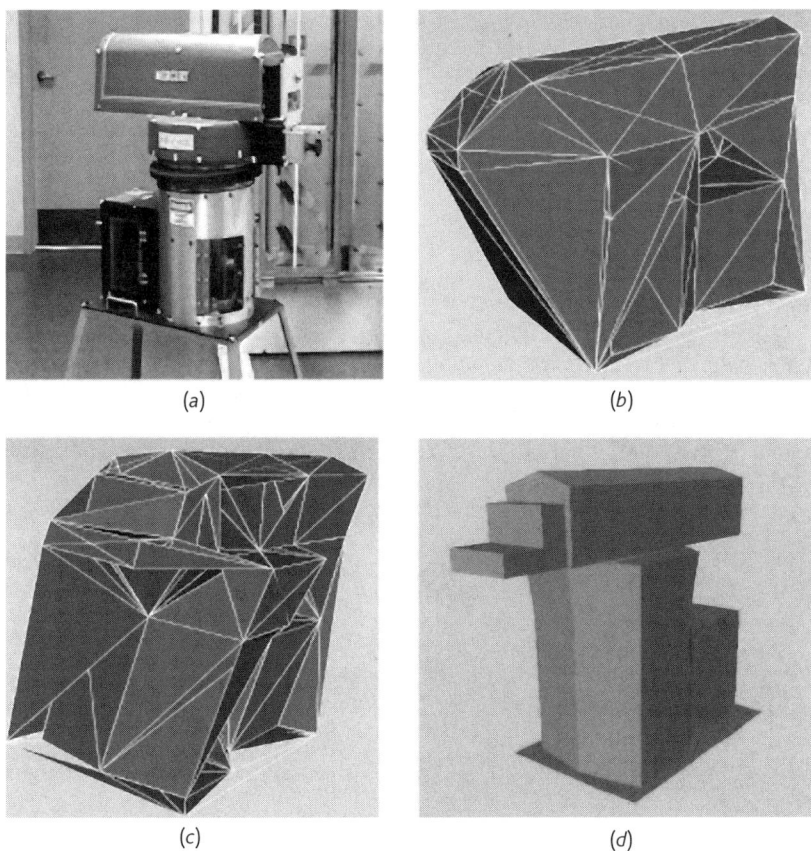

(*a*)   (*b*)

(*c*)   (*d*)

Figure 5.23 Object recovered with our methodology: (*a*) original object; (*b*) intermediate shape from 1 viewpoint; (*c*) intermediate shape from 7 viewpoints; (*d*) final shape (17 viewpoints). (Also shown in the color section.)

FURTHER READING

Additional information about the computational geometry concepts presented in this chapter can be found in Preparata and Shamos (1985) and de Berg et al. (1997). The same concepts, with an emphasis on their computer implementation in C language, were described in detail in O'Rourke (1994). The incremental flip algorithm for three-dimensional Delaunay triangulation was described first by Joe (1988). Edelsbrunner and Shah (1992) have improved the performance of this algorithm by adding the capability of randomized point insertion. Boissonnat's technique is due to Boissonnat (1984). The concept of α-shapes has been proposed by Edelsbrunner and Mucke (1994).

REFERENCES

Boissonnat, J. D., "Geometric Structures for Three-Dimensional Shape Representation," *ACM Transactions On Graphics,* Vol. 3, No. 4, pp. 266–286, 1984.

De Berg, M., *Computational Geometry: Algorithms and Applications,* Springer-Verlag, New York, 1997.

Edelsbrunner, H., and E. P. Mucke, "Three-Dimensional Alpha-Shapes", *ACM Transactions on Graphics,* Vol. 13, No. 1, pp. 43–72, 1994.

Edelsbrunner, H., and N. R. Shah, "Incremental Topological Flipping Works for Regular Triangulations," *Proceedings of the Eighth Annual Symposium on Computational Geometry,* 1992, pp. 43–52.

Joe, B., "Three-Dimensional Triangulations from Local Transformations," *SIAM Journal on Scientific and Statistical Computing,* Vol. 10, No. 4, pp. 718–741, 1988.

O'Rourke, J., *Computational Geometry in C,* Cambridge University Press, New York, 1994.

Preparata, F. P., and M. I. Shamos, *Computational Geometry: An Introduction,* Springer-Verlag, New York, 1985.

6

HYBRID TRACKING FOR MANUFACTURING SYSTEMS AUTOMATION

6.1 INTRODUCTION

The task of a tracker in VR applications is to report the position and orientation of a user's head and hand. Accordingly, the VR system updates the perspective display to make it consistent with the user's viewpoint. There are multiple types of tracking systems used in VR: magnetic, optical, mechanical, acoustic, and inertial. Each of these trackers has advantages and disadvantages. For example, magnetic trackers have no line-of-sight constraints (i.e., are not affected by objects blocking a straight line from signal transmitter to receiver), but their accuracy decreases dramatically with increase in distance from the transmitter and is also influenced by metallic objects in the neighborhood. Optical trackers are very fast and accurate and are also immune to magnetic interference, but their use is restricted by the line-of-sight constraint. Mechanical trackers, based on linkages, are very accurate, but their work is severely restricted within a small range volume (determined by the geometry of linkages). Acoustic (ultrasonic) trackers are relatively cheap and accurate, but they also have limited range and line-of-sight restriction. The most widely used trackers for VR applications have been magnetic trackers.

On the other hand, extended or long-range tracking effectiveness is crucial for the automation of manufacturing systems. Autonomous navigation of mobile robots and material handling equipment (such as automated guided vehicles or forklifts) is often a prerequisite for automation of manufacturing systems. To achieve autonomous navigation of such components of manufacturing systems, they need to know at any time where they are located within the environment, with respect to a global coordinate system. Similarly, in virtual reality–aided manufacturing systems design and maintenance applications, it is necessary to capture the motion of human participants to replicate it on avatars within virtual environments (VEs) representing specific manufacturing systems. This motion often has to be captured over a longer range than the ranges of current tracking systems for virtual reality (VR) applications. Currently, extended-range trackers are employed for tracking mobile robots. The most widely used long-range tracking systems in robotics are active beacon systems. The biggest disadvantage of active beacon navigation systems is the line-of-sight constraint (the line of sight is the straight line between the transmitter and the receiver of the navigation signal — a laser beam, for example). There might be instances when tracking a certain part (such as the end-effector) of a robot is necessary, and this part cannot be visible to the tracking system unless multiple beacons are placed within the motion environment, thus increasing substantially the cost of the tracking system.

The considerations above suggest combining some of the advantages offered by individual tracking systems to design a hybrid tracker for autonomous navigation in real manufacturing environments and human motion in virtual environments. In this chapter we conceptualize and develop a prototype long-range hybrid tracker based on a combination of a laser tracker and a magnetic tracker and apply the concept to extended-range human motion tracking on the factory floor. This easily portable system not only utilizes the strengths of a laser tracker in tracking mobile objects over long ranges in large environments such as a manufacturing shop floor and the strength of a magnetic tracker to compensate for violation of line-of-sight constraint, but also reduces the overall cost by reducing the number of expensive beacons required by the laser tracker. The hybrid tracker assists in developing real-time synchronization of human head and hand motion in a manufacturing environment with those of an avatar in a virtual manufacturing environment.

The hybrid tracker described in this chapter is based on a combination between a laser tracker and a magnetic tracker. The laser tracker used is an active beacon system called CONAC™, manufactured by MTI Research, Inc., and the magnetic tracker employed is called MotionStar®, manufactured by Ascension Technology Corporation. The laser tracker has the advantage of enabling accurate tracking of position and orientation over long ranges (the system we use has a maximum range of 100 m, but through serialization of many such systems, unlimited range can be obtained). The magnetic tracker enables tracking of multiple parts of an object without problems due to occlusion. Moreover, when the line-of-sight constraint for the laser tracker is temporarily violated, the magnetic tracker can compensate for it. This is an important advantage since it reduces the number of beacons or landmarks that have to be placed within the environment, thus reducing substantially the cost of tracking systems. Details are given in the following section.

Due to their contribution to end-to-end latency of VR systems, tracking systems have received a great deal of attention among VR research community. Hybrid tracking has been explored mostly in the area of augmented reality (AR), where accurate registration between real environment and virtual objects superimposed on it is critical. Most of the tracking in AR is being performed with the aid of videocameras, by tracking fiducial marks, placed at known locations in the environment, using computer vision techniques. Despite the accuracy of these techniques, they are slow (due to the necessity of searching the marks by scanning the images pixel by pixel), their range is limited by the placement of the fiducial marks, and they are not robust to occlusions of the fiducial marks. One type of hybrid tracker combines computer vision-based tracking with inertial tracking. Since it is well known that inertial trackers exhibit drift with time (their errors increase over time), their output is corrected by using vision-based tracking. Another option is to combine Ascension magnetic trackers with a passive image-based system that observes known fiduciary marks in the real world. The magnetic tracker measurements help in reducing the search area of the fiducials in two-dimensional images captured by head-mounted cameras, thus reducing the latency of the hybrid tracker. Combinations of inertial and optical technologies have also been proposed. None of these approaches addresses the problem of tracking motion in large environments such as a factory floor. For this kind of application, active beacon systems are very suitable, due to their accuracy and extended range, but due to the line-of-sight constraint, usually a large number of beacons have to be mounted on the factory floor. We overcome this disadvantage by using a magnetic tracker in combination with a laser tracker.

6.2. DESCRIPTION OF THE HYBRID TRACKING SYSTEM AND GENERIC METHODOLOGY FOR MOTION TRACKING

As mentioned before, our hybrid tracker for motion tracking is a combination of a laser tracker and a magnetic one. The laser tracker provides high accuracy and a high update rate for high ranges (0 to 100 m), but its use is restricted by the line-of-sight constraint. On the other hand, the magnetic tracker does not require line of sight but is accurate only within small working volumes. The laser tracker is based on triangulation of laser signals emitted by two beacons and received by one or more position transponders (PTs) attached to the moving object. One PT can report only the position with respect to one beacon, so to retrieve the orientation, one has to employ at least three PTs mounted rigidly on a special fixture. The advantage of employing the MotionStar magnetic tracker in combination with a laser tracker is its suitability for applications with frequent occlusions between transmitter and receiver. MotionStar uses pulsed dc magnetic fields instead of ac magnetic fields (which are used by Polhemus, Inc. magnetic trackers and older versions of Ascension Technologies trackers). Dc fields are significantly less susceptible to metallic distortion than are ac fields. However, dc-based magnetic trackers are susceptible to interference with magnetic fields generated by ferromagnetic objects (such as computer monitors or dc motors). Even though it is difficult upfront to estimate the probability of encountering such objects during a motion sequence, it is reasonable to assume that in most cases the wearer of a magnetic tracker will not be in the immediate proximity of ferromagnetic objects that would catastrophically affect the tracker's output. Overall, by weighing its advantages, MotionStar remains a reliable magnetic tracker for motion capture in manufacturing environments. By using a Kalman filter (see Appendix A3) to minimize the external effects on its performance, reasonable results can be obtained, as will be seen later. The advantages of incorporating a magnetic tracker into our hybrid tracker are as follows:

- It can track multiple targets without worrying about occlusions between transmitter and receivers.
- Since its behavior is not influenced by a line-of-sight constraint, the magnetic tracker can be used as a backup when the line of sight between laser tracker's beacons and PTs is occluded temporarily. This enables reduction in the number of beacons (landmarks) used for motion tracking, thus reducing the cost of the tracking system.

Typically, magnetic tracker's receivers are placed on components whose motion trajectories have to be captured and cannot be "seen" at all times by beacons of the laser tracker. The PTs of the laser tracker are mounted on the moving objects, in a location that is always visible to the transmitting beacons. The positions of the tracked components (i.e., the components equipped with a MotionStar receiver) are reported either with respect to a beacon's coordinate system or to a coordinate system attached to the motion environment, termed a world-coordinate system (WCS; in this case the transformation between WCS and beacon coordinate system is known a priori). For this purpose, one of the MotionStar receivers is rigidly attached to the PTs, so the transformation between this receiver and PTs is invariant as the tracked object moves. The MotionStar transmitter is also placed on the moving object. Thus the magnetic tracker will report position and orientation only within small

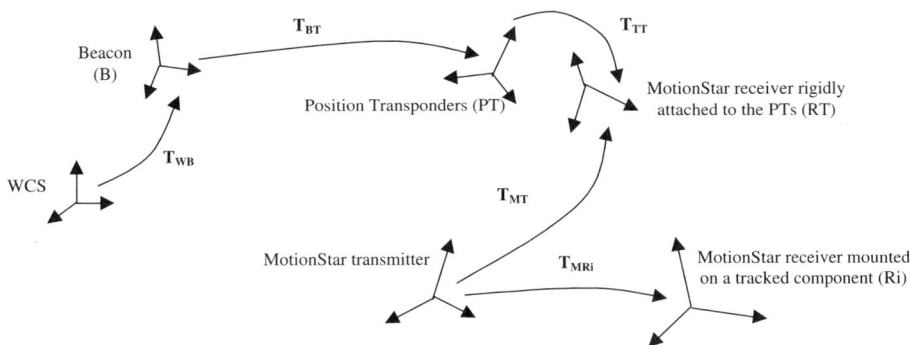

Figure 6.1 Geometry of the hybrid tracker.

working volumes. Two questions arise: How accurate are the tracker outputs when the magnetic tracker aids in circumventing line-of-sight problems, and for how long can violation of the line-of-sight constraint be tolerated so that the position estimates fall within acceptable accuracy limits? These questions are addressed through an example in Section 6.6.

The generic geometry of our hybrid tracker is depicted in Figure 6.1. In this figure, a case in which the positions of the tracked components are reported with respect to a WCS is illustrated. Note that only one beacon (B) of the laser tracking system is shown in Figure 6.1. In reality, the laser tracker has two beacons, but the position is reported with respect to a coordinate system associated with one of the beacons, so to simplify the figure, only this beacon is shown. The notations employed in Figure 6.1 are summarized in Table 6.1. In the figure only one single-tracked component (denoted Ri) is shown, for the

TABLE 6.1 EXPLANATION OF NOTATION IN FIGURE 6.1

Notation	Meaning
T_{WB}	Transformation matrix from WCS to the coordinate system associated with the laser tracker (attached to one of the beacons); known a priori
PT	Position transponders coordinate system; coincident with the center of the circle circumscribing the three PTs (hereafter, when we mention position transponders, we refer to this particular point associated with them)
T_{BT}	Coordinate transformation matrix between beacon (B) and PT
RT	MotionStar receiver rigidly attached to PTs
T_{TT}	Invariant transformation matrix between PT coordinate system and magnetic receiver rigidly attached to PTs (RT)
T_{MT}	Coordinate transformation matrix between MotionStar transmitter and RT
Ri	MotionStar receiver mounted on a tracked component
T_{MRi}	Coordinate transformation matrix between MotionStar transmitter and Ri (not used, but defined for the purpose of clarity)

purpose of clarity. Our hybrid tracker can track as many components as MotionStar allows (Ascension Technology claims that it can track up to 40 targets).

Let the position of the tracked component Ri, with respect to WCS, be represented by the vector $\mathbf{x} = (x,y,z)^T$ and with respect to the MotionStar transmitter be given by $\mathbf{x}_m = (x_m, y_m, z_m)^T$. As can be seen from Figure 6.1, $\mathbf{x}$ and $\mathbf{x}_m$ can be related by the following equation:

$$\mathbf{x} = \mathbf{T}_{MT}^{-1} \mathbf{T}_{TT} \mathbf{T}_{BT} \mathbf{T}_{WB} \mathbf{x}_m \qquad (6.1)$$

The vector $\mathbf{x}_m$ is measured by the receiver Ri with respect to the MotionStar transmitter. So equation (6.1) is the basic equation for tracking a component within WCS. For tracking the object globally (as a whole), only PT is used; therefore, the magnetic tracker is not needed (unless the line-of-sight constraint is violated).

6.3 HYBRID TRACKER PRECALIBRATION

To compute the position of a tracked component with respect to WCS, one needs the transformation between PT coordinate system and the coordinate system associated with the MotionStar receiver that is rigidly attached to the PTs (labeled RT in Figure 6.1). This transformation is labeled $\mathbf{T}_{TT}$ in Figure 6.1 and is invariant as the PT–RT ensemble moves. In order to compute $\mathbf{T}_{TT}$, precalibration of the hybrid tracker is performed before starting the motion tracking process. The geometry associated with precalibration is depicted in Figure 6.2. The notations used in Figure 6.2 are summarized in Table 6.2 for a generic case as well as for an application that demonstrates the use of our hybrid tracker (human motion capture).

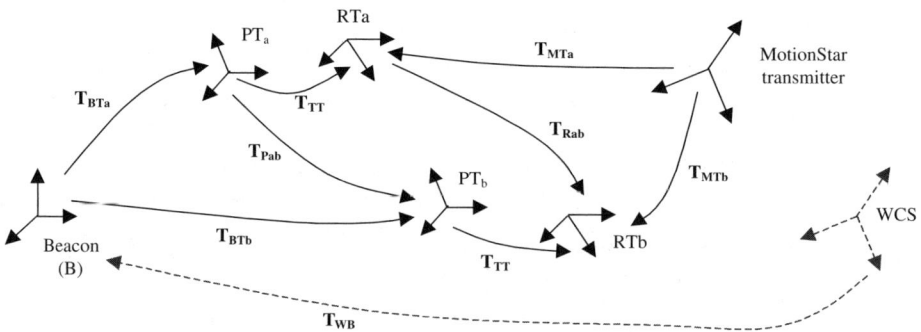

Figure 6.2 Hybrid tracker precalibration setup.

TABLE 6.2   EXPLANATION OF NOTATION IN FIGURE 6.2

Notation	Meaning	
	Generic	*Human Motion*
$PT_{a,b}$	PT coordinate system in positions labeled a and b, respectively	Same (PTs are head-mounted)
$T_{BTa,b}$	Coordinate transformation matrix between beacon coordinate system (B) and PT coordinate system in positions labeled a and b, respectively	Same
T_{Pab}	Transformation undertaken by PTs from position a to position b (w.r.t. B)	Same
T_{Rab}	Transformation undertaken by MotionStar receiver RT from position a to position b (w.r.t. MotionStar transmitter)	Same
WCS/T_{WB}	World-coordinate system/transformation between WCS and beacon (B)—used for the purpose of registering real environments with virtual environments	
T_{TT}	Invariant transformation between PT and RT (parameter to be calibrated)	Same
$RT_{a,b}$	RT in positions labeled a and b, respectively	Head-mounted magnetic receiver in positions labeled a and b, respectively
$T_{MTa,b}$	Coordinate transformation matrix between MotionStar transmitter and RT in positions labeled a and b, respectively	

Let

$$\mathbf{T}_{Pab} = \mathbf{T}_{BTb}\mathbf{T}_{BTa}^{-1} \tag{6.2}$$

and let

$$\mathbf{T}_{Rab} = \mathbf{T}_{MTb}\mathbf{T}_{MTa}^{-1} \tag{6.3}$$

From Figure 6.2, the following equation can be written:

$$\mathbf{T}_{Rab} = \mathbf{T}_{TT}\mathbf{T}_{Pab}\mathbf{T}_{TT}^{-1} \tag{6.4}$$

Equation (6.4) follows from the fact that the transformations involved form a closed loop. From equation (6.4) it follows that

$$\mathbf{T}_{Rab}\mathbf{T}_{TT} = \mathbf{T}_{TT}\mathbf{T}_{Pab} \tag{6.5}$$

from which $\mathbf{T}_{TT}$ is computed. [Equation (6.5) is an equation of the form $\mathbf{AX} = \mathbf{XB}$; the same type of equation has been encountered in Chapter 4.]

6.4 VIOLATION OF THE LINE-OF-SIGHT CONSTRAINT

To track all components with respect to WCS, the transformation between beacon (B) and PT coordinate system has to be known and is given by the laser tracker. To recover this transformation, all three PTs have to be visible at any time by both beacons of the laser tracker. In tracking mobile robots on the factory floor using active beacon systems, beacons are usually placed at optimal locations throughout the environment. This can easily be done when the paths are predefined or are expected to take place in well-known areas, but also increases the cost of tracking systems. When a tracker is used to capture unpredictable motion (such as human motion), one cannot design a priori an optimal configuration of beacons to prevent violation of the line-of-sight constraint. To get around this problem, we can use the magnetic tracker (specifically the receiver attached to PTs–RT in Figure 6.1) to backup the system when the line of sight of the laser tracker is temporarily occluded. Consider Figure 6.2 again. Let us assume that the tracked component moves from position a to position b. In position a, all three PTs are visible, and therefore the transformation $\mathbf{T}_{BTa}$ is reported correctly. In position b, at least one PT is occluded. In this case, the transformation $\mathbf{T}_{BTb}$ can be recovered from the previous estimate of the PTs position and orientation ($\mathbf{T}_{BTa}$) and the relative motion undertaken by magnetic receiver RT (denoted as $\mathbf{T}_{Rab}$) by the following equation:

$$\mathbf{T}_{BTb} = \mathbf{T}_{TT}^{-1}\mathbf{T}_{Rab}\mathbf{T}_{TT}\mathbf{T}_{BTa} \qquad (6.6)$$

In equation (6.6), $\mathbf{T}_{Rab}$ is measured with respect to the MotionStar transmitter and the assumption that the transmitter does not move while RT moves from a to b is made. However, this is not the case. In practice, MotionStar transmitter is mounted onto the tracked object component least susceptible to change its position regularly over small time intervals. We therefore assume that the MotionStar transmitter remains fixed over such time intervals (on the order of milliseconds). The potential violation of this assumption is considered while designing a Kalman filter that deals with line-of-sight constraint violations, by scaling up the measurement noise uncertainty.

6.5 OPERATING THE HYBRID TRACKER

When retrieving position and orientation information by fusing data provided by two or more sensors, typically the assumption is made that measurements are available simultaneously from all sensors. In reality, this is almost never the case, due to different update rates of the various sensors. In our case, measurements from the laser and magnetic trackers are fed to a 300-MHz Pentium PC via serial cables and from there to an SGI workstation that performs all the calculations for position and orientation estimates. Since tracking is initiated only

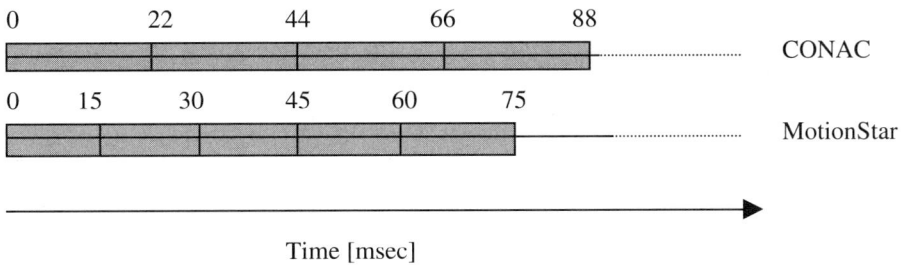

Figure 6.3 Temporal diagram of the hybrid tracker measurements.

when the first data packet arrives from both sensors, communication over-head is not relevant for the time increment between two consecutive measure-ments. The interval between two measurements of the laser tracker is 22 mil-liseconds (ms). When using a single receiver, the update rate of MotionStar tracker is 5 ms and increases by the same amount as a new receiver is added. The temporal diagram shown in Figure 6.3 depicts the succession of measure-ment packets as those arrive to the SGI workstation when using three receivers of the magnetic tracker (in this case the update rate is 15 ms).

As can be seen from Figure 6.3, measurements arriving from the laser and magnetic trackers are not synchronous. This introduces an error in estimating the true position of a tracked component, since it is not possible to collect a measurement from both trackers at exactly the same moment in time. The fact that there is no constant offset between readings complicates the prob-lem. Due to the small temporal difference between measurements collected from the two sensors and due to the fact that the expected number of mag-netic sensors typically used in our applications is between two and six, the errors are not expected to be significant by comparison with the errors inflicted by the noise in the measurements. Consider the case shown in Figure 6.3. In the current stage of our hybrid tracker, if in-between two successive readings from the laser tracker there is only one reading from the magnetic tracker, this one is considered in the calculations. If two or more readings appear, these are first averaged to obtain a more realistic estimate.

The position of a tracked component is estimated through a Kalman filter (Appendix A3). In our case, two Kalman filters are used alternately, depending on whether or not the line-of-sight constraint of the laser tracker is violated. When there is occlusion between the beacon and any PT, the laser tracker stops sending data to the interface module (each PT has its own interface mod-ule in order to increase update rate). This case is tested by monitoring the time interval Δt_l elapsed from the previous measurement. If Δt_l exceeds 26 ms, the line-of-sight constraint is considered violated (recall that the update rate of the laser tracker is 22 ms and we allow 4 ms for possible communication glitches). In this case, the system switches to the alternative Kalman filter, which uses the same state and measurement models but has a larger initial error covari-

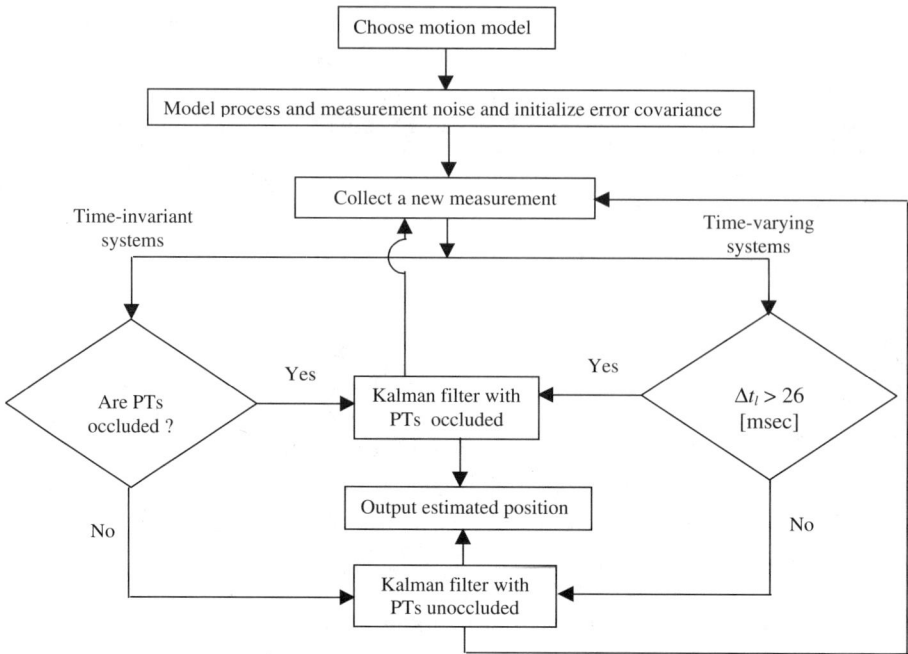

Figure 6.4 Generic methodology for motion tracking.

ance and different measurement noise model due to the fact that the accuracy of the magnetic receiver RT (that backs up the laser tracker) is expected to be lower than that of the laser tracker. When the line of sight between beacons and PTs is free, the laser tracker starts outputting measurements automatically and a switch to the regular Kalman filter is performed. Filtering is resumed with the predicted state variables and error covariance given by the backup filter, instead of the same values before occlusion of the line of sight (we found out that this approach is more appropriate because the motion estimation is smoother and the amount of jitter is reduced). The only change is that direct laser tracker measurement is used instead of equation (6.6). The generic methodology of operating the hybrid tracker when capturing motion can be summarized as shown in Figure 6.4.

6.6 APPLICATION TO HUMAN MOTION

Human motion is an example of using the hybrid tracker with time-varying systems. (An example of applying the tracker to time-invariant systems is the telemetry-based depth recovery procedure described in Chapter 4.) Capturing human motion in manufacturing environments in order to be replicated in VEs is a challenging task. In VR applications, typically the head and hand of a user are tracked in order to update the perspective. To achieve realistic human motion in VEs, more components of a human body have to be tracked (such as torso and joints). To illustrate the application of our hybrid tracker to human motion capture, we limit ourselves to tracking only the head and hand of a human on a factory floor, replicated by an avatar in a VE representing the real factory floor. VE is a priori registered with the real environment.

The three PTs of the laser tracker and one receiver of the magnetic tracker (RT in Figure 6.1) are rigidly mounted on a fixture with the shape of a hat, mounted on the user's head. The MotionStar transmitter is placed in a back-pack, located on the back of the user, or when motion takes places within a small volume, it can be placed in a fixed position, close to the human operator. The user's hand is tracked by means of a magnetic receiver, attached to the wrist, based on which the hand position with respect to WCS can be computed. The laser tracker gives the head position.

The measurements performed by the tracker over time are noisy. It is reasonable to assume that the process of position estimation is driven by normally distributed noise. To estimate the coordinates of hand position optimally with respect to WCS in a noisy environment, we design a Kalman filter, as mentioned in Section 6.5. The precise dynamic model of hand and head motion is unknown, but the Kalman filter can provide very good results even for this kind of application. Also, the Kalman filter has been found to perform well even when the assumptions of normal distribution of noise representing the uncertainties in the measurements in the model are violated.

The use of a Kalman filter for motion capture requires a motion model. Unfortunately, it is almost impossible to obtain an accurate model for the hand and/or head motion. To get around this problem, different models to approximate head and/or hand motion are used. One approximation is the *position–velocity* (PV) *model,* which assumes that motion takes place at constant velocity and models acceleration as white noise. The PV model can be enhanced by adding constant acceleration. Another approximation model assumes that head rotations are infrequent and that angular speed and angular acceleration are nonzero only during infrequent change in viewing direction. These assumptions lead to the choice of an integrated Gauss–Markov process to model the head movement. All these approximations provide satisfactory results, with occasional overshoot when sudden change of direction or velocity occurs. We have used the PV with constant acceleration model to approximate hand and head motion. The Kalman filter for hand and head tracking is described briefly in Appendix A4.

To determine the performance of the hybrid tracker, the magnetic receiver that records the hand position is positioned initially at some known world locations in order to determine the process and measurement noise covariance matrices $\mathbf{Q}$ and $\mathbf{R}$, respectively. Measurements are collected and the filter is run off-line. The error is the difference between the estimated and actual positions of the tracker. A cost function is defined as the sum of the squared errors at each time step. Through minimization of the cost function, matrices $\mathbf{Q}$ and $\mathbf{R}$ are computed. The Kalman filter error covariance matrix $\mathbf{P}$ is assumed to be diagonal. The diagonal elements of $\mathbf{P}$ are initialized to some large values (two for the elements corresponding to position, 50 for velocity, and 60 for acceleration). The elements corresponding to velocity and acceleration are initialized to higher values than the ones corresponding to position because motion tracking starts with the user being stationary.

When tracking human motion, one does not have available ground truth data, since it is impossible to predict the movement path exactly. To assess the accuracy and consistency of our Kalman filter model, we have captured motion sequences with frequent changes of direction and monitor the difference between Kalman filter predictions and real tracker measurements. Metallic objects were present close to the movement path to illustrate a relatively insignificant impact on our model. The results are shown in Figures 6.5 and 6.6, where it is seen that there is a reasonable consistency between Kalman filter predictions and real tracker measurements, with slight overshoot or undershoot when a sudden change of direction occurs. The type of motion depicted in Figures 6.5 and 6.6 takes place in very unfavorable conditions for our motion model. Typically, we do not expect such frequent changes of direction, and therefore the overshoot or undershoot will be reduced. Additional experiments provided similar results. The maximum overshoot encountered is about 7 cm. Figure 6.6 depicts a special case, in which the laser tracker is occluded for approximately 9 s. This occlusion happens to coincide with a change in direction of motion. When such situations occur, the measurement noise level is scaled up to reflect the additional uncertainty in position estimation when using the relative motion of the magnetic tracker [equation (6.6)] instead of the laser tracker [equation (6.1)]. As can be seen, even though it is still at an acceptable level, the amount of overshoot in this case is larger than the typical overshoot when changing direction of motion, as shown in Figure 6.5.

The results presented in this section show the consistency of our Kalman filter and show that the motion model we have chosen gives sufficiently accurate results even in unfavorable cases encountered in human motion. Also, as was shown in Figure 6.6, the magnetic tracker can effectively back up the laser tracker when it is occluded for an amount of time (9 s in our case). We do not recommend using the magnetic tracker to compensate for the violation of line-of-sight constraint for more than 10 s (which is sufficient in most cases). The presence of metallic objects definitely affects the accuracy of the tracker, but not to an unacceptable extent. Based on the results obtained, we conclude that our hybrid tracker can be used effectively for extended-range motion capture in manufacturing environments, for applications that can accept a maximum error of 5 to 7 cm in position estimation.

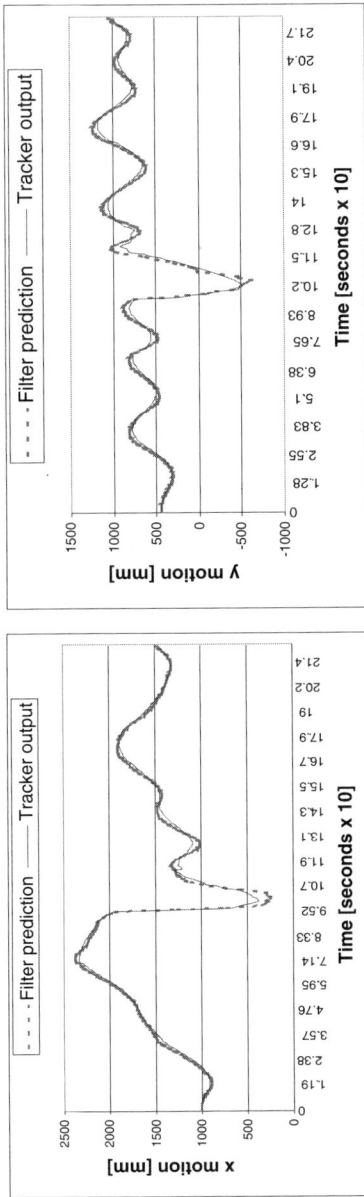

Figure 6.5 Kalman filter predictions (x and y directions of motion) versus real tracker output: first sequence.

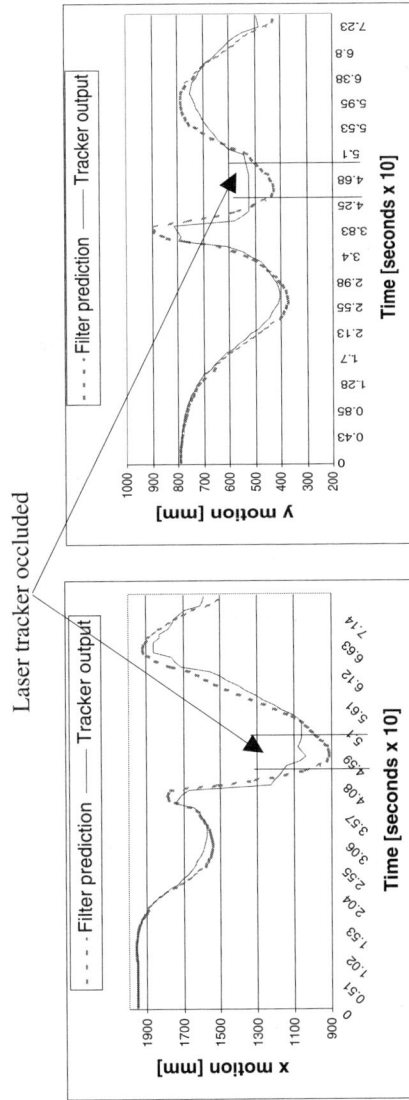

Figure 6.6 Kalman filter predictions (x and y directions of motion) versus real tracker output: second sequence.

Figure 6.7 Human operating in a manufacturing system whose motion is replicated in a VE, a priori registered with the real system (also shown in the color section).

In Figure 6.7 (also shown in the color section) an operator portrayed as an avatar is illustrated operating in a virtual manufacturing environment (the virtual environment is the same as the one shown in Figure 4.54). The human operator (setting up a machine tool) is being tracked with our hybrid tracker, and his motion is being replicated within the VE, a priori registered with the real environment. The example shown in Figure 6.7 illustrates the potential application of our hybrid tracker to monitoring and training human operators in virtual manufacturing systems.

Possible future improvements can be performed in the following areas:

- *A better dynamic model.* Even though the dynamic model we have used for human motion capture enables reasonable accuracy, a better dynamic model can still decrease the errors in position estimation. At this time, no mathematical model is available that can describe the nature of human motion precisely. Future research might be able to develop such models.

- *Elimination of errors due to different update rates of trackers used as components of the hybrid tracker.* Recent developments in this area suggest that any incomplete measurement can be incorporated into a Kalman filter as it arrives, without waiting for the complete information required to estimate the state variables. The estimate is recorded only when complete information is available. We plan to adapt this methodology to our hybrid tracker, by processing separately the information from magnetic and laser trackers and combining them subsequently. This approach can speed-up filter prediction, thus reducing the overall latency and improving the accuracy of state estimates.

FURTHER READING

For more information about tracking systems used in VR, the reader is referred to the surveys of Ferrin (1991) and Meyer et al. (1992). Hybrid tracking in the context of augmented reality has been the topic of extensive research. Representative approaches have been described in Azuma (1993, 1995), Durlach and Mavor (1995), State et al. (1996), and Neumann et al. (1998). The CONAC laser tracking system has been described in MacLeod and Chiarella (1993) and Borenstein et al. (1996). In fact, the latter contains an exhaustive review of all active beacon positioning systems. Approximation of the human motion model has been tackled in Liang et al. (1991), Friedmann et al. (1992), and Welch (1996). Kalman filtering literature was mentioned at the end of Chapter 4.

REFERENCES

Ascension Technologies Inc., "MotionStar User Manual," available from *http://www.ascension-tech.com.*

Azuma, R., "Tracking Requirements for Augmented Reality," *Communications of the ACM,* Vol. 36, No. 7, pp. 50–51, 1993.

Azuma, R.T., "Predictive Tracking for Augmented Reality," Ph.D. dissertation, University of North Carolina at Chapel Hill, Department of Computer Science Technical Report TR95-007, 1995.

Borenstein, J., H. R. Everett, and L. Feng, *Where Am I? Sensors and Methods for Mobile Robot Positioning,* The University of Michigan, prepared for the Oak Ridge National Laboratory and U.S. Department of Energy, 1996.

CONAC™ Laser Tracking Systems, *http://www.mtir.com/tracdesc.html.*

Durlach, N., and A. Mavor, eds., *Virtual Reality: Scientific and Technological Challenges,* Report of the Committee on Virtual Reality Research and Development to the National Research Council, National Academy Press, Washington, DC, 1995.

Ferrin, F. J., "Survey of Helmet Tracking Technologies," Proceedings SPIE *Vol. 1456: Large Screen Projection, Avionic and Helmet-Mounted Displays,* pp. 86–94, 1991.

Friedmann, M., T. Starner, and A. Pentland, "Device Synchronization Using an Optimal Linear Filter", *Proceedings of the 1992 Symposium on Interactive Graphics,* Cambridge, MA, pp. 57–62, 1992.

Liang, J., C. Shaw, and M. Green, "On Temporal-Spatial Realism in the Virtual Reality Environment", *Proceedings of the ACM Symposium on User Interface Software and Technology,* Hilton Head, SC, pp. 19–25, 1991.

MacLeod, E., and M. Chiarella, "Navigation and Control Breakthrough for Automated Mobility", *Proceedings of the 1993 SPIE Conference of Mobile Robots,* Boston, MA, pp. 57–68, 1993.

Meyer, K., H. L. Applewhite, and F. Biocca, "A Survey of Position Trackers," *Presence,* Vol. 1, No. 2, pp. 173–200, 1992.

Neumann, U., S. You, Y. Cho, J. Lee, and J. Park, "Augmented Reality Tracking in Natural Environments", *Proceedings of the First International Workshop on Augmented Reality,* San Francisco, CA, 1998.

State, A., G. Hirota, D. T. Chen, W. F. Garrett, and M. Livingston, "Superior Augmented Reality Registration by Integrating Landmark Tracking and Magnetic Tracking", *Proceedings of ACM SIGGRAPH '96,* New Orleans, LA, 1996.

Welch, G., *SCAAT: Incremental Tracking with Incomplete Information,* University of North Carolina at Chapel Hill, Department of Computer Science Technical Report TR96-051, 1996.

EXERCISES

A forklift carrying parts on a factory floor is being tracked by a tracking system for the purpose of replicating its movement in a virtual environment. Even though the movement of the forklift is continuous, when displaying it in a virtual environment it becomes discrete due to frame rate constraints. Assuming a frame rate of 30 frames/s, the interval between two recordings of the forklift position is approximately 0.03 s.

We also assume that the forklift motion is uniform, so the equations that describe it are as follows:

$$x_{k+1} = x_k + v_{xk}(t_{k+1} - t_k)$$

where $x_{k, k+1}$ are the positions of the forklift along the x axis of the coordinate system relative to which its motion is tracked (named World Coordinate System — WCS), at time steps t_k and t_{k+1}, respectively, and $v_{xk, k+1}$ are the velocities of the forklift at the same time steps. The difference $t_{k+1} - t_k$ is constant and is equal to 0.03 s. The same equations are valid for position and velocity along y and z axes of WCS.

The tracker measures only the position of the forklift (x_k, y_k, z_k) at time step t_k. Let the vector of state variables be the vector formed by the three position and the three velocity elements: $(x_k, y_k, z_k, v_{xk}, v_{yk}, v_{zk})^T$. The vector of measurements is: $\mathbf{z}_k = (x_k, y_k, z_k)^T$.

6.1. Find the matrices ϕ_k and $\mathbf{H}_k$ that describe the process and measurement equations on which Kalman filtering is based.

6.2. If we record the sequence of positions shown below, perform the Kalman filtering for the recorded measurements and plot the positions of the forklift. Also plot the error covariances corresponding to x_k and v_{xk} (only for these components of the state vector) and find after how much time they tend to stabilize to a minimum (in other words, find the time elapsed until the convergence of the Kalman filter). Initialize the position elements of the state vector with the position at $t = 0$ and consider the forklift stationary in this position. The covariance matrices are also given below.

Hint: Consider $\hat{\mathbf{x}}_0^- = (155.43, 155.36, 255.35, 0, 0, 0)^T$ and treat the recording at $t = 0$ as the first measurement also, in order to make the Kalman filter converge faster (in order to nullify the effects of the large initial error covariance).

```
Record motion sequence:

                    x       y       z

    t=0:     155.43 155.36 255.35

    t=0.03:  157.3   157.4   258.25

    t=0.06:  159.67 158.35 260

    t=0.09:  163.28 164.2   266.3

    t=0.12:  170.44 168.85 270.5

    Note: x, y and z are in [cm] and t in [sec].
```

$$
\mathbf{Q} = \text{const.} = \begin{bmatrix}
1.06e-11 & 0 & 0 & 0.8e-9 & 0 & 0 \\
0 & 1.06e-11 & 0 & 0 & 0.8e-9 & 0 \\
0 & 0 & 0.2e-8 & 0 & 0 & 1.8e-7 \\
0.8e-9 & 0 & 0 & 0.8e-7 & 0 & 0 \\
0 & 0 & 0.8e-9 & 0 & 0.8e-7 & 0 \\
0 & 0 & 1.8e-7 & 0 & 0 & 1.8e-5
\end{bmatrix}
$$

$$
\mathbf{R} = \text{const.} = \begin{bmatrix}
0.01 & 0 & 0 \\
0 & 0.01 & 0 \\
0 & 0 & 0.01
\end{bmatrix}
$$

$$
\mathbf{P}_0^- = \begin{bmatrix}
2 & 0 & 0 & 0 & 0 & 0 \\
0 & 2 & 0 & 0 & 0 & 0 \\
0 & 0 & 2 & 0 & 0 & 0 \\
0 & 0 & 0 & 50 & 0 & 0 \\
0 & 0 & 0 & 0 & 50 & 0 \\
0 & 0 & 0 & 0 & 0 & 50
\end{bmatrix}
$$

7

EXACT COLLISION DETECTION

7.1 INTRODUCTION

The objective of this chapter is to discuss techniques that add validation capability in a VR manufacturing simulator instead of just using VR manufacturing simulator for visualization. Collision detection is an important problem in this regard. The problem of collision detection has been studied extensively in robotics, computational geometry, and computer graphics. The goal in robotics has been mainly the planning of collision-free paths between obstacles. This differs from the requirements in virtual manufacturing (VM), where the motion is subject to dynamic constraints or external forces and cannot typically be expressed as a closed-form function of time. The emphasis in computational geometry has been on theoretically efficient intersection detection algorithms. Most of them are restricted to a static instance of the problem and are nontrivial to implement. None of these algorithms adequately address collision detection in a VE, which requires performance at interactive rates for thousands of pairwise tests between object features. Collision detection is a phase in contact modeling, described next.

Contact Modeling

The two main phases of contact modeling are collision detection and collision response. The timing and location of collision is determined in the *collision detection phase*. The physically correct behavior issues of the colliding objects are addressed in the *collision response phase*. Collision response can also cover issues where a collision detection event is used to trigger other events. The collision detection phase can be treated as independent of collision response as long as we can assume perfectly rigid bodies. Only the collision detection phase is addressed in this chapter.

Collision Detection

In most VM applications, perfectly rigid bodies are assumed. For deformable bodies, such as in automobile crash testing, proprietary finite-element modeling solutions have been employed for this transient phase of contact modeling. Because of the computational expense these are not suitable for real-time interactive VM simulation.

7.2 GENERAL TECHNIQUES FOR COLLISION DETECTION

Most collision detection algorithms make two main categories of assumption:

1. *Geometric assumptions:* for example, a collision test is restricted to convex objects and applications where topology information is not used during a collision check (i.e., objects are treated as polygon soups).
2. *Motion assumptions:* for example, object position and velocity are known a priori.

Previous attempts at contact modeling can be classified into analytical and geometric techniques.

7.2.1 Analytical Techniques

A popular technique in this class is the potential field–based approach. Objects are surrounded with repulsive vector fields. The bounds of influence of a vector field around an object are determined prior to generation of the repulsive fields by enclosing each object with a surrounding volume which provides some security space between the outer surface of the object and the envelope. A repulsive vector is assigned to the surrounding volume so that approaching objects are prevented from penetrating the object. This technique is powerful. However, it does not guarantee a solution to the collision problem. In addition, due to the use of object envelopes as representatives of objects during collision processing, this approach is not suitable for applications where a high degree of accuracy is essential.

An interesting approach based on interval arithmetic and recursive subdivision is also popular. Simulation environment is described as a collection of point-set volumes where complex objects are modeled as combinations of primitive volumes using various operators (e.g., intersection, union). The decision as to whether objects collide is made based on the presence of points occupied by more than one object. This technique presents an efficient way to identify and refine regions that potentially contain collisions. However, the fact that object motion must be known a priori eliminates this approach for interactive applications. Another technique assumes a high degree of geometric coherence between successive frames. The decision as to whether collision between two objects occurred is based on the existence of a separating plane such that each object lies in a different half-space of the plane. A contact force model and a characteristic function (expressed as a function of time) are employed in the decision-making process. The characteristic function is used as a measure of the distance, while its first derivative with respect to time is used as a measure of the relative velocity between approaching objects. The strongest restriction on this technique is its applicability to convex objects only. We have not encountered the use of this technique in interactive VR applications.

7.2.2 Geometric Techniques

Whereas collision detection in analytic techniques has been treated primarily as a dynamic problem assuming an a priori knowledge about object position and velocity, a geometric approach treats collision detection as a static problem, meaning that objects are considered stationary during the analysis, which corresponds to one frame in the sequence. Once the objects are frozen, various techniques based primarily on computational geometry are employed to detect interferences between objects. By itself, a geometric approach does not require knowledge of the motion of an object. However, some approaches have used a concept of motion prediction to rectify some weaknesses that have arisen in geometrical modeling of collision detection. The two important weaknesses associated with most geometrical techniques are:

1. *Fixed-time-step weakness.* Since the collision detection process is applied at equally spaced time instances, it is possible that a collision occurs between two successive time instances. In such extreme cases, some collisions may remain undetected. In case two small objects are approaching one another, or a small object is approaching a thin object, one object can pass completely through another between collision checks. A classic example is a bullet approaching a thin wall. The bullet speed can be chosen such that it passes through the wall without the collision being detected! Also, if collision occurs between two successive collision checks, objects penetrate, and penetration rather than contact is detected in the next check.

2. *All pairs weakness.* Since no assumptions about the current and the next positions of an object are made, each part (feature) of an object's surface has an equal chance to collide with other objects' features. Accordingly, we have to check every pair of features (one feature from each approaching object) at every time instance. It is obvious that every algorithm that contains all-pair check admits quadratic complexity [$O(n^2)$].

The first geometric technique is *space partitioning.* There are many space-partitioning techniques. The concept of *binary space partition tree* (BSP-tree) has been used to improve the efficiency of an interference check. Normally, a uniform spatial subdivision scheme is used. Collision is reported if more than one object occupies the same unit cell. An efficient hashing scheme is used to speed up computation. Another partitioning approach is based on an *octree scheme.* This approach can be seen as a nonuniform spatial subdivision scheme where the size of the unit cell in a region depends upon the occupancy of the region, meaning that regions which contain many objects have smaller unit cells than regions with fewer objects. The common weakness of space partitioning techniques is their unsatisfactory speed.

An improved version of the partitioning scheme is introduced in the form of hierarchical bounding volumes. The best bounding volume is a convex hull, but it is computationally expensive for real-time needs. This approach assumes that objects are surrounded by some bounding volume (e.g., box, sphere). The practicality of a bounding volume arises from the fact that for a bounded aspect ratio and scale factor, the number of the bounding volume intersections is asymptotically proportional to $n + k$, where n is the number of objects and k is the number of intersecting objects. The term *aspect ratio* for an object is the volume of the smallest enclosing sphere divided by the volume of the largest enclosing sphere, and the *scale factor* for an object collection is the volume of the largest enclosing sphere divided by the volume of a smallest enclosing sphere.

There are many ways of developing the bounding volumes, one of which is to use *oriented bounding boxes* (OBBs). An OBB is a bounding box in an orientation that has the tightest enclosing box for a given object (Figure 7.1). There is computational cost involved in determining the direction of the tightest fit, but OBBs provide the most accurate box approximation for an object. An OBB tree is used to partition an object into smaller and smaller boxes progressively. An application of OBB is in a technique known as RAPID, described later.

A related concept is *k*-discrete orientation polytope or *k*-DOP, where *k* is the number of orientations that are checked to determine the bound-

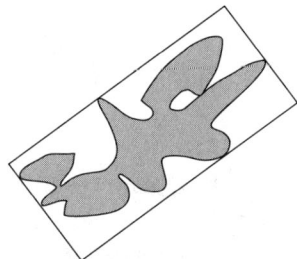

Figure 7.1 OBB.

ing box. Theoretically, k is infinity for an OBB. Normally, k is set to 6 or 14 to reduce the computational cost in determining a bounding box that is reasonably tight and accurate. A 6-DOP is also known as an axis-aligned bounding box (AABB) and is a popular choice (such as in I-COLLIDE, described later). The six orientations are $+x$, $-x$, $+y$, $-y$, $+z$, and $-z$ (i.e., the six cardinal axis directions). In 14-DOP, these six directions are further refined by adding eight more directions. These eight directions are along

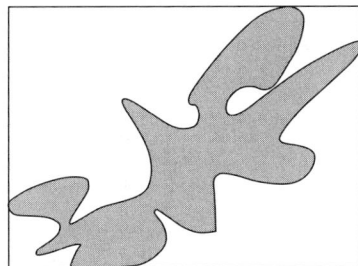

Figure 7.2 AABB or 6-DOP.

the center of each octant of a three dimensional coordinate system starting from the origin. The idea of a k-DOP is to reduce the computational effort needed in OBBs. An illustration of AABB or 6-DOP is provided in Figure 7.2.

These bounding volume techniques are especially important for the development of real-time collision detection algorithms because the bounding volumes have been used successfully as a preliminary test for the interference. The fact that the absence of bounding volumes interference guarantees the absence of collision between corresponding objects has been used to reduce significantly the number of features that need to be checked. To date, the best results have been reported in a software known as I-COLLIDE. A brief insight into this approach is provided next.

Large-scale environments consist of stationary as well as moving objects. Let there be n moving objects and m stationary objects. Each of the n moving objects can collide with the other moving objects, as well as with the stationary ones. Keeping track of $\binom{n}{2} + nm$ pairs of objects at every time step can become time consuming, as n and m get large. To achieve interactive rates, this number is reduced by pruning multibody pairs before performing pairwise collision tests. Sorting is the key to the pruning approach. Each object is surrounded by a three-dimensional bounding volume. These bounding volumes are sorted in three dimensions to determine which pairs are overlapping. Exact pairwise collision tests are performed on these remaining pairs. However, it is not intuitively obvious how to sort objects in three dimensions. A dimension reduction approach is used. If two bodies collide in a three-dimensional space, their orthogonal projections onto the xy, yz, and xz planes and x, y, and z axes must overlap. Based on this observation, axis-aligned bounding boxes are chosen as bounding volumes. These bounding boxes are then projected efficiently onto a lower dimension, and the sort is performed on these lower-dimensional structures.

This approach is different from typical space partitioning approaches used to reduce the number of pairs. A space partitioning approach puts considerable effort into choosing good partition sizes. But there is no partition size that prunes out object pairs as ideally as does testing for bounding box overlaps. Partitioning schemes may work well for environments where n is small compared to m, but object sorting works well whether n is small or large.

Many collision detection algorithms have used bounding boxes, spheres, ellipses, and so on, to rule out collisions between objects that are far apart. I-COLLIDE uses bounding box overlaps to trigger the *exact collision detection* algorithm. Two types of axis-aligned bounding boxes are considered: fixed-sized bounding cubes (fixed cubes) and dynamically resized rectangular bounding boxes (dynamic boxes). The size of the fixed cube is computed to be large enough to contain the object at any orientation. This axis-aligned cube is defined by a center and a radius. Fixed cubes are easy to recompute as objects move, making them well suited to dynamic environments. If an object is nearly spherical, the fixed cube fits it well. The center and radius of the fixed cube are calculated as a preprocessing step. At each time step, as the object moves, the cube position and orientation are recomputed.

The size of the rectangular bounding box for the dynamic boxes is computed to be the tightest axis-aligned box containing the object at a particular orientation. It is defined by its minimum and maximum x, y, and z coordinates. As an object moves, its minima and maxima are recomputed, taking into account the object's orientation. For oblong objects, rectangular boxes fit better than cubes, resulting in fewer overlaps. This is advantageous as long as few of the objects are moving, as in a walkthrough environment. In such an environment, the savings gained by the reduced number of pairwise collision detection tests outweigh the cost of computing the dynamically resized boxes. As a precomputation, each object's initial minima and maxima are determined along each axis. It is assumed that the objects are convex.

The one-dimensional sweep and prune algorithm begins by projecting each three-dimensional bounding box onto the x, y, and z axes. Because the bounding boxes are axis-aligned, projecting them onto the coordinate axes results in intervals (see Figure 7.3). Overlaps among these intervals are of interest, because a pair of bounding boxes can overlap if and only if their intervals overlap in all three dimensions. The two-dimensional intersection algorithm, on the other hand, begins by projecting each three-dimensional axis-aligned bounding box onto any two of the xy, xz, and yz planes. Each of these projections is a rectangle in two dimensions. Typically, there are fewer overlaps of these two dimensional rectangles than of the one dimensional intervals used by the sweep and prune technique.

To summarize, the first phase of I-COLLIDE consists of candidate feature reduction using a fixed-size box or a dynamically resized box. This results in a reduction from $O(n^2)$ to $O(n + m)$. The second phase consists of exact collision on the candidate features chosen. The closest features of two temporally coherent moving polyhedra is based on Voronoi regions and has an expected

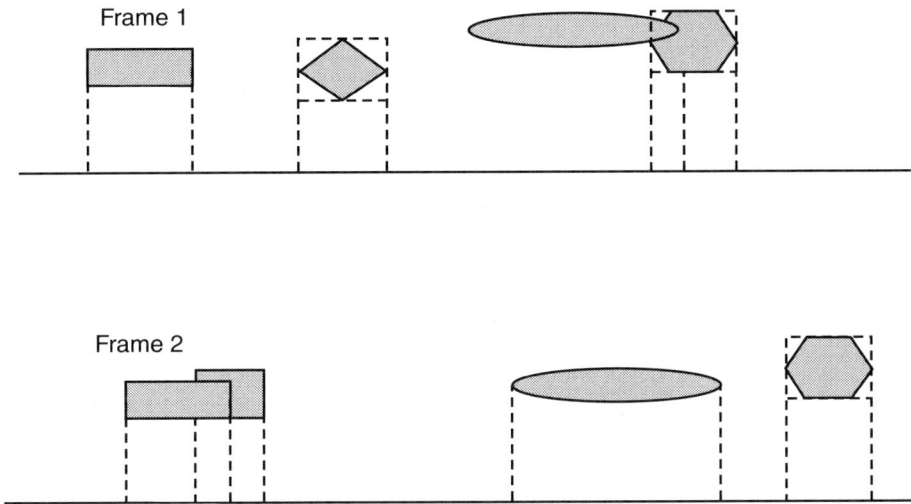

Figure 7.3 Projections of axis-aligned bounding boxes onto coordinate axes and overlaps.

constant time if polytopes are not moving swiftly. The Voronoi region of a feature is the set of points closer to that feature than to any other feature in the set. Voronoi regions form a partition of space outside the polytope. A data structure for each feature is conceived and it is called a *cell*. Each cell has pointer to neighboring cells sharing a constraint plane (CP) defined by Voronoi constraints. The algorithm starts with a candidate pair of features from a pair of polytopes and checks for closest points. For convex polytopes, this is a local test. For nonconvex polytopes, the algorithm enters into a cyclical loop when it enters from external to internal Voronoi regions. This is one of the reasons why I-COLLIDE works only for convex polytopes. For convex polytopes, the local test takes the following form: If either feature fails a test, keep trying to examine the neighboring feature. The Euclidean distance between feature pairs must always decrease for nonpenetration if the polytopes are convex. A simple example is provided (see Figure 7.4). Check if V_b is in cell 1? The answer is no. Next from cell 1, go to CP, then go to cell 2. Ask the question: Is V_b in cell 2? The answer is again no, but cell 2 is the closest Voronoi region. Hence the search stops. Next, a switch is made to cells of B. Then check if the nearest point P_a lies in B's cell. In this way one eventually arrives at the closest feature pair.

Geometric coherence between two successive interference checks has been exploited, but the strongest restriction is its limitation to convex objects only (or convex hulls of nonconvex objects). This limitation restricts its use for validation of manufacturing simulation in VR. Although progress has been reported in developing software entitled RAPID (Robust and Accurate Polygon Interference Detection) for collision detection involving nonconvex objects, we

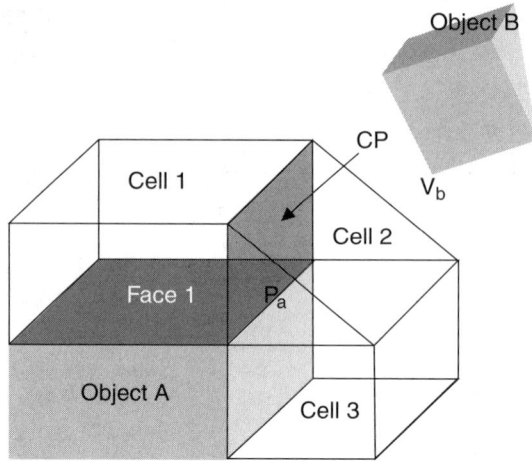

Figure 7.4 Closest features of object *A* with respect to object *B* defined by Voronoi regions.

are not aware of any methods that handle the more complex issues in manufacturing collision: namely, near-miss detection and detection of collision inside complex regions such as holes that are so important in manufacturing process validation.

Decomposing nonconvex objects to convex objects is a NP-hard problem, so this possibility is also ruled out. Another possibility would be to approach collision detection before the VR scene is created, which assumes building models as assemblies of convex pieces. This adds a new level of complexity to model building, which is already one of the main bottlenecks in developing VR systems, and the approach cannot be applied to existing databases. Hence this possibility is also ruled out. The fact that none of the potential global techniques have been found suitable for solving the exact collision detection problem and the fact that the versatility of the data sets (e.g., models with sharp edges, isosurfaces in CFD and medicine, mathematically defined manifolds) make it even more improbable for a potential global solution have led to our quest for an appropriate local method.

7.3 SPECIALIZED LOCAL COLLISION DETECTION TECHNIQUE FOR VIRTUAL MANUFACTURING

The key notion in our approach is the creation of virtual objects (VOs) associated with specific scene objects and testing collisions between VOs instead of testing collisions between the original scene objects. One can treat VOs as representatives of specific scene objects in collision detection. The term VO is

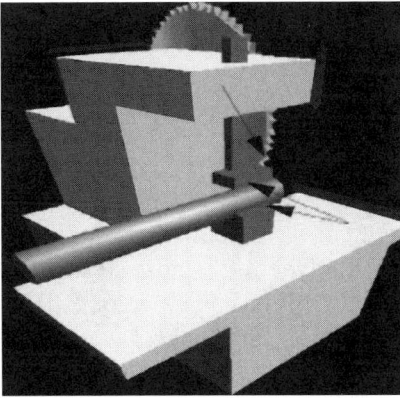

Figure 7.5 Virtual cutter.   **Figure 7.6** Virtual lathe.

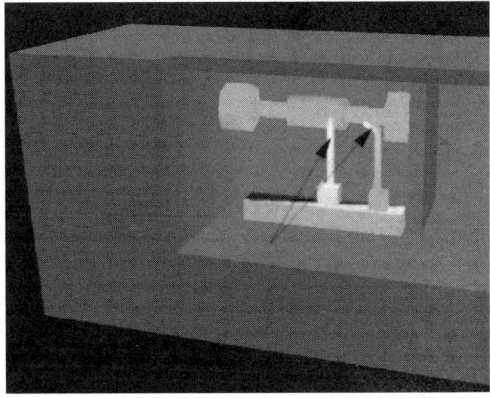

used since VOs are present only for collision-detection computational process. The rationale behind our approach is that not every object feature (face, edge, or vertex) is of interest for collision detection. In the case of a virtual cutter machine (Figure 7.5), regions that contain features of interest for collision detection are indicated with arrows. One can see that a majority of features are not significant for the virtual cutting, so they are not of interest for collision detection. Also, some object features can never reach each other. In the case of a virtual lathe (Figure 7.6) it is obvious that the tools (indicated with arrow lines) can never collide. Once this observation is made, the objective becomes quite clear — encapsulate candidate features into VOs.

The key contribution is a methodology for fast creation of convex objects for a class of frequently encountered nonconvex manufacturing objects that can be used in any existing collision detection approach (such as I-COLLIDE). The methodology is an alternative for the   nondeterministic (NP-hard) non-convex object decomposition problem. The methodology described is best suited for interactive simulation and animation applications where high accuracy of object contact modeling is required. Some of the research areas where this technique can make an immediate impact are:

- *Virtual assembly* applications, where a lack of computationally efficient exact collision detection algorithms has discouraged researchers from introducing collision detection into complex parts assembly simulation. In some applications, the technique can improve the level of interactivity significantly by eliminating the need for predefined constraint relationships.
- More reliable and efficient validation capabilities in *mobile robot simulation* applications, instead of existing computationally expensive space subdivision techniques.

- Simulation of manufacturing processes where accurate modeling of *near-miss detection* is essential. Examples of such processes include *robotic painting, robotic welding,* and *numerically controlled machining operations.*

Before introducing terminology and assumptions, some background in computational geometry and topology is presented. A three-dimensional polyhedron is defined as a finite, connected set of features called facets (or faces), edges, and vertices. Manifold polyhedron P is a polyhedron whose surface δP is a 2-manifold, which assumes that each point on δP has an ϵ-neighborhood that is homeomorphic to a half-ball. A polyhedron is simple if no two facets have only one vertex in common. A polyhedron is defined to be of genus n if it contains n holes (handles). An edge of a manifold polyhedron is called a *notch* if the dihedral (interior) angle between its incident facets is a reflex angle (exceeds π radians). A polyhedron with a notch (AB) is shown in Figure 7.7. A manifold polyhedron is defined to be nonconvex if it contains at least one notch.

Let P be a simple manifold polyhedron of arbitrary genus and let δP be a surface of the polyhedron. A patch of P is defined as a polygonal mesh which is a subset of δP. An example of a patch (indicated by the arrow) is shown in Figure 7.8. Vertices encountered on a patch are divided into two main categories: simple and boundary, and two subcategories: interior edge vertex and corner.

A vertex is called *simple* if it is bounded by a cycle of polygons. Each edge that is incident to a simple vertex has exactly two adjacent facets. A border vertex lies on a border of a patch. Consequently, some of its incident edges belong to only one facet. To find interior edges, a threshold is set for the dihedral angle formed by facets incident to an edge. If a dihedral angle associated with

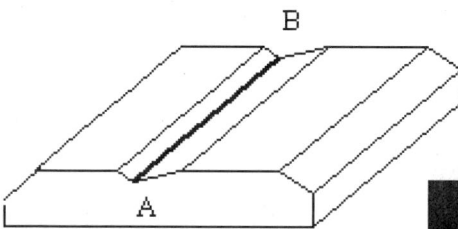

Figure 7.7 Polyhedron with a notch.

Figure 7.8 Patch.

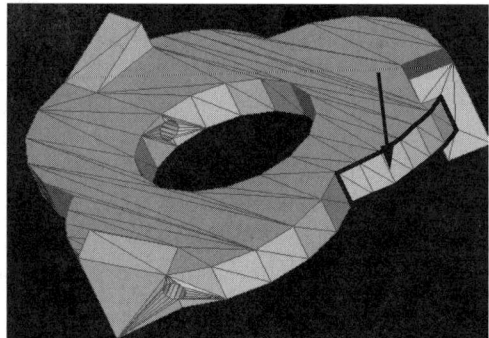

an edge exceeds a threshold, it is classified as an *internal edge*. For example, the threshold value for the mechanical parts used in the evaluation reported below is 80°. A vertex is said to be an *interior edge vertex* if it has exactly two incident interior edges. In case one, if three or more interior edges are incident to a simple vertex, it is called a *corner vertex*. Let P be a planar polygon. The boundary of P can be split into polygonal chains. A polygonal chain is said to be *monotone* with respect to a line L if every line that is normal to L intersects the chain in at most one point. A polygon is said to be monotone with respect to the line L if its boundary can be split into two polygonal chains where both chains are monotone with respect to L. A vertex is defined to be *reflex* if the dihedral angle between its incident edges is a reflex angle. Cusp C is defined to be a reflex vertex whose adjacent vertices are either both at or above, or both at or below, C with respect to the direction of L. A polygon that is split into two polygonal chains C_1 and C_2 (dashed lines denote separation between chains) is shown in Figure 7.9. The polygon is monotone with respect to the horizontal line since both chains are monotone.

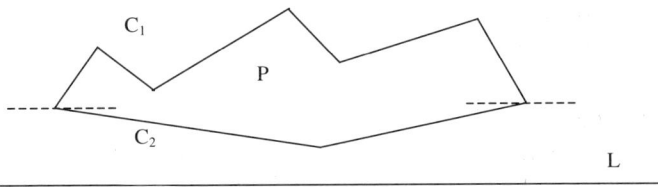

Figure 7.9 Polygon monotone with respect to horizontal line.

Let S be an n-dimensional set of points. A point V which belongs to the set is defined to be an extremum of the set with respect to an oriented line L if S's intersection with a small enough sphere (or circle in case of a two-dimensional problem) centered at V lies entirely in one of the two closed half-spaces defined by the plane (or line in case of a two-dimensional problem) normal to L that passes through V. A set with three extremum points (a, b, and c) is shown in Figure 7.10.

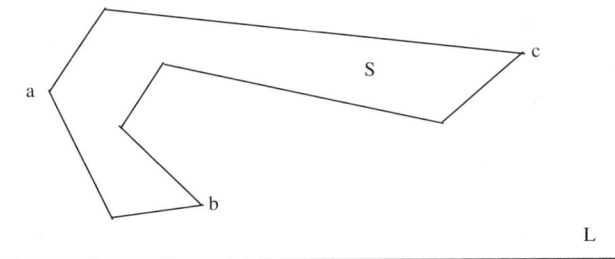

Figure 7.10 Extremum points.

A patch of a polyhedron is said to be *convex* if it lies on a boundary of its convex hull and the interiors of both the patch and the convex hull lie on the same side of the boundary with respect to each of the facets of the patch. The *convex hull* of a set of points is the smallest convex domain that contains the set. A patch is called *monotone* with respect to a plane if there is one-to-one mapping between vertices on the patch and the normal projections of the vertices on the plane. Sufficient conditions for a patch to be convex are: 1) the interior of a patch (a boundary is not included) is free of the polyhedron's notches, and 2) a patch is monotone with respect to a plane onto which it projects into a complex polygon.

The approach assumes simple manifold polyhedra of arbitrary genus (i.e., number of holes) with a triangulated boundary. For objects available in CAD data format, triangulation is now done using features already embedded in many CAD software packages (such as ProEngineer, Alias). For objects that are not available in CAD data format, triangulation can be done using well-known algorithms (see Chapter 5 for an example of such an algorithm). Nonmanifold objects are ignored because they either represent artificial objects with no real counterparts whose sole purpose is algorithm explanation, or the nonmanifold characteristics are a result of degeneracies that can be detected and then the nonmanifold object can be decomposed into a set of manifold objects.

The input into the algorithm consists of a set of triangle meshes which can be created inside a CAD environment (e.g., ProEngineer). The input can be created interactively inside a virtual environment as well. After the input has been defined, the algorithm consists of (1) removal of initial dihedral angles whose incident facets (triangles) contribute insignificantly to local geometry and topology, (2) resolving holes, (3) disassembling meshes along notches and patching all remaining non-convex meshes into convex ones, and (4) creation of virtual objects. The methodology is designed for objects with well-defined sharp corners and sharp edges. The steps are as follows:

1. Initial notch removal
   a. Characterization of local geometry and topology
   b. Resolving candidate dihedral angles
   c. Local retriangulation
2. Resolving holes
   a. Creation of hole's cap
   b. Interior hole tiling
   c. Creation of virtual sensor that activates collision detection inside a hole
3. Mesh disassembling
   a. Identification of connected notches
   b. Initial decomposition along closed notch polylines

  c. Cutting along remaining notches
  d. Additional cutting with respect to local coordinate system (LCS)
4. Virtual object creation
  a. Projecting a convex patch onto the LCS plane
  b. Fence-off phase (connecting boundary of a convex patch with its projection boundary)
  c. Attaching virtual objects to original objects

Initial Notch Removal

In the first phase, the goal is to resolve isolated notches whose incident facets (triangles) contribute insignificantly to local geometry and topology. Here one is referring primarily to facets whose area is insignificant with respect to the local topology (e.g., facets whose area is at most 1/20 of the area of each of the surrounding facets). Several factors can contribute to the appearance of these notches in the data set. Factors frequently encountered are:

- *Incorrect CAD model.* Since high-resolution visualization tools are available inside CAD packages, these errors are relatively easy to detect.
- *Errors in CAD triangulation modules.* Once we obtain triangular meshes using tools embedded into CAD software, it is very difficult to detect errors by visual inspection. To capture these errors, geometrical processing tools need to be used.

Since the number of virtual objects that are created at the end of the algorithm depends on the number of notches, by resolving such notches, one significantly reduces the number of virtual objects and consequently, relaxes the computational process. Once selected, the notches are marked appropriately in the database and removed from consideration in downstream processes. To visualize the object without marked notches, one has to perform a local retriangulation to secure mesh continuity. This process can also be seen as a local rectification.

Resolving Holes

In the second phase, the strategy is again to resolve a certain number of notches to relax the third phase and to reduce the number of virtual objects. The key observation here is that holes are sources of notch accumulation. Consequently, the goal is to separate accumulation areas and to treat them locally. The first step in this phase is to identify sections of the mesh corresponding to holes. To recognize hole areas, we make use of vertex classification, explained earlier. All interior vertices having at least one incident notch are identified. An example in which these vertices are recorded is shown in Figure 7.11. The bold arrow points to the vertex and the other arrow points to the notch. Once these vertices are identified, the next step is to locate open-

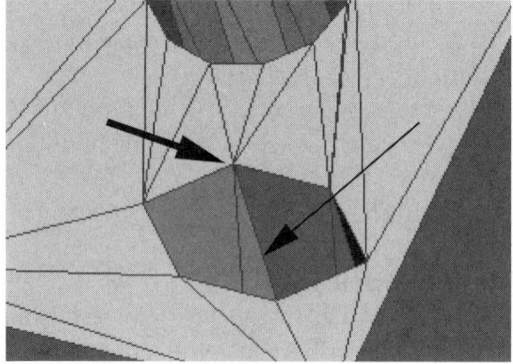

Figure 7.11 Hole opening.

ings of the holes. A hole opening can be seen as a closed edge set (loop) that satisfies the following two conditions:

1. Each vertex on the loop is interior vertex with at least one notch.
2. Each edge on the loop is an edge whose dihedral angle is a feature angle.

To recognize blind holes, we have to identify vertices that belong to hole bottoms. Vertices having at least three incident notches are classified as *bottom vertices*. An example where these vertices are recorded is shown in Figure 7.12. The bold arrow points to the vertex while the other arrows point to the notches. Once these vertices are identified, the next step is to search for connected edge sets associated with holes' bottoms. These sets are similar to opening sets in a sense that all vertices on the sets are vertices with at least three incident notches and all edges are notches. To identify sections of the mesh corresponding to the holes, we search for correspondence between opening and bottom loops. The loops are paired so that each vertex on one loop is connected with vertices on the mate where at least one connecting edge must be a notch.

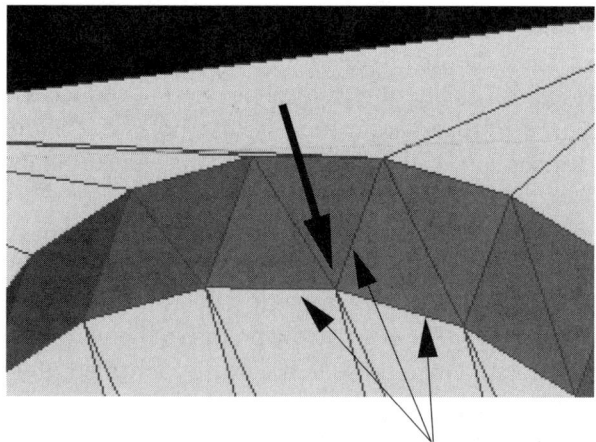

Figure 7.12 Hole bottom.

The next step is to disassemble the hole mesh along notches and interior edges into convex patches. A set of tiles is created where each patch lies on exactly one tile and covers (tiles) the hole interior. In the case of the belt drive (Figure 7.13), the tiles (identified with a bold arrow) are shown in Figure 7.14. Since the original object will be included in the scene as an object free of holes, one has to patch the hole openings. Now the hole opening is covered and the interior is tiled, but one is still not able to check for a collision detection inside the hole. To overcome this challenge, one creates a virtual sensor, a cylindrical object, which retains, to the extent possible, the shape of the hole cover. The term *sensor* refers to an imaginary hole cover (of user-defined thickness) and to the fact that collision detection with this hole cover can be used to trigger a collision detection process for the hole that it covers. The outer surface of the sensor lies slightly above the hole cover, so the approaching object will first hit the sensor. In the case of the hole shown in Figure 7.13, the sensor is shown in Figure 7.15. Once the collision between an approaching object and the sensor is detected, collision detection between the approaching object and the object that contains the hole is deactivated and the interior of the hole can be explored.

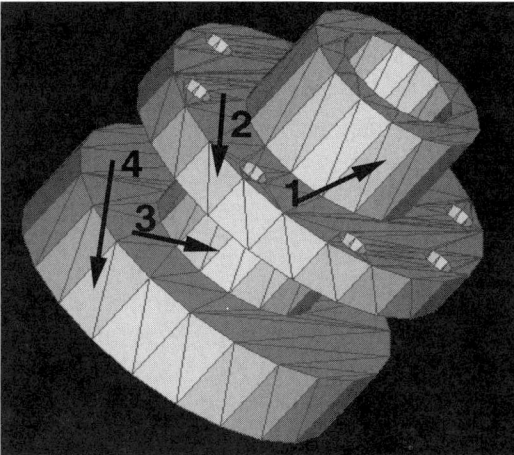

Figure 7.13 Belt drive.

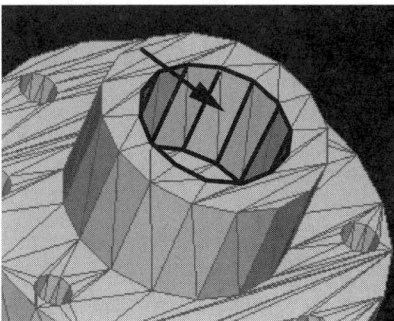

Figure 7.14 Tiles.

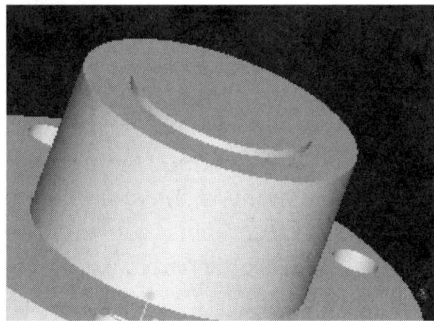

Figure 7.15 Sensor.

Mesh Disassembling

The third phase begins with the search for sets of connected notches. Geometrically, each set is represented as a closed notch polyline. Since all the edge incidences are explicitly stored, the sets can be found in linear time. The mesh is disassembled along the polylines, leaving two meshes for each polyline. The next step is to patch the openings in both meshes. This phase is especially efficient for families of objects frequently encountered in manufacturing processes characterized by the presence of protrusion features (e.g., shafts). The next step is to cut all meshes along remaining notches. The last step in the third phase ensures that all patches created by mesh decompositions are convex. An obvious approach to secure convexity would be to use constrained Delaunay triangulation, but because of its computational complexity, difficult implementation, and lack of convergence guarantees, an alternative approach based on recent advances in computational geometry is used. A local coordinate system (LCS) is assigned to each mesh, consisting of facets adjacent to remaining notches so that LCS's xy plane is initially perpendicular to the dominant facet direction. The dominant facet direction is computed as

as $\vec{d} = \sum_{i=1}^{k} A_i \vec{n}_i / \sum_{i=1}^{k} \vec{A}_i$, where k denotes the number of facets on the mesh while n_i and A_i denote a facet normal and area, respectively. Additional requirements are that no facet is perpendicular to the y axis and no edge is normal to the x axis. The latter constraints can be enforced by rotating the LCS. Once LCS is properly oriented, the last step of the third phase can be performed. Additional cuts are made along the edges whose incident vertices are extremum points with respect to the y axis.

Virtual Objects Creation

In the final phase, the patches are first projected onto the corresponding xy planes. Note that the third phase secures that each patch projects onto the xy plane into a convex polygon. This step is visualized in Figure 7.16. The next step is to create virtual objects by connecting vertices that lie on the boundary of a patch with corresponding vertices that lie on the projection boundary of a patch. After the fence and the projection are triangulated, a VO is ready to be added to the scene graph. A wireframe representation of the VO created from the patch shown Figure 7.16 is presented in Figure 7.17. Once all VOs are created, a hierarchy for each object and the scene hierarchy need to be specified. An *object hierarchy* usually contains subobjects obtained by the initial decomposition and VOs obtained by further decomposition of subobject boundaries. An object hierarchy is shown in Figure 7.18. In the case of the belt drive, the object hierarchy contains four subobjects (pointed to by arrows in Figure 7.13), 16 virtual objects attached to the first subobject (14 tiles, a sensor, and an additional virtual object attached to the hole bottom), and 96 virtual objects attached to the second subobject (80 tiles and 16 sensors).

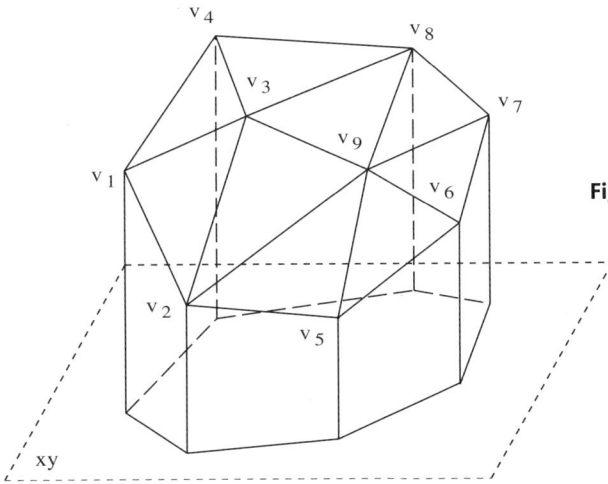

Figure 7.16 Patch projection.

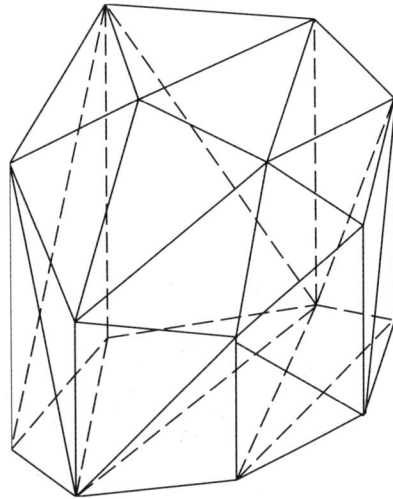

Figure 7.17 Triangulated virtual object.

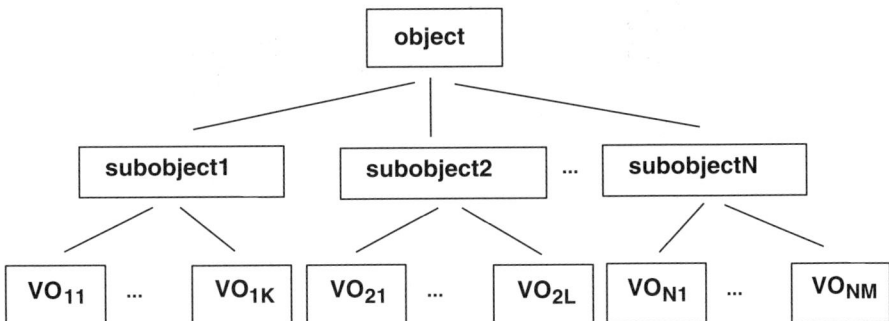

Figure 7.18 Object hierarchy.

Implementation

The algorithm has been tested on a set of nonconvex objects that have well-defined corners and sharp edges. Most of the objects have more than one hole. Each object was checked for the presence of nonmanifold characteristics. The search was restricted for nonmanifold characteristics to search for geometric degeneracies related to edges. The objects that had any edge with more than two incident facets were rejected. Since CAD representation of each object was assumed available, triangulation was performed inside a CAD environment (ProEngineer). CAVE and ImmersaDesk virtual reality platforms have been used for the implementation. In general, input is represented by bordered meshes, where each mesh contains certain object features (facets, edges, vertices) whose integration into a collision check is requested by the simulation objective.

The coordinate measuring process was selected as a testbed application. This was due primarily to the importance of exact collision detection for accurate modeling of the coordinate measuring process. A virtual coordinate measuring machine is shown in Figure 7.19. In a real coordinate measuring process, the coordinates of a point on the object boundary are measured by touching the point by a probe. Consequently, all points (vertices) that lie on the object boundary whose coordinates need to be measured become features of interest for collision detection. Collision detection enables a virtual probe to stop as soon as it touches the object boundary. Collision detection is visualized by changing the color of the probe. Audio effects can easily be integrated into the virtual measuring process so that collisions can be reported not only by

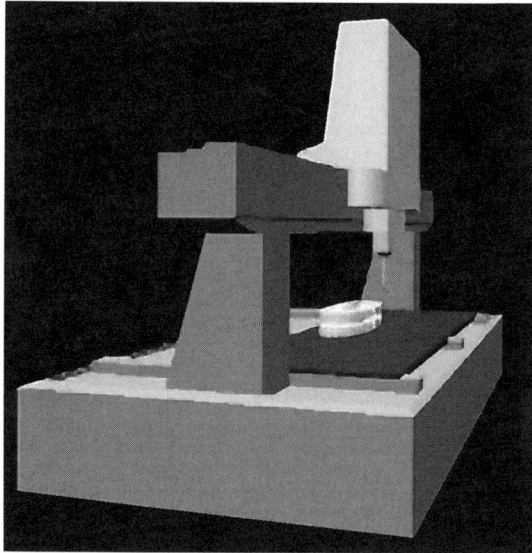

Figure 7.19 Virtual coordinate measuring machine.

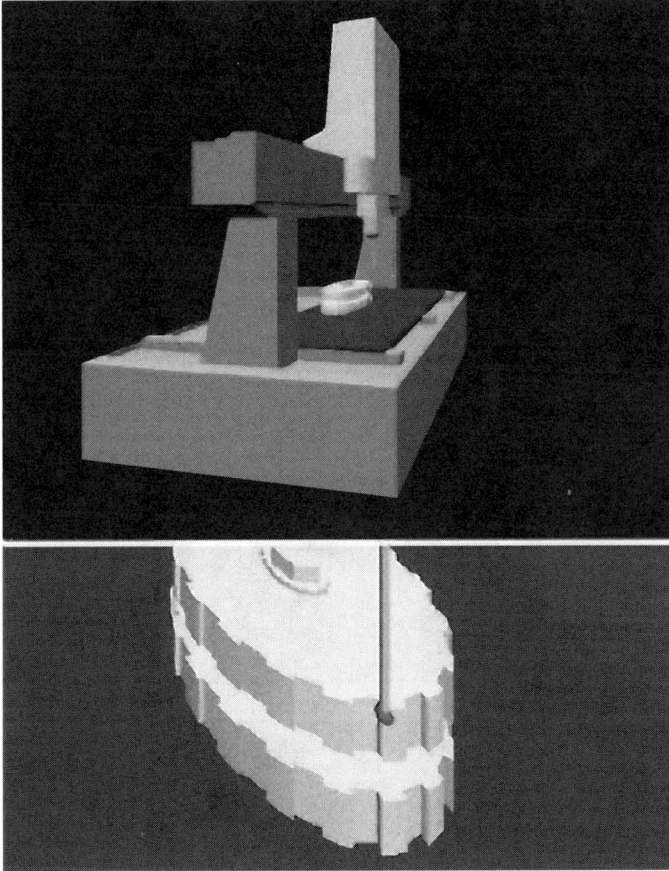

Figure 7.20 Virtual probe and collision detection (also shown in the color section).

changing probe color (turning the probe red) but also by emitting an appropriate sound. In Figure 7.20 (also shown in the color section), the relevant portions of Figure 7.19 are zoomed in to illustrate the exact collision detection with the probe.

The software libraries used in the implementation consist of a CAVE™ library for visualization and navigation inside the VR environment, Silicon Graphics IRIS Performer™ for scene design and update, and I-COLLIDE for collision detection between convex objects. To ease loading of the scene objects and integration of collision detection routines, ProEngineer implementation of triangulated IRIS Inventor™ file format was used. VRML file format was also used to facilitate retrieval of scene hierarchy information, since no scene loader inside the CAVE library for objects created in ProEngineer was available.

To illustrate the efficiency of the methodology, its performance is measured using a set of 15 mechanical parts. The complexity of the experimental set ranged from 426 to1386 edges, 284 to 924 facets, 48 to 310 notches and 0 to 9 genus, where notches and genus attributes are indicative of the degree of part nonconvexity. All of these tests were run on a Silicon Graphics Indy workstation. The collision detection was possible in realtime on a Silicon Graphics Indy workstation, which is indicative of its usefulness in other more or less powerful VR environments. The parts tested are presented in Figure 7.21. To

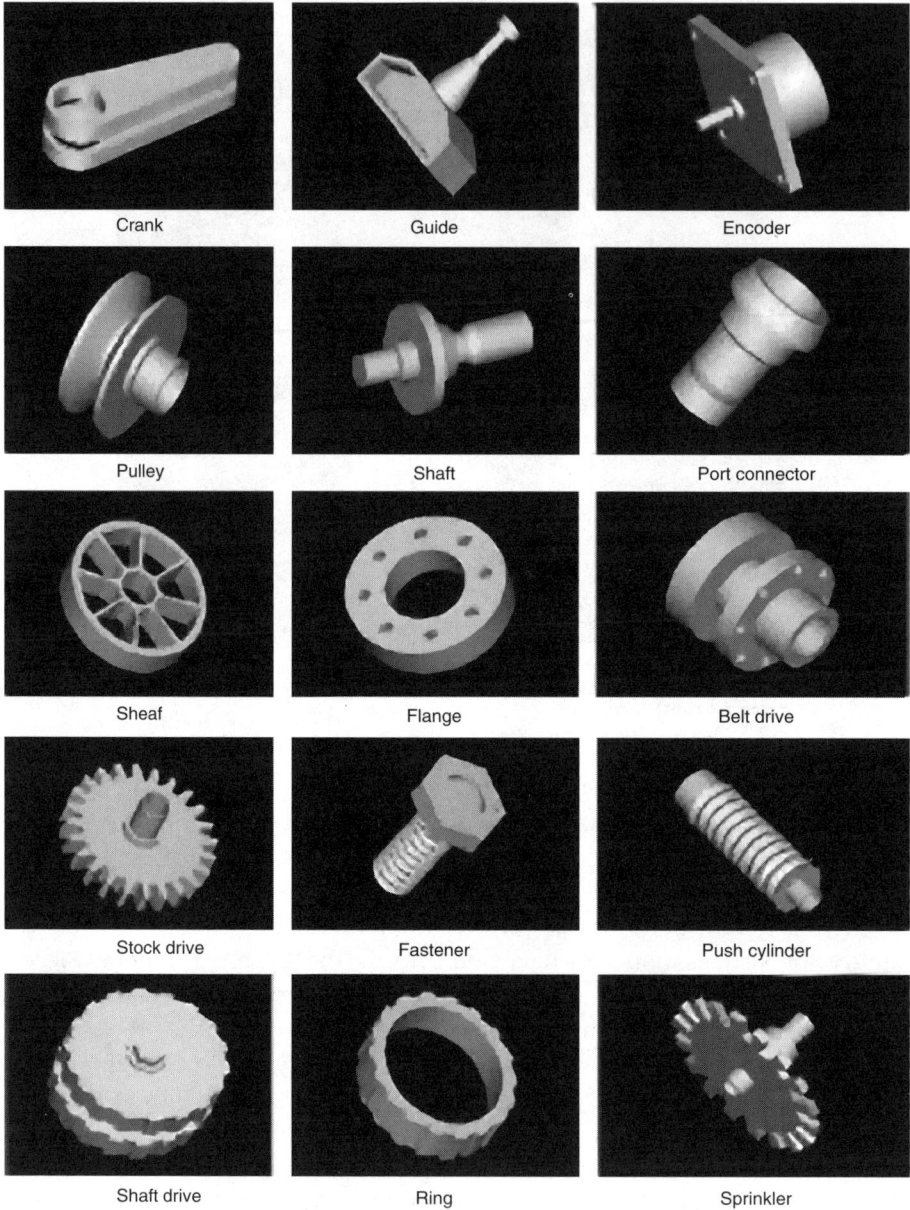

Crank	Guide	Encoder
Pulley	Shaft	Port connector
Sheaf	Flange	Belt drive
Stock drive	Fastener	Push cylinder
Shaft drive	Ring	Sprinkler

Figure 7.21 Objects used in experiments.

show that the methodology is truly practical, none of its routines are excluded from the time measurement process. The reported cost of the methodology covers the entire process, including triangulation of the virtual objects.

As expected, parts where most of the notches belong to closed notch polylines (e.g., crank, guide, and shaft) admit the lowest execution time. The increase in time is not unreasonable for more complex models. In the sheaf and flange examples, the increase in time is due largely to increased genus. One should also remember that this time occurs in preprocessing, before the simulation starts.

The evaluation is summarized in Table 7.1. It is important to remember that most of the virtual objects created are simple convex objects with as few as 12 facets. In the sheaf, flange, and belt drive examples, the increase in the number of virtual objects created is due to increased genus. Most of the virtual objects generated here are tiles. As can be seen from the shaft drive example, a more complex shape (1386 edges, 924 facets) admits more virtual objects even in the absence of holes. Most important, our methodology performed favorably when it was applied in VR applications where collision detection was not available previously. No performance degradation was noted when collision detection using virtual objects was added.

Note that the method does not work in real time, but it is a rapid modeling technique (e.g., VO computation for moderately complex parts takes less than 18 s). For an assembly with 50 parts it may take about 15 min. However, one should note that if the VO computation preprocessing overhead is removed, the method performs adequately at real-time interactive speeds.

TABLE 7.1 EXPERIMENTAL RESULTS

Part	Edges	Facets	Notches	Genus	Number of Virtual Objects	Times
Crank	426	284	48	2	35	0.82
Guide	606	404	64	1	12	1.07
Encoder	768	512	80	4	52	2.74
Pulley	1002	668	130	1	21	6.05
Shaft	480	320	64	0	6	0.76
Port connector	840	560	144	2	36	4.41
Sheaf	744	496	94	9	113	15.94
Flange	780	520	98	9	117	17.37
Belt drive	1062	708	156	9	116	11.54
Stock drive	930	620	96	1	89	7.27
Fastener	1380	920	310	9	108	9.42
Push cylinder	1170	780	190	1	27	6.57
Shaft drive	1386	924	140	0	115	9.46
Ring	672	448	58	1	79	7.95
Sprinkler	798	532	90	1	70	3.91

FURTHER READING

The material presented in this chapter is based largely on the work of Tesic and Banerjee (1999). The I-COLLIDE algorithm is due to Cohen et al. (1995). The convex decomposition algorithm described here has been developed by Chazelle and Pallios (1997). A good recent description of collision research can be found in Klosowski et al. (1998).

REFERENCES

Chazelle, B., and L. Pallios, "Decomposing the Boundary of a Non-convex Polyhedron," *Algorithmica,* Vol. 17, pp. 245–265, 1997.

Cohen, J. D., M. C. Lin, D. Manocha, and M. K. Ponamgi, "I-COLLIDE: An interactive and Exact Collision Detection System for Large-Scale Environments," *Proceedings of the ACM Interactive 3D Graphics Conference,* pp. 112–120, 1995.

Klosowski, J. T., M. Held, J. S. B. Mitchell, H. Sowizral, and K. Zikan, "Efficient Collision Detection Using Bounding Volume Hierarchies of K-DOPS," *IEEE Transactions on Visualization and Computer Graphics,* Vol. 4, No. 1, pp. 21–36, 1998.

Tesic, R., and P. Banerjee, "Exact Collision Detection Using Virtual Objects in Virtual Reality Modeling of a Manufacturing Process," *Journal of Manufacturing Systems,* Vol. 18, No. 5, pp. 367–376, 1999.

CHAPTER

8

MOTION MODELING

8.1 INTRODUCTION

The role of motion modeling in a virtual manufacturing (VM) simulator is to facilitate the user's description of motion and to enable the user to have precise motion control at real-time interactive simulation speeds. Whereas in a real environment motion characteristics are based on the driving mechanism (e.g., servomechanism), in VM the motion fundamentals are kinematic in nature. To achieve physically correct simulation, appropriate dynamic constraints have to be imposed. The connection of kinematics and dynamics is a difficult topic. Some augmented reality interfaces attempt to address this issue. Motion modeling thus refers to the kinematic phase only. It specifies parameters such as position, velocity, and orientation of a rigid object for every frame. A well-ordered frame series (i.e., temporal consistency) is assumed. Most VR manufacturing applications can be classified into two scenarios from a motion modeling standpoint:

Figure 8.1 Octahedral Hexapod machine installed at NIST machining circle–diamond–square test part. *Source:* photo courtesy of NIST.

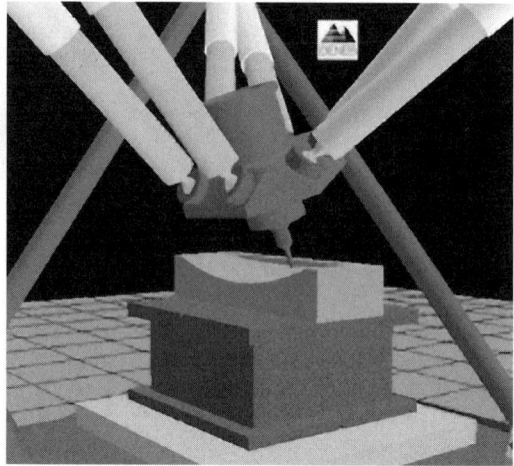

Figure 8.2 Deneb TeleGrip simulation of Octahedral Hexapod machining composite panel forming die. With part on riser block, no joint limits are exceeded. (Part geometry supplied by NASA Johnson Space Flight Center.) *Source:* photo courtesy of NIST.

1. *Motion parameters for every frame are generated by real or synthetic controllers.* An example of a real controller is hardware-in-the-loop VR simulation, such as the hexapod machine model developed at National Institute of Standard and Technology (NIST) (Figure 8.1) using Deneb's TELEGRIP (Figure 8.2). Examples of synthetic controllers abound in training examples using VR.

2. *Motion parameters are either entirely or partially unknown.* The vast majority of VR applications fall in this category. In such a case the trajectory specification is usually uncoupled from parameter specification process. This second scenario is addressed mainly in this part; hence the next topic is trajectory specification.

8.2 TRAJECTORY SPECIFICATION

Due to incomplete or unavailable trajectory data, trajectory specification tools are frequently used in a VR manufacturing simulator. Trajectory modeling is strongly coupled with the application being considered. A good understanding and strict enforcement of curve continuity is critical for accurate simulation. A detailed explanation of the importance of various forms of curve continuity is presented later. Here we demonstrate the significance of the continuity using as an example a numerically controlled milling process in which discontinuity or inappropriate continuity can cause a leap in the milling tool and damage it. In addition to continuity, a choice of parametric curve representation is very important for accurate trajectory specification. In general, as far as a VM simulator is concerned, parametric representation where trajectory passes through a given set of points is more useful than parametric representation where the trajectory passes through the vicinity of the points.

Since Bezier curves and surfaces have been used heavily in computer graphics and have improved constantly due to significant research effort devoted to their development, we decided to assume a Bezier parametric representation. However, both motion control and trajectory specification algorithms presented in this chapter can easily be modified for use with another parametric scheme.

The trajectory specification approach that is designed is useful for applications in VR manufacturing, where motion of an object is constrained by a set of locations in space where the moving object must arrive in time and where values of motion parameters other than position (e.g., velocity, orientation) must be strictly enforced. Curve continuity, interpolation, and local control are the three main requirements that are considered. Each of these requirements is addressed next.

1. Parametric curve continuity needs to be enforced if one wants to use a VM simulator as a validation tool. In general, parametric curve representation is not as intuitive as a functional representation; hence it needs to be studied closely. There are two types of continuities in parametric space:

   a. *Parametric continuity.* A composite parametric curve is considered to have an nth-order parametric continuity (C^n) if the first n derivative vector pairs agree where the curve segments meet. For example, for C^1 continuity, the tangent vectors at joins are equal; that is, their direction and magnitudes are equal.

   b. *Geometric continuity.* This is a relaxed form of parametric continuity. For example, a composite parametric curve is considered to have first-order geometric continuity (G^1) if the slopes are equal at joins, meaning that their directions are the same but their magnitudes may be unequal. For example, consider two parametric curve segments that meet at point (8,4):

$$p(u) = \{8u, 4u\} \qquad 0 \le u \le 1$$
$$l(u) = \{4(u+2), 2(u+2)\} \qquad 0 \le u \le 1$$

The first derivatives are $p'(u) = (8,4)$ and $l'(u) = (4,2)$. The directions are the same but the magnitudes are different.

Failure to enforce parametric continuity in VM simulator can cause damage to actual manufacturing processes in the physical domain when the simulation specifications are implemented.

2. An interpolation requirement in this context refers to a curve that models trajectory passes through a given set of control points with limited curve oscillations between consecutive control points. Polynomial interpolation techniques (such as Lagrange, Gauss, Newton, and divided differences) are not suitable for trajectory modeling because they fail to

provide a path that has no oscillating segments. A Bezier composite curve is a good choice to enforce interpolation requirement. Following are some of its useful properties:

- A Bezier curve is uniquely defined by an ordered sequence of control vertices known as a *control polygon.*
- The curve passes through the points of the control polygon.
- The curve satisfies the convex hull property; that is, the curve remains inside the convex hull formed by control vertices.
- The shape of the curve is related intuitively to the shape of the control polygon, meaning that the curve closely follows its control polygon.
- The curve has linear precision capability. If the control polygon vertices are uniformly distributed on the line joining its endpoints, the curve generated is a straight line. This enables integration of commonly encountered straight-line segments into the Bezier curve–modeled trajectory.
- The endpoint tangent directions are uniquely determined by the slopes of the corresponding arms of the control polygon. This enables easy specification of both start and end conditions which are mostly used for the specification of motion parameters at locations corresponding to the interpolating points.

Note that Bezier curves are more appropriate here than B-splines, which have been used extensively for corner detection (see Chapter 4) because the trajectory needs to pass through all the control points. Recall that in a B-spline there are four control points, which influence interpolation, but not all are actually interpolated (i.e., the curve does not necessarily pass through all of them). In a Bezier curve there are two endpoints and two other points that control the endpoint tangent vectors.

3. The local control requirement assumes that a change in path specification has a local effect, meaning that only neighboring composite curve segments are affected by the change. An example of a change in path specification may be a change in the set of points that need to be interpolated. Polynomial techniques fail to meet this requirement because the shape of the curve is influenced by each data point. This requirement is difficult to impose, especially if continuity and interpolation constraints are also imposed. The best way is to use higher- degree curves, but here also trade-offs are involved because these curves are more difficult to manage.

8.3 TRAJECTORY MODELING

Input into the trajectory model is an ordered set of locations that a moving object must pass through during the simulation loop. Any path planning technique can be used to increase efficiency in generation of the input set, if desired. The trajectory model makes no assumption related to the methodology used to generate the input set.

To relax the composite curve generation process, breakpoints where only C^0 continuity is enforced are identified. These points correspond mostly to intermediate locations where a moving object rests before it resumes motion. The breakpoints are used to divide the original input into subsets of points that need to be interpolated. Interpolation subsets are further divided into two groups with respect to the level of continuity requested. A subset that is subject to C^2 continuity enforcement is used for fourth-degree composite Bezier curve generation, and a subset subject to C^1 continuity enforcement is used for generation of third-degree composite Bezier curves. The methodology is summarized in Figure 8.3.

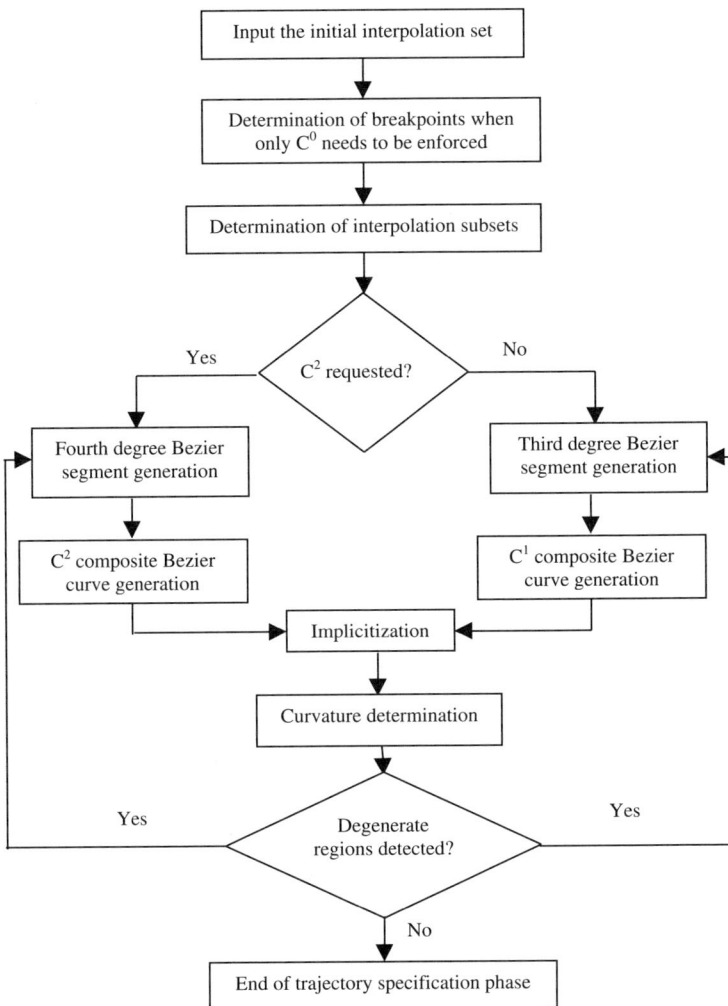

Figure 8.3 Trajectory modeling methodology.

A Bernstein polynomial form of Bezier curve generation is chosen. A Bezier curve can be expressed by

$$P(u) = \sum_{i=0}^{n} P_i B_i^n(u) \tag{8.1}$$

where *Bernstein polynomials* are defined by

$$B_i^n(u) = \frac{n!}{i!(n-1)!} u^i (1-u)^{n-1} \tag{8.2}$$

The following property of Bernstein polynomials is used to illustrate the linear precision property of a Bezier curve:

$$\sum_{i=0}^{n} \frac{i}{n} B_i^n(u) = u \tag{8.3}$$

The expression for a third-degree Bezier curve generated by a control polygon (P_0, P_1, P_2, P_3) is

$$P(u) = P_0(1-u)^3 + P_1 3u(1-u)^2 + P_2 3u^2(1-u) + P_3 u^3 \tag{8.4}$$

The expression for a fourth-degree Bezier curve, which is used when C^2 continuity is needed, generated by a control polygon $(P_0, P_1, P_2, P_3, P_4)$, is

$$P(u) = P_0(1-u)^4 + P_1 4u(1-u)^3 + P_2 6u^2(1-u)^2 + P_3 4u^3(1-u) + P_4 u^4 \tag{8.5}$$

Composite Bezier curves are used in modeling complex trajectories. The trajectories are constructed from a set of Bezier curves where the last control vertex of the ith composite curve segment coincides with the first control vertex of the $(i + 1)$th composite curve segment. To avoid the appearance of an inflection point inside a curve segment, an attempt is made to maintain a convex shape in the control polygon. In some cases it is not possible to satisfy both convexity and continuity constraints. In case of conflict, the convexity constraint has a lower priority than the continuity constraint. To enforce the appropriate level of curve continuity, certain restrictions need to be applied. Since the shape of a Bezier curve depends on the shape of its control polygon, every constraint is expressed in terms of control polygon vertices. The derivative of the Bernstein polynomial and the derivative of the Bezier curve obtained from it are

$$\frac{d}{du} B_i^n(u) = n\left[B_{i-1}^{n-1}(u) - B_i^{n-1}(u) \right] \tag{8.6}$$

$$\frac{d}{du} P^n(u) = n \sum_{i=0}^{n-1} (P_{i+1} - P_i) B_i^{n-1}(u) \tag{8.7}$$

The Bezier curve derivative is used to find an appropriate position of control vertices to enforce appropriate continuity for consecutive segments in a composite Bezier curve. Assuming that the sequences $(P_0^1, P_1^1, \ldots, P_n^1,)$ and $(P_0^2, P_1^2, \ldots, P_n^2,)$ denote control polygons corresponding to two composite Bezier curve segments joining at $P_n^1 (P_n^1 = P_0^2)$, Δ_1 and Δ_2 denote lengths of parameter domains for the curve segments, and P_i^j denotes a position vector. The following condition should be satisfied if C^1 and C^2 continuity are desired:

$$\Delta_1 (P_n^1 - P_{n-1}^1) = \Delta_2 (P_1^2 - P_0^2) \tag{8.8}$$

To enforce C^2 continuity, the following condition must be satisfied:

$$\frac{\Delta_1}{\Delta_2} P_2^2 - \frac{\Delta_1 + \Delta_2}{\Delta_2} P_1^2 = \frac{\Delta_2}{\Delta_1} P_{n-2}^1 - \frac{\Delta_1 + \Delta_2}{\Delta_1} P_{n-1}^1 \tag{8.9}$$

The constraints above become easier to handle if one chooses $\Delta_1 = \Delta_2 = 1$. Implicit forms of the parametric representations are determined for:

- Convenience in curvature determination, which is helpful in resolving degenerate regions such as high-curvature regions, self-intersections, and cusps
- Simplifying the problem of curve intersections to help in analytical detection of obstacles between moving objects
- Ease of determination whether a given location lies on the trajectory to facilitate many aspects of VM simulation, such as analytical detection of obstacles, making subdivision of composite curves easier, and determination of collision-free routes as feedback to path planning algorithms

In implicitization, a parametrically defined curve of the form $x = x(t)/w(t)$ and $y = y(t)/w(t)$ is converted into its implicit form, $f(x,y) = 0$. Here x and y represent the Cartesian coordinates of points on a curve, and $x(t)$ and $y(t)$ are polynomials in t. $w(t)$ is a conversion factor, which can be assumed as 1 for simplicity in some cases. It is always possible to find an implicit representation of a parametric curve, but the reverse process is not guaranteed. In this implementation, implicitization is done using elimination theory. Note that calculating curvature in parametric form may cause numerical instabilities. In such a case, an approximation called *pseudocurvature* can be used for detection of degeneracies if the curve is specified in parametric form. Pseudocurvature is defined as $\sqrt{\left(\dfrac{d^2 x}{dt^2}\right)^2 + \left(\dfrac{d^2 y}{dt^2}\right)}$. A distinct advantage of the implicit form of a parametric curve is the ability to provide exact curvature which does not exhibit numerical instabilities. This curvature for the implicit form of curve representation is given by $\dfrac{y''}{(1 + y'^2)^{3/2}}$. Curvature information can be used to

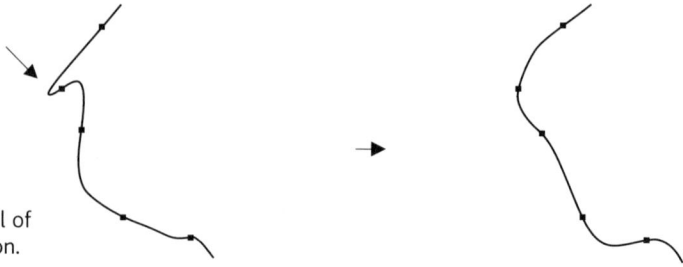

Figure 8.4 Removal of high-curvature region.

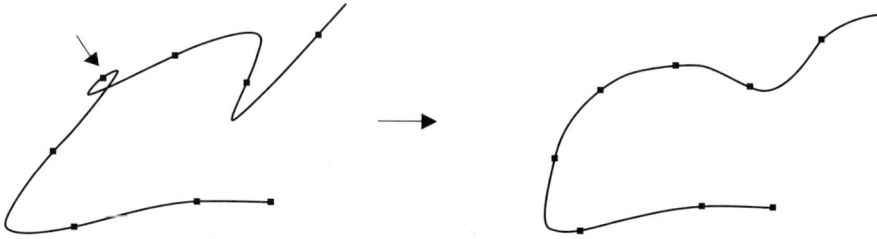

Figure 8.5 Removal of self-intersection.

detect potential degeneracies on the trajectory. The common ones are high-curvature regions where, for example, high curvature and limited steering angle can prevent a moving object from negotiating a curve, and self-intersections can, for example, cause collisions among moving objects. In case of degeneracy, first an attempt is made to remove it without affecting curve continuity by moving free vertices of the control polygon. In cases where changing the shape of a control polygon does not offer sufficient flexibility for removal of a degeneracy and where degeneracies can be removed at the expense of C^2 continuity, one can use a lower-degree composite Bezier curve in trajectory specification. Two examples where degeneracies are removed by replacing a fourth-degree with third-degree composite Bezier curve are shown in Figures 8.4 and 8.5. Notice, however, that strict enforcement of the degeneracy removal constraint is not enforced.

8.4 DETERMINATION OF MOTION PARAMETERS

Motion parameters are either entirely or partially unknown. As has already been indicated, in a vast majority of VM applications, motion parameters are either entirely or partially unknown. In such a case the trajectory specification is usually uncoupled from the motion parameter specification process. Having addressed this process, attention is now directed toward the determination of motion parameters. Efficient and accurate tools for determination of motion simulation parameters are important for both macroscopic and microscopic VM simulators. Whereas in macroscopic simulators the aim is to capture human expectation, in microscopic simulators the aim is to secure sufficient accuracy of motion modeling.

The most efficient way for high-level motion control during simulation is to derive closed-form analytical solutions to motion parameters. Such analytical solutions are easier to handle using a functional representation of motion parameters. Unfortunately, a functional representation approach is not flexible enough to facilitate modeling of complex surfaces and trajectories. On the other hand, a parametric representation (splines) of curves and surfaces that has been widely used in computer graphics is more useful for motion control in VE. Due to the fact that the mapping between parametric and Cartesian spaces is not linear, in most cases it is not possible to obtain closed-form solutions to motion parameter specifications. Since the increasing complexity in surface and trajectory modeling in computer-aided design makes the use of splines necessary, a technique that will enable users to retain the necessary level of a motion control in the parametric space would substantially increase both correctness and degree of realism in VM simulator. Most techniques developed so far rely either on computationally expensive numerical analysis methods or on inaccurate simplification models. An efficient technique based on analytical and iterative procedures is presented. In the first phase of the technique, an analytically derived curvature of a parametric curve supplies an initial guess to the second phase, which is an iterative derivation of certain motion parameters. This two-pass approach secures fast convergence and user-controllable accuracy.

Finding appropriate input parameters that secure realistic and correct motion simulation is difficult. Traditional animation has relied on keyframing, anticipation, timing, staging, arcs, and exaggeration. Although an acceptable level of motion control is obtained, the dependence of such traditional animation on the user's action has almost eliminated its use in manufacturing applications. Considerable interest in developing automatic techniques for determination of motion simulation parameters has resulted in developing various computer-assisted techniques, which can be classified as *interactive techniques*, where the user remains in the loop for fine-tuning and making selection among computer-generated intermediate alternatives, and *automatic techniques*, where motion simulation parameters are determined according to a user-defined objective function. Although introducing automatic techniques into a difficult simulation parameter setting is appealing, it comes with a price. Increased automation normally comes at the expense of control. Since the importance of control in a parameter setting for simulation in VM environment is widely acknowledged, an effort is needed to increase the level of control while keeping the advantages of automatic techniques.

A method is designed to increase the level of control in parameter setting while keeping the advantages of the user's ability to set parameters based on user-defined objectives. In general, an ordered sequence of moving object displacements during the simulation loop reflects the object's speed and acceleration. In the case of parametric trajectory modeling, an intuitive approach would be implicit arc length parametrization, which unfortunately, is extremely difficult to construct. Due to nonlinear mapping between parametric and Cartesian space, analytical solutions are also difficult. An obvious thought would be to use numerical analysis. In general, a numerical model would try to find the desired value of parameter u by solving the equation

$$l = \int_{u_0}^{u} \sqrt{[X'(u)^2 + Y'(u)^2 + Z'(u)^2]}\,du \tag{8.10}$$

where l is the arc length and u_0 is the parameter value of the reference point. Standard numerical techniques such as Newton–Raphson can be used. However, for time-critical interframe calculations, such techniques fail to converge within a given time. Local approximation techniques work only for certain curves. The method designed here is an efficient iterative procedure for locating points on a curve with desired spacing. It makes no assumptions related to curve shape and parametrization.

A trajectory model and motion specifications are input to the system. A trajectory model can be obtained by using the technique described previously. If another technique is used for trajectory modeling, implicitization has to be performed if necessary. If motion specification is not complete, missing parameters need to be estimated. This usually happens in macroscopic motion simulation such as various traffic simulations. Since it proved useful in practice, it is assumed that the speed of the moving object is proportional to the reciprocal value of the curvature. Once the motion parameter specification is complete, one needs to calculate the sequence of displacements for each moving object that involve the object motion parameters. This displacement is proportional to speed. The coefficients for speed estimation and the displacement depend on the nature of the application and desired frame rate.

The next step is to determine acceptable lower and upper bounds of the displacement. The bounds that reflect the accuracy requested are used as stopping conditions in the algorithm. The initial guess can be a critical step in the iterative process. Previous parameter increment du_{i-1} and previous curvature k_{i-1} are both used to calculate the initial value of parameter increment du_i:

$$du_i = \frac{k_{i-1}}{k_i}\,du_{i-1} \tag{8.11}$$

where i represents the frame index. The rationale for including only one parameter is the strong correlation between the curvature and the arc length of the parametric curve. Examples include circles where uniform spacing of sample points corresponds to uniform curvature. The value of the parameter increment is adjusted using the following expression if the initial value does not provide the desired distance (e.g., $D_{lower} < D < D_{upper}$):

$$du_i = m\,du_{i-1}\cdot 2^{-n} \tag{8.12}$$

where n is the iteration count and m is a coefficient, which is set to $+1$ if the parameter increment exceeds the upper bound and is set to -1 if it falls below the lower bound. The methodology is summarized in Figure 8.6.

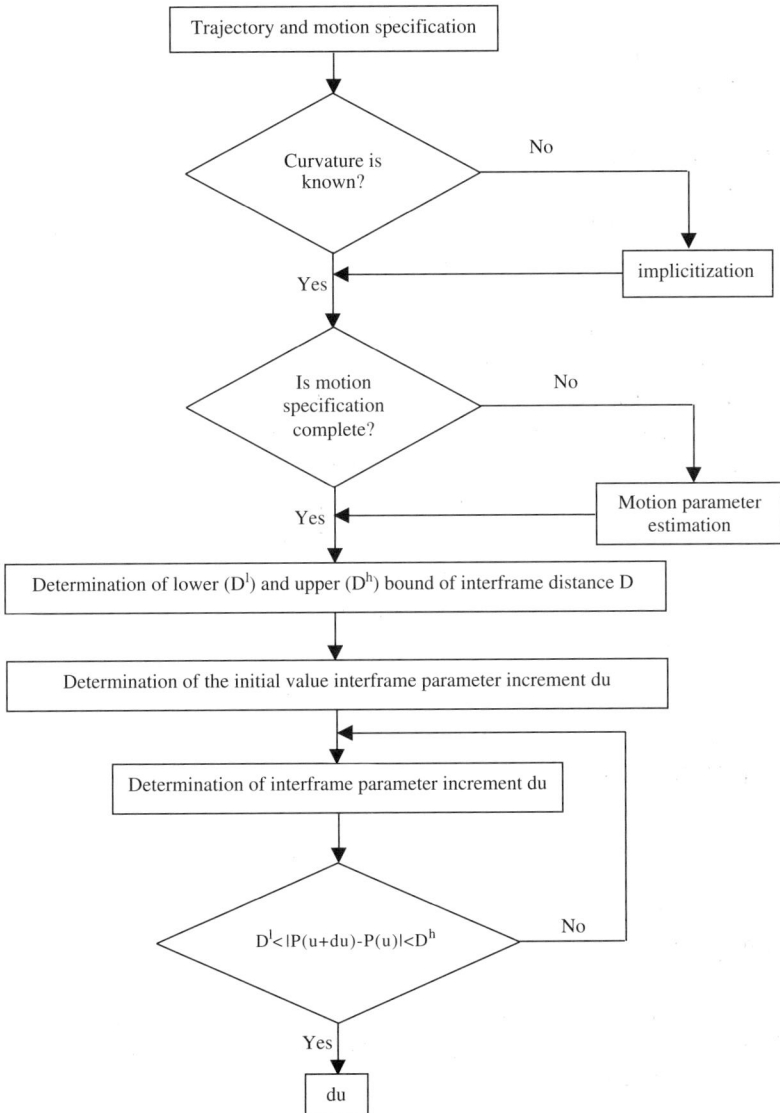

Figure 8.6 Motion parameter determination.

Experiments with Uniform Motion

The methodology is experimented with using a set of third-degree Bezier curves. In each case the goal is to come up with an appropriate set of sampling points on the curve, which would reflect uniform motion along the trajectories. To assess the accuracy, a Euclidean distance between two consecutive points along the curve is calculated. This is a good estimate of the distance between two points because a large sampling set of 1500 points is used. The desired precision is specified by setting lower (D_{lower}) and upper (D_{upper}) bounds on interframe distance. The number of iterations necessary to achieve desired accuracy is recorded for each frame and maximum and average number of iterations are found. The results show fast convergence in all cases even in cases where high precision (0.005%) is employed.

Experiments with Nonuniform Motion

Two scenarios are tested: (1) acceleration does not change significantly between two consecutive frames, and (2) acceleration changes randomly in a given interval. For each of the two scenarios, 100 randomly generated Bezier curves are used. For each curve a sampling set of 1000 points are randomly generated. In the first scenario, initial acceleration a_i is randomly selected from the interval [0.01, 0.099] and coherence is secured by linearly changing acceleration in the interval $[a_i, 2a_i]$. In the second scenario, a_i for each frame was randomly selected from the interval [0.01, 0.099]. To assess the accuracy of the algorithm, it is assumed that the Euclidean distance between two consecutive points is a good estimate of the distance between the two points, measured along the curve. Since a large sampling set of 1500 points was used, this assumption is believed to be reasonably accurate. The results show fast convergence even in cases where high precision (0.0005%) was requested. For more information on methodology and experiments, see "Further Reading."

FURTHER READING

This chapter is based largely on the work of Tesic (1999).

REFERENCE

Tesic, R., "Collision Detection and Motion Generation for Virtual Manufacturing Simulator," Ph.D. dissertation, Department of Mechanical Engineering, University of Illinois at Chicago, 1999.

9

TELECOLLABORATIVE VIRTUAL MANUFACTURING ARCHITECTURE

9.1 VIRTUAL MANUFACTURING LATTICE DATA STRUCTURE

9.1.1 Scenegraph Limitations

A *scenegraph* is the most common abstract data structure used to represent objects in a VE. Objects are represented as nodes in the scenegraph and the relationship between the objects and the environment is described by the edges. The scenegraph is a hierarchical structure that describes the position of each node in the environment and the dependency of the node with respect to its parent nodes. All the software used for creating virtual environments use some form to describe the scenegraph. For example, VRML uses a text-based approach to describe the scenegraph. IRIS Performer™, on the other hand, utilizes special data types to create its internal representation of the scenegraph.

The scenegraph contains the nodes' geometry, material, and other physical attributes and the relative displacement and orientation with respect to its parent. Simulated operations in the virtual environment can be performed by manipulating the scenegraph structure appropriately. For example, moving an object in the VE can be achieved by updating the transformation matrix by the displacement. When the scene is updated, the object appears in the new location obtained from the updated transformation matrix. The scenegraph has callback features that are used to handle and service specific events. VRML provides structures such as TouchSensor and TimeSensor to perform actions based on an event occurrence. For example, the click of a wand button on a doorknob simulates touch and leads to a callback to open the door by rotating it along the axis of the door hinge. The doorknob object has a callback routine that checks for the wand button click event and accordingly, services the event by invoking the module to open or rotate the door.

Some of the scenegraph limitations are as follows:

1. Often, there is a lack of complete information related to an event to be serviced and the corresponding action by the callback routine. For example, suppose that the click of the wand button on an object leads to the callback to move the object to another location. The action of moving to the location depends on the path to be taken. If the path is unknown or if there exist multiple paths to the destination, it leads to a situation that requires special handling.

2. Another limitation of the scenegraph is the inability to handle the status of an object and performing instantaneous queries. For example, it is necessary to check the status of a machine before allocating a job to the machine. The machine needs to be in a proper state to be able to process the job. The scenegraph does not have a mechanism to maintain the state of the objects in the VE.

There are two types of tasks in virtual manufacturing:

a. *Interactive tasks.* In this type there is a high degree of interaction between the user and the objects in the VE. An assembly task in VE is an example of an interactive task. The user is actively involved in the process (e.g., the user grasps objects using an interactive device, such as a wand or a data glove, and assembles them together). The user can navigate around the objects to a desired location and execute the task.

b. *Automated tasks.* The interaction between the objects in the VE and the user is minimal. The user retains control of the VE. A robotic work cell is an example of an automated task. The robot is in charge of assigning jobs to machining centers and transporting the jobs between work cells. The user can navigate in the VE, examine the tasks being executed, start task execution and pause, and make decisions in emergency situations.

These tasks are currently specified either by programming or by a menu system. Programming in a VR environment is a difficult task, and it involves the expertise of a programmer to program the tasks. The menu system is often limited to performing a set of predefined tasks and it becomes difficult to handle situations outside the predefined set of tasks. The scenegraph limitations apply to interactive and automated tasks. In interactive mode, the environment must be capable of imposing restrictions on the user's actions by a rule enforcement mechanism. In the automated mode, the tasks are being executed through a driver module that provides data at some time interval for running the application. The environment must be capable of providing rule enforcement or status information data to the driver module so that it can control the interface data in a meaningful way.

3. The scenegraph in its present form is not able to apply the rule enforcement.

The foregoing three limitations on the scenegraph have led to the creation of a behavioral layer architecture. It consists of a virtual manufacturing lattice (VML) structure, an object library structure, and a virtual manufacturing script (VMS). The VML, coupled with the object library, is aimed at supplementing the scenegraph structure to facilitate VR-based simulation of design, planning, and control applications in manufacturing. The VML structure aims at encapsulating the object features needed for the design, planning and control applications and provides an interface with external modules to drive simulation in the virtual environment. The object library is used for efficient storage and retrieval of the prototypes from and to the VML structure, respectively. The VMS is a scripting mechanism, which has a grammarlike structure for users to express the tasks in the CVE. The VMS operators manipulate the VML in order to simulate the manufacturing operations in a CVE. Figure 9.1 shows the relationship of the scenegraph to the CVE, and Figure 9.2 illustrates the overall

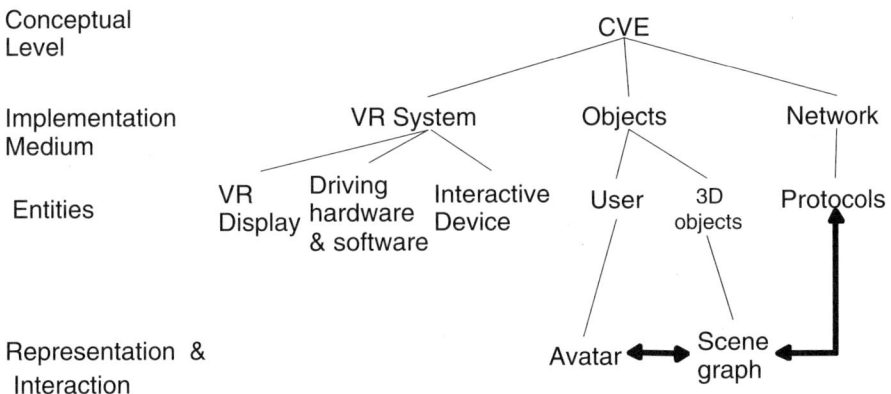

Figure 9.1 Relationship of scene graph to CVE.

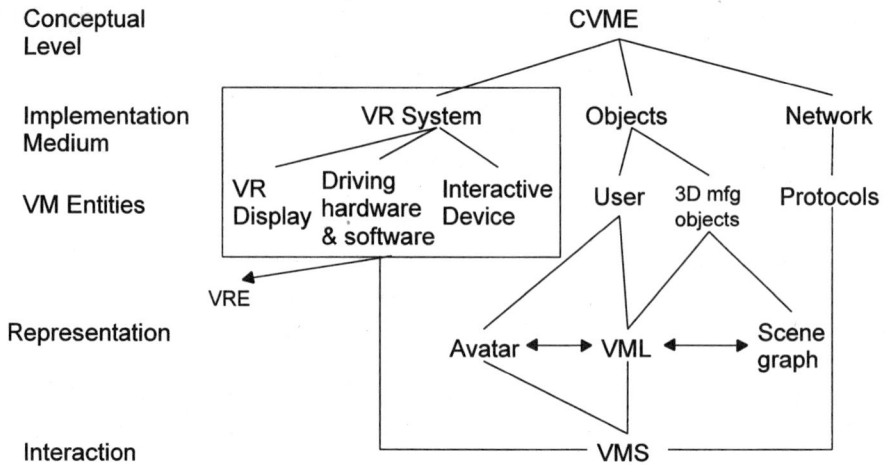

Figure 9.2 Collaborative virtual manufacturing environment (CVME) incorporating behavioral layer architecture.

block diagram of a collaborative virtual manufacturing environment (CVME) incorporating behavioral layer architecture. The following section describes the elements (VML, object library, and VMS) of the behavioral layer architecture.

9.1.2 VML Structure and Object Library

The methodology to address limitations in the scenegraph structure led to the idea of VML structure coupled with object library structure, which is aimed at supplementing scenegraph structure. This would facilitate VR-based simulation of design, planning, and control applications in the manufacturing domain. The VML is a lattice structure used to represent convex as well as nonconvex nondeformable solid objects in the VE. The composition of an object can be described most efficiently using a hierarchical relationship between all its components. All the objects and their hierarchy of components that are present in the VE can be combined in the form of a lattice structure, as shown in Figure 9.3. The lattice is a multiple inheritance structure that can be used for representing component reusability. A 4-tuple that encapsulates the properties of the objects describes each node in the lattice.

The scenegraph needs to be updated at a regular interval on a "need to display" basis. The "need to display" attributes are the properties that are useful in the graphical display and the user needs to be aware of the changes by visual means. Scenegraph updates are performed by having callback routines to update the scenegraph nodes whenever the nodes in the VML are updated. The inverse process also needs to be satisfied to maintain continuity and logical synchronization between the scenegraph and VML. Any action on an

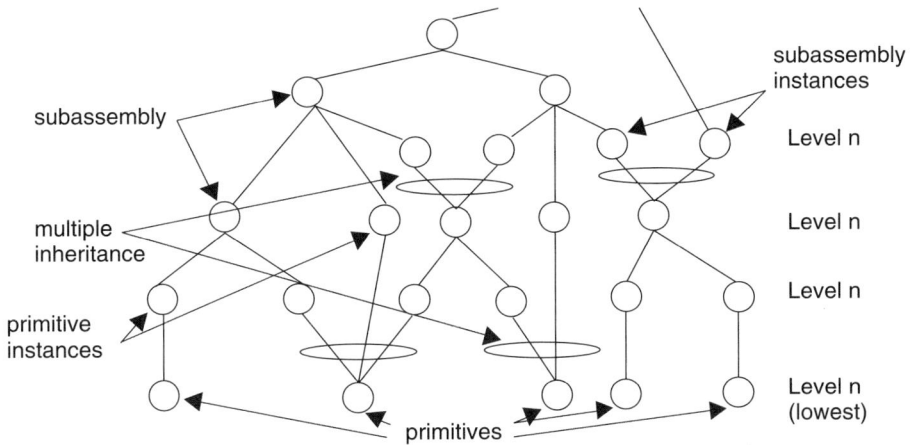

Figure 9.3 Virtual manufacturing lattice structure.

object in the VE gets "recorded" in the corresponding node of the scenegraph. For example, interactive movement of an object in the VE by a user results in a change in the coordinates of the node in the scenegraph. The change needs to be propagated to the corresponding node in the VML for further processing by other external modules, such as a collision detection algorithm, controller module, and so on. The update from the scenegraph to the VML is also initiated via callback routines whenever there is a change in the scenegraph node.

Conceptually, the lattice is a multiple inheritance structure. The management of multiple inheritance starts getting difficult with increase in the number of parents for an object. During implementation of the lattice structure, an object library structure coupled with multiple instantiation methods is used to reduce the multiple inheritance structure of the lattice to a single inheritance structure that reduces the lattice to a tree structure. An object library is created to store the primitives efficiently. A primitive is an object that is used more than once in the virtual environment. A typical example of a primitive is a washer or a rivet in an assembly; they are used at multiple places when two objects are being joined. The object library stores a copy of the primitive, and instances of the primitives are created every time a new primitive is required. The main function of the object library structure is proper indexing and faster storage and retrieval of primitives. A hash table implementation of the object library is used in this case. A hash function is used to encode and decode the key for an object primitive. The key is then used to store or retrieve the object primitive and create an instance of the object to be used in the lattice. The choice of implementation medium for the object library depends on the nature of the CVR application. Some other implementation strategies include

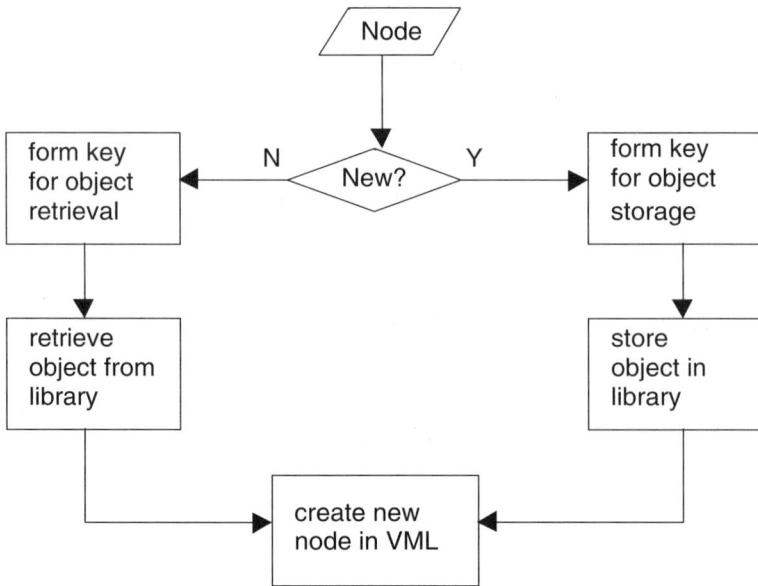

Figure 9.4 Storage and retrieval of nodes while creating VML.

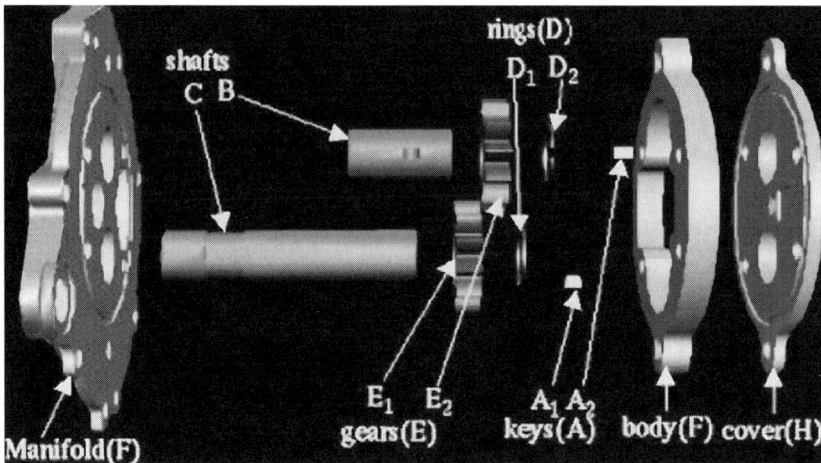

Figure 9.5 Exploded view of a portion of a gear pump assembly.

an array or a database interface. The flowchart shown in Figure 9.4 illustrates the use of object library in the storage and retrieval process of primitives.

To compare VML with scenegraph, we can use the example shown in Figure 9.5, an exploded view of a gear pump assembly. The VML structure for the assembly is shown in Figure 9.6, and Figure 9.7 illustrates the limitations of scenegraph in this case by not being able to enforce the assembly rules required.

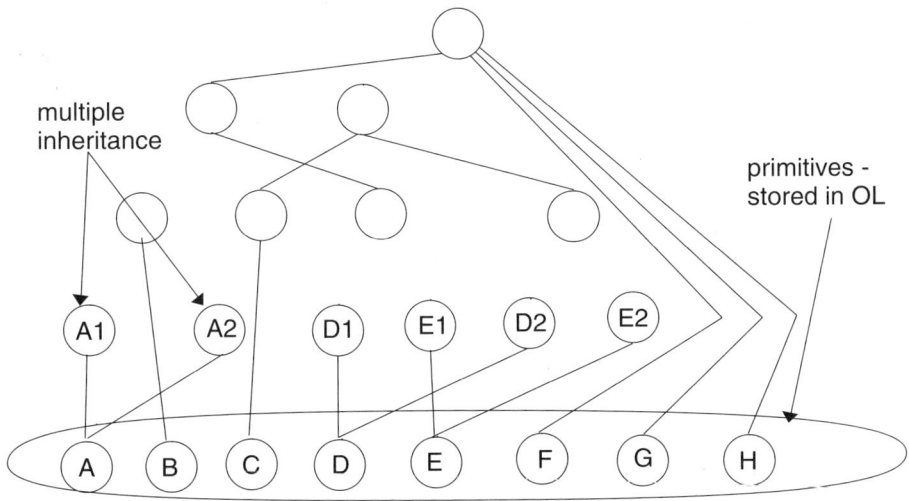

Figure 9.6 VML structure for the gear pump assembly example.

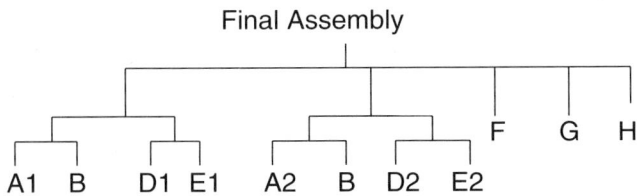

Figure 9.7 Example of scenegraph limitation: It cannot enforce the assembly rules.

9.2 THE FOUR-TUPLE NODE STRUCTURE

The properties of objects are encapsulated at the node level in the VML structure using a 4-tuple $<C,R,T,E>$ structure. These include the geometric data, relationships with other objects, status information, and other data essential for simulating some of the object behavior in the virtual world.

1. **C:** defines the component structure of the node. It describes the object geometry, material information, and texture details. Geometric data include vertex, normal, and face composition. Material information includes specular, diffusive and ambient lighting, transparency, shini-

ness, and reflective and refractive index. A member in the component set can be created either as instances from the generic object library part definition or by performing elementary set operations (union, intersection, and difference) on two or more components belonging to the set, which are treated as parts, in the virtual manufacturing environment. Manipulating the part or parts in the component set **C** produces changes in physical properties during the virtual manufacturing session.

2. **R:** defines the precedence relationship and the hierarchy between the components. The precedence relationship specifies the parent–child relationship between objects. Using the precedence relationship it is possible to determine the object composition. The precedence relationship can be exported from the three-dimensional modeler used to model the components and can be read by the hierarchy loader module. The precedence relationship can also be set in an interactive design session in the VE where the users are designing a new entity. The precedence relationship rules are enforced to check illegal operations, the steps that are considered infeasible in a physical environment due to the laws governing the processes.

A temporal logic framework is used here to represent the precedence relationships. Propositional temporal logic is used for the temporal logic framework. The framework has been adapted to be used in the lattice structure for virtual manufacturing. The task of implementing the precedence relationship using the temporal logic framework in the virtual lattice is performed at the node level. Functions are used to represent the precedence relationship that either a Boolean value [TRUE/FALSE] depending on whether the condition has been satisfied or returns the relationship that needs to be satisfied. The temporal operator P is represented by a function *argument$_2$-precede* (*argument$_1$*). For instance, P_1 P P_2 is represented as P_2-*precede* (P_1). The temporal operator U is represented by a function *argument$_2$-until* (*argument$_1$*). Normally, the U operator is combined with an $\neg$ operator to indicate a *not until* scenario, which is then represented by function *argument$_1$-not_until* (*argument$_2$*). For instance, $\neg P_2 U P_1$ represented functionally as P_2-*not_until* (P_1) signifies NO PROCESSING of task formula P_2 UNTIL P_1 is DONE. The $\wedge$ and $\vee$ operators are represented by keywords *and* and *or,* respectively. The operators are used in prefix form in this implementation, but they can be used in other forms as well without loss of continuity. For instance, the notation $(p_1 \wedge p_2)$ P p_3 is expressed in the functional form as p_3-*precede* ([*and,* p_1, p_2]). The other temporal operators, $\square$ and $\lozenge$, are represented similarly by functions *always* (*argument*) and *eventually* (*argument*). These are more useful in expanding an assembly sequence expressed in state form and mapping them according to the *precede*() and *not_until*() functions for every node in the lattice. A *check*() function is provided to get the response on the status of the node. For example, p_1-*check* () checks the event control list of node p_1 and returns either TRUE or FALSE, based on the state it currently is in. It returns TRUE if the state has been encountered.

3. **T:** defines the trajectory information of a component. The trajectory information contains links to the path that a component can follow while moving. These can be updated from an external motion-planning algorithm or from an external control algorithm at run-time, and the virtual environment can refresh from the trajectory data to simulate motion in the virtual world. The trajectory information can also be linked to a collision detection algorithm to determine accurate collision detection and to provide feedback to the control mechanism for computing new trajectory data. These are plug-in modules, and appropriate ones can be used depending on the circumstances and availability and the computational power of the VR system. The **T** element of the 4-tuple serves as the link with the externally used modules and the lattice structure.

4. **E:** defines the event control list for a component for performing planning and control. The behavior of discrete event systems (DESs), such as manufacturing systems, are modeled naturally by finite-state automata. A controlled automata-based approach is used to describe the event control list. During simulation, the event control list is used to track the state of the objects. **E** is a nondeterministic finite automaton which can be represented by a 5-tuple $\mathbf{E} = (Q, \Sigma, \delta, q_0, F)$, where:

   a. Q is a finite set of states.
   b. Σ is a finite set of permissible input parameters, transitions, or event labels.
   c. δ, referred to as the *state transition function,* is a mapping from $Q \times \Sigma$ to $\mathcal{P}(Q)$ which dictates the behavior of the finite-state control.
   d. q_0 in Q is the initial state of the finite-state control.
   e, $F \subseteq Q$ is the set of final or marker states.

The precedence relationship rules are enforced by the VMS using the event control list of objects. If a node in the VML has not reached the desired state, appropriate action is initiated according to the logic of the script specified. The event control list can be used, for example, along with the precedence relationship rules to indicate any user-attempted illegal operations in the virtual environment. Figure 9.8 illustrates an example of precedence relationship and its VML implementation.

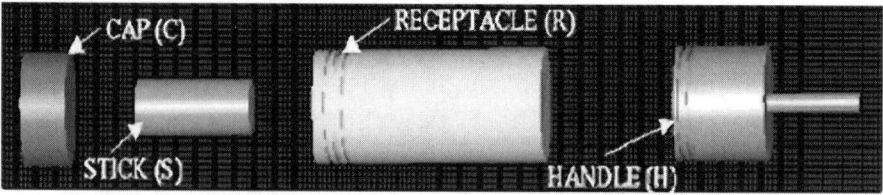

$$p_2 = R \lor H; \quad p_3 = S \lor p_2; \quad p_1 = C \lor p_3$$

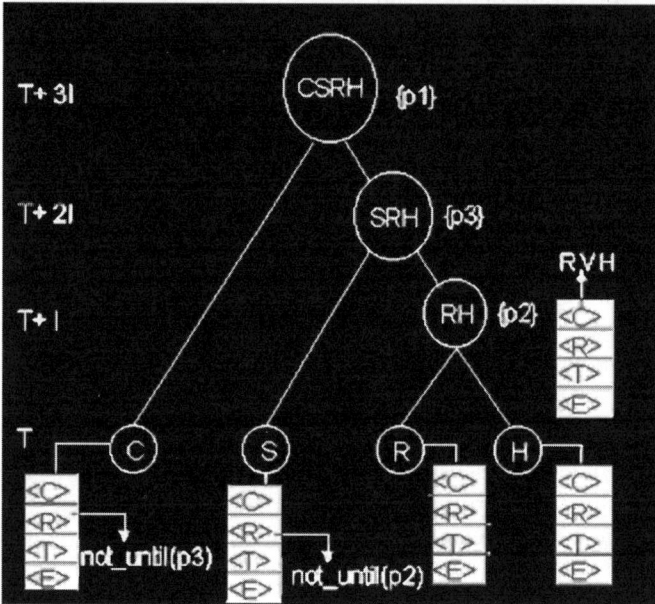

Figure 9.8 Example of precedence relationship and its VML implementation.

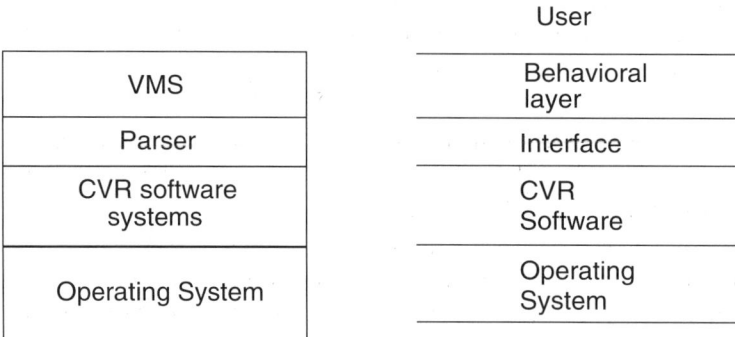

VMS	User
	Behavioral layer
Parser	Interface
CVR software systems	CVR Software
Operating System	Operating System

Figure 9.9 VMS organization.

9.3 VIRTUAL MANUFACTURING SCRIPT

9.3.1 The Need for VMS

- Numerical control (NC)
  — Not a machining method
  — Machine control concept
  — Led the foundations for computer-aided manufacturing (CAM)

- Processing language (APT)
  — Facilitated programming of NC tools

- VML with its 4-tuple node structure
  — NOT a tool for performing virtual manufacturing operations

- VMS
  — Similar in spirit to a NC processing language in VM
  — Provides a basis for performing telecollaborative VM operations
  — Has a vocabulary of operators

The success of having a telecollaborative VE for synthesizing manufacturing operations depends on the ease and effectiveness of specifying the tasks by designers or users. High-level programs or a menu-driven system often perform the tasks in a CVE. It is difficult and time consuming to code specific programs for performing the tasks. The menu system is often unsuitable to perform tasks, especially when certain tasks need to be performed or visualized continuously, without interruption. Also, the menu system has to be tailored for specific operations. Many designers and users in manufacturing are familiar with numerically controlled (NC) programming concepts and processing languages. The VMS described here is similar in spirit to a NC processing language for the virtual manufacturing domain. The VMS provides a basis for performing telecollaborative virtual manufacturing operations and has a vocabulary of operators for synthesizing the operations. *The VMS does not attempt to create an entirely new or different VE on any platform. The VMS is a layer above the CVR software systems where the users can specify the tasks in a simplified and organized manner.* This is shown in Figure 9.9. The script has been developed to be included with existing VR environments running on a number of different computer platforms for performing virtual manufacturing operations.

9.3.2 VMS Classification

The VMS has a vocabulary of operators that can be classified into four primary categories: *event, status, telecollaborative,* and *plug-in*. Each of these categories has operators for performing related operations in the virtual manufacturing domain. The operators are at a higher level and they need to be parsed, and corresponding VR software-dependent functional code needs to be generated to perform task-specific manufacturing operations. A set of data structures is defined and is instantiated.

1. V_OBJECT: defines the object handle
2. V_POINT: defines a point in three dimensions
3. V_ORIENTATION: defines object orientation in three dimensions
4. V_USER: defines a person or user in CVE
5. V_PROCESS: structure to group all operators for a process
6. V_STATE: enumeration of all the possible states of V_OBJECT

Event Operators

A set of event operators is used to specify normal manufacturing functional tasks in CVE. These are generic operators that are parsed to generate the underlying (platform-dependent) code to simulate the tasks in CVE. Some of these operators are synchronized together when two or more objects interact with one another to perform a related task. A few of the operators are listed below.

1. V_RUN(): to perform a virtual manufacturing operation
2. V_STOP(): to stop the current operation initiated by a V_RUN() operator
3. V_LOAD(): to perform any preprocessing actions before a V_RUN()
4. V_UNLOAD(): to perform any postprocessing actions after a V_RUN()
5. V_BREAKDOWN(): to simulate a breakdown of an object (e.g., an equipment)
6. V_WAIT(): to make a process wait for an event indicated by a V_STATE
7. V_INTERRUPT(): to allow forcible interruption of current operation; often used to simulate breakdowns
8. V_SYNCHRONIZE(): to synchronize two or more tasks
9. V_CREATE(): to create a new node in VML representing completion of a task

Query Operators

The *query operators* are used to inquire on the current status of an object in the CVE. The *event operators* rely on the output from query operators to ensure that the object precedence relationships in the lattice structure are satisfied and operations are performed in a feasible sequence.

1. V_STATE(): generic query operator to inquire as to the status of an object

Telecollaboration Operators

The set of *telecollaboration operators* is meant for geographically separated users to interact with each other through the virtual environment. These operators are used for initiating users to enter and exit the ongoing virtual manufacturing session and create a suitable form for representing them in the VE. Generic operators to implement user interaction are also included in the set.

1. V_OPEN(): to initialize the telecollaboration environment and establish connection with the remote host
2. V_CLOSE(): to close the connection with remote hosts at the end of the telecollaborative session
3. V_ADD_USER(): to enable a user to join the telecollaborative session
4. V_DEL_USER(): to enable a user to leave the telecollaboration environment
5. V_INDICATE(): to perform user interaction in the environment
6. V_TRANSMIT(): to send initial loading data across to a new user in the environment
7. V_INTERACTIVE(): to indicate that the user is taking over control of the operations
8. V_AUTOMATIC(): to indicate that the CVE is taking over control of the operations

Plug-in Operators

The set of *plug-in operators* is similar to the plug-and-play plug-in operators found for various data types in World Wide Web browsers. There is a set of operations that is generic and is performed for a wide range of operations. For example, on a factory floor, people move from point to point, as do a conveyor, a forklift truck, and even tools mounted on machine. In a traffic simulation model, vehicles as well as pedestrians move from point to point. But the "movement" for the various objects is different in nature. So a specific *movement operator* needs to be loaded to perform movement on a specific type of object. Such operations are categorized in the plug-in operators. The plug-in operators are usually external modules that provide a data stream to simulate specific types of object behavior. Examples of plug-in operators include:

1. V_MOVE()
2. V_CHECK_COLLISION()

9.3.3 VMS Description

The VMS is a collection of commands to be executed in the CVE, expressed in terms of the operators described above and a few other keywords. Each command is expressed in a new line. The command is parsed in the same order, and the script parser generates the underlying code. Normally, each command is treated as an individual event and serviced as a callback. The parser does not wait for completion of the preceding command unless the command being processed explicitly needs to wait for the completion of a previous task. The underlying code enforces the precedence relationship rules specified for a node in the VML prior to task execution and updates the status information of the node that is created or modified.

```
V_PROCESS(P1): V_WAIT(P0) - implies that the process P1
must wait until the process P0 is complete.
```

The parser generates the underlying code for explicit wait expressed in terms of pseudocode:

```
while (P0.status != 'Y') // keep checking until P0 is
complete
    wait (10 s)
trigger(P1) // trigger an event that will initiate the
callback to execute process P1
```

The callback will have the code to perform any additional precedence checks that might be required. If all the conditions are satisfied, the actions in the process are performed, followed by the postprocessing step where status of the objects is updated.

A process itself can be expressed in terms of a set of commands: for example,

```
V_PROCESS(P1): V_WAIT(P0)
{
O2.V_MOVE() TOWARD O3 UNTIL O2.V_CHECK_COLLISION(O3) = "Y"
O2.V_RUN()
}
```

The process P1 asks object O2 to move in the direction of O3 and keep moving until it collides with O3. A pseudoversion of the underlying code generated by the parser is as follows:

```
    {
    perform precedence check on O2
    use the trajectory data of O2 to retrieve the travel
    path
    use the plug-in data of O2 to load the movement algo-
    rithm
    use the plug-in data of O2 to load the collision
    detection algorithm
    initiate movement of O2
    {
        move at a constant speed
        detect obstruction in path
        recompute trajectory if path is obstructed
        stop movement on collision with O3
```

```
    transmit application-specific data to all connected
    hosts
    {
            user movement data
    }
}
postprocessing: update status of O2
}
```

The (platform-independent) pseudocode above serves as input for the underlying (platform-dependent) CVR software on which a user decides to implement our behavioral layer architecture.

The VMS can operate in either automatic or interactive mode, and a copy of the script resides locally at each user's host. In the automatic mode, complete control is given to the VMS operators, and users can observe the tasks being performed. This mode is useful in training applications where the user needs to study a manufacturing process. There is a minimal amount of control available to the user in the form of the interactive device whereby the user can cause the automated process to pause in order to study an intermediate step in the operation. However, the user can freely move or navigate around the VR environment. The interactive mode provides control to users to perform interactive tasks in the CVE. The script is also specified in the interactive mode and provides the intent of the manufacturing process. In the event of a network bottleneck, control passes over to the automated VMS execution, and tasks are executed from the local copy of the VMS. Control passes back to the interactive mode when the required bandwidth is available. This transfer process is monitored by the network monitor module, which is described in a later section.

9.4 EXAMPLE: AUTOMATED TASK EXECUTION USING VML–VMS

An assembly example is described here to illustrate the working of the VML and VMS. The six objects shown in Figure 9.10 are to be assembled into a single assembly. The objects were modeled using ProEngineer and exported as VRML objects to be used in the CVE. There are four tasks that are performed in the assembly, which is specified in the reference above:

1. p_1: insert task, connecting L1 and B giving subassembly P1
2. p_2: insert task, connecting L3, SD and P1 giving subassembly P2
3. p_3: insert task, connecting L2 and P2 giving subassembly P3
4. p_4: insert task, connecting HP and P3 giving final assembly P4

This example is used to illustrate the automatic mode of VMS and the precedence rules enforcement mechanism using the precedence relationship

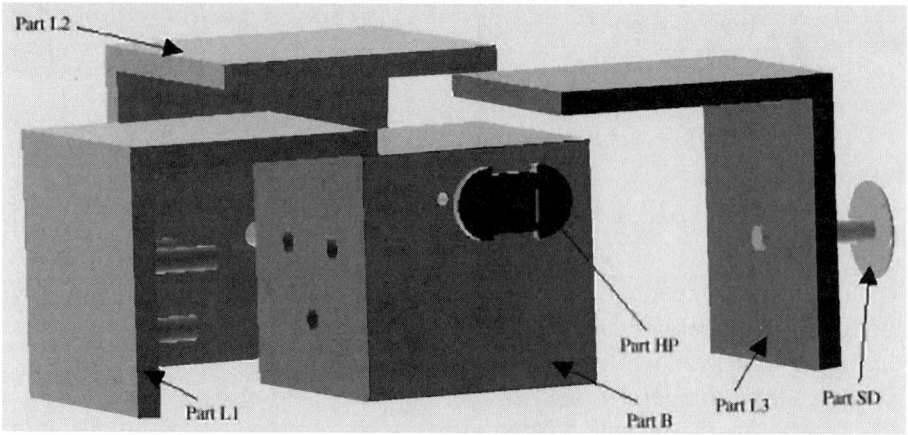

Figure 9.10 Assembly example.

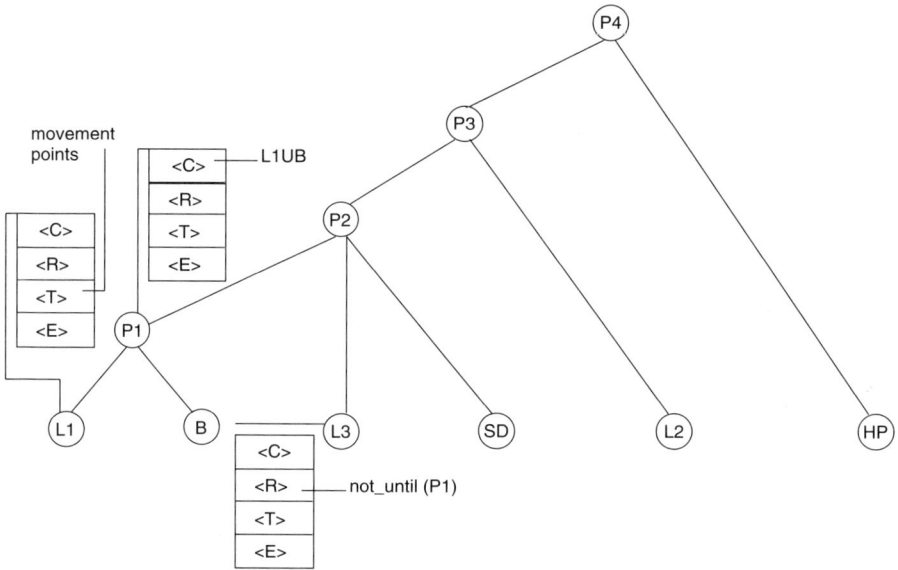

Figure 9.11 Portions of the VML structure of the assembly VMS example.

(R) element and the event control list (E) of the 4-tuple node structure in the VML. The assembly sequence to perform the task is given by $\sigma = (p_1\, P\, p_2 \wedge p_2\, P\, p_3 \wedge p_3\, P\, p_4)$, where P is the precedence relationship, indicating *precede*. The VML structure of the assembly is shown in Figure 9.11. Each node in the lattice structure maintains its status information. A check is performed to see whether the node has been created, which corresponds to the completion of each of the task listed above. The sequence can be expressed in terms of the VMS to depict the assembly tasks automatically for users in the CVE:

```
V_OBJECT VO1="L1",VO2="B",VO3="L3",VO4="SD",VO5="L2",
VO6="HP"
V_OBJECT VOP1="P1",VOP2="P2",VOP3="P3",VOP="P4"
```

The next set of commands read in the geometry data for the primitives:

```
VO1.V_LOAD("L1.VRML") - read in-part L1 geometry
VO2.V_LOAD("B.VRML")
VO3.V_LOAD("L3.VRML")
VO4.V_LOAD("SD.VRML")
VO5.V_LOAD("L2.VRML")
VO6.V_LOAD("HP.VRML")
```

Initially, the subassemblies do not exist. So setting their status to "N":

```
VOP1.V_STATE(N)
VOP2.V_STATE(N)
VOP3.V_STATE(N)
VOP4.V_STATE(N)
```

The following set of commands are used to add the local user, establish connection with a remote host, add a remote user to the environment, and transmit the current VML to the remote location:

```
V_OPEN("hostname.domain")
V_ADD_USER("user1@localhost")
V_ADD_USER("user2@hostname.domain")
V_TRANSMIT()
```

The following command specifies the automatic execution mode:

```
V_AUTOMATIC()
```

The following sets of commands describe the four assembly processes along with the dependencies:

```
V_PROCESS(P1):
{
    VO1.V_MOVE() TOWARDS VO2 UNTIL
    VO1.V_CHECK_COLLISION(VO2) = "Y"
    VO1.V_RUN()
    VOP1.V_CREATE() ON COMPLETION
}
```

The syntax ON COMPLETION indicates that the status of VOP1 should be updated on completion of VO1.V_RUN():

```
V_PROCESS(P2): V_WAIT(VOP1)
{
    VO3.V_MOVE() TOWARD VOP1
    UNTIL VO3.V_CHECK_COLLISION(VOP1) = "Y"
    VO4.V_MOVE() TOWARD VOP1 UNTIL
    VO3.V_CHECK_COLLISION(VOP1) = "Y"
    VO3.V_RUN()
    VO4.V_RUN()
    VOP2.V_CREATE() ON COMPLETION ALL
}
```

The syntax ON COMPLETION ALL indicates that the status of VOP2 should be updated on completion of VO3.V_RUN() and VO4.V_RUN():

```
V_PROCESS(P3): V_WAIT(VOP2)
{
    VO5.V_MOVE() TOWARD VOP2
    UNTIL VO5.V_CHECK_COLLISION(VOP2) = "Y"
    VO5.V_RUN()
    VOP3.V_CREATE() ON COMPLETION
}
V_PROCESS(P4): V_WAIT(VOP3)
{
    VO6.V_MOVE() TOWARD VOP3 UNTIL
    VO6.V_CHECK_COLLISION(VOP3) = "Y"
    VO6.V_RUN()
    VOP4.V_CREATE() ON COMPLETION
}
V_DEL_USER("user2@hostname.domain")
V_DEL_USER("user1@localhost")
V_CLOSE()
```

The parser parses the script and generates the underlying code. The collaborative virtual environment software CAVERNSoft and the CAVE™ virtual reality platform were used to simulate the assembly operations from the script.

9.5 EXAMPLE: INTERACTIVE TASK EXECUTION USING VML–VMS

We now construct an example to illustrate the interactive mode of VMS. One of the best examples involves telecollaboration using different networks. This example targets two main objectives: (1) to demonstrate the use of VMS and the role of the 4-tuple $< C,R,T,E >$ of the VML during interactive tasks, and (2) to study the effect of network factors during the telecollaboration process. The latter objective is an important issue for the performance of a CVE and is an integral part of an effective telecollaborative virtual environment. A brief description of the network parameters is provided here to give a background of the network management aspect of the CVME. The description below first describes the experiment plan (network planning) and follows it up by a detailed experiment.

9.5.1 Network Planning

A key component of an effective telecollaborative virtual environment is the network through which the data transfer occurs. The availability of a reliable and high-bandwidth network is a basic requirement for satisfactory performance and perceptual continuity. Often, the notion of persistence or continuity is affected due to a lag in the system known as *latency*. Latency is the delay between performing an action and seeing the result of that action in VR. Latency can affect a system from a variety of sources. The latency sources in VR are system latency and network latency. The use of a network to perform collaborative tasks gives rise to network latency. *Network latency* is the time it takes to get a unit of information from source to destination through a network. A study of an acceptable network latency in CVE demonstrates that humans tolerate a maximum of 200 ms of network latency for real-time coordination. Networks also exhibit variable latencies, resulting in display frame jitter, which means that data packets do not arrive at a fixed delay. The issue here is to reduce network latency and jitter so that it increases the interactivity and dynamic nature of the CVE and provides consistency among virtual environments.

9.5.2 Bandwidth Study

Bandwidth is the available network capacity for data transfer. The available bandwidth varies widely depending on the network medium. Common network media include modem, Ethernet, Asynchronous Transfer Mode (ATM), and Integrated Services Digital Network (ISDN). Modems normally use telephone lines, and current data transfer rates are typically up to 56 kilobits per second (kbps). Many of the current CVEs communicate over the Ethernet using a T1 or a T3 connection, which is limited in the local area network. ISDN replaces the existing analog telephone lines with reliable high-speed and flexible digital connections. ISDN uses two channels, bearer channel (B channel) and data channel (D channel) for voice and data and can be configured at 128 kbps and 16 or 64 kbps, respectively, depending on the service type. ATM is a high-speed fiber optic broadband network that has typical speeds of up to 155 Mbps, or up to 622 Mbps depending on the configuration.

In a CVE, there are different types of data packets that need to be sent through the network to the participating hosts. These include tracker data packets, audio data packets, video data packets, and application-specific data packets. Each of these data packets is of different size and varies in importance. For example, it does not have a significant effect on the CVE if intermittent audio or tracker data packets are lost, but the transmission speed of these packets is of primary concern. Depending on the nature of the application, it is often necessary to avoid packet loss for application-specific data packets. In the former case, UDP is the most suitable protocol and TCP is the preferred protocol for the latter. Also, the smaller the packet size, the better is the performance in terms of continuity and latency.

An experiment involving two users was set up to study network latency. A manufacturing facility model developed by Searle was used, and users were asked to pick an object and move along a path and place the object at the destination specified. The users were represented as avatars in the CVE and communicated with each other using audio feedback through the network. Their head and wand (a three-dimensional mouse) positions were tracked and the data were sent across the network to the other host, where the CVE would update the location of the avatar. A VMS was prepared for the operation and placed locally at each host. A portion of the VMS is provided here to illustrate some of the features.

```
V_OPEN("hostname.domain")
V_ADD_USER("user1@localhost")
V_ADD_USER("user2@hostname.domain")
V_TRANSMIT()
V_INTERACTIVE() [indicating interactive mode]
V_PROCESS(P1): BY user1
{
    VO1.V_CHECK_COLLISION()
    VO1.V_MOVE()
}
V_PROCESS(P2): BY user2
{
    VO2.V_CHECK_COLLISION()
    VO2.V_MOVE()
}
```

There are two processes specified by the script: P1 and P2. Process P1 is performed by `user1` and P2 is performed by `user2`. Here the underlying code for the process P1 involves attaching an external (plug-in) collision detection algorithm as a callback to the object VO1 and extracting the trajectory information from the T component of the 4-tuple of node VO1 in the VML. When the users are moving the objects interactively in the CVE, the collision detection algorithm warns of any imminent collision and guides the user through the environment. The processes are specified even during the interactive mode of operation. This way the processes can be executed if the users decide to switch over to the automatic mode.

Two different network mediums were used: Ethernet and ISDN connecting two O2 workstations running IRIX 6.3. CAVERNSoft and the CAVE simulator were used as the medium for the users and the script to execute. The code is more than 10 megabytes in size. (Due to size and proprietary reasons, contact the authors for details regarding the code.) Three different-sized data packets were transferred in the experiment. Application-specific data size of 27 bytes, tracker data of 55 bytes, and audio data were split into multiple packets of 110 bytes each. Data transfer was carried out using TCP and UDP protocols separately, and another program was written to study the effect of sending the same data sets using Ping. Ping is a tool for network testing, measurement, and management. It imposes considerable load on the network and is normally used for manual network fault isolation. Ping sends out data packets to the remote host at regular intervals and provides latency statistics in terms of minimum delay, maximum delay, average delay, and the percentage loss. Individual latency data were recorded for TCP and UDP protocols, and the average and standard deviations of the latency data were computed. Three different sampling rates — 15 frames per seconds (fps), 30 fps, and 60 fps — were used.

The results from the experiments are shown in Tables 9.1 and 9.2. The experimental results indicate the obvious fact that the network latency for ISDN is higher than Ethernet. This is consistent with the fact that Ethernet has a higher bandwidth than ISDN. Moreover, the latency increases beyond the threshold value of 100 ms for a sampling rate of 60 fps for all three network protocols while using ISDN. All the available bandwidth is used up by the system, and the network is saturated giving rise to jitter. Saturation may be reached earlier, depending on the state of the network and available bandwidth. The performance of a telecollaborative VE is affected adversely if the network latency is above the threshold value.

TABLE 9.1 LATENCY USING ETHERNET WITH DIFFERENT PACKET SIZE AND SAMPLING RATE

Size (bytes)	Sampling (fps)	TCP Mean (ms)	TCP Std. Dev. (ms)	UDP Mean (ms)	UDP Std. Dev. (ms)	UDP Loss (%)	Ping Min. (ms)	Ping Avg. (ms)	Ping Max. (ms)	Ping Loss (%)
27	15	1.349	0.334	1.056	0.350	0.0	1	1	1	0
55	15	1.872	1.017	1.517	0.403	0.0	1	1	10	0
110	15	1.380	0.307	1.216	0.385	0.0	1	1	7	0
27	30	1.122	0.253	1.053	0.278	0.0	1	1	2	0
55	30	1.579	0.580	1.079	0.304	0.0	1	1	3	0
110	30	1.301	0.446	1.181	0.374	0.0	1	1	3	0
27	60	1.076	0.201	1.043	0.315	0.0	1	1	2	0
55	60	1.159	0.436	1.075	0.346	0.0	1	1	9	0
110	60	1.258	0.301	1.996	0.861	0.0	1	1	4	0

TABLE 9.2 LATENCY USING ISDN WITH DIFFERENT PACKET SIZE AND SAMPLING RATE

Size (bytes)	Sampling (fps)	TCP Mean (ms)	TCP Std. Dev. (ms)	UDP Mean (ms)	UDP Std. Dev. (ms)	UDP Loss (%)	Ping Min. (ms)	Ping Avg. (ms)	Ping Max. (ms)	Ping Loss (%)
27	15	42.642	19.384	39.101	5.369	0.011	44	260	424	59
55	15	43.551	3.873	40.395	6.237	0.016	75	332	483	71
110	15	50.310	26.248	45.687	3.133	0.001	92	457	676	83
27	30	45.865	15.861	42.717	2.226	0.0	103	269	390	64
55	30	47.308	17.159	42.848	7.312	0.024	63	331	485	70
110	30	50.346	13.969	47.378	3.459	0.002	78	452	634	82
27	60	10803.584	5626.444	134.562	84.078	0.171	50	251	384	64
55	60	6801.107	1977.888	175.979	97.482	0.146	65	310	477	71
110	60	13665.358	2406.148	156.723	146.30	0.399	73	458	663	86

FURTHER READING

The material presented in this chapter is based primarily on the work of Banerjee (1999).

REFERENCE

Banerjee, A., "A Behavioral Layer Architecture to Intergrate Tellecollaborative Virtual Manufacturing Operations," Ph.D. dissertation, University of Illinois at Chicago, 1999.

CHAPTER

10

SPECIALIZED ROOM AIRFLOW DESIGN USING COMPUTATIONAL FLUID DYNAMICS AND VIRTUAL REALITY

10.1 INTRODUCTION

Removal of airborne particles from an industrial indoor environment is an important issue in specialized manufacturing. Examples are microelectronics manufacturing rooms, hospital isolation and operating rooms, clean rooms, and rooms of pharmaceutical and chemical industrial plants. The quality of the indoor environment within which the manufacturing operation is carried on is important. For example, an extremely clean indoor air is necessary for many high-precision manufacturing operations. In other toxic indoor environments, contaminant removal may be required to protect the health of workers. Air pollutants exert a toxic effect on health through inhalation and a direct effect on the respiratory system.

The most effective and used method for removing indoor airborne particle is using mechanical ventilation. Standards and recommendations have been issued to improve the indoor manufacturing conditions and to prevent hazards to manufacturing plant workers by societies or governmental agencies such as

ASHRAE or CDC. Unfortunately, they are always developed for an entire category of manufacturing rooms and, in many cases, cannot guarantee the desired air cleanliness level. This is why new methodologies are needed to allow building engineers to easily design the plant layout and the ventilation system so as to obtain an effective contaminant removal. In fact, tackling this problem during the design phase is much cheaper than improving the particle removal in an already existing manufacturing room. In this chapter, computational fluid dynamics (CFD) and virtual reality (VR) technologies have been used to develop a new method for easy and quick design of an efficient contamination control system for the specialized manufacturing industry.

10.2 VENTILATION IN THE MANUFACTURING INDUSTRY

In this section we give examples of common applications in which an efficient ventilation is needed due to manufacturing requirements: clean rooms and industrial process air-conditioning, and in the pharmaceutical industry and, in general, the manufacturing industry where hazardous or dust components are emitted.

Clean rooms are enclosures where airborne particulate matter is controlled to within exceedingly fine prescribed limits. Such areas are typically used for the manufacture or assembly of electronic or mechanical components to high specification and fine tolerances (e.g., transistors). Sources of contaminants are the supply air, personnel, clothing, test specimens or products, building construction elements, and furnishing.

Process air-conditioning is the maintenance of an artificial environment for a process or a product; the provision of comfort conditions for personnel is usually of secondary importance. Industrial air-conditioning may be necessary to meet specific requirements of air cleanliness: this should consider the effects of minute dirt or dust deposits on the manufactured product (e.g., electronic components), as well as requirements for prevention of other air contaminants entering the conditioned space.

Control of the industrial environment is necessary to maintain the efficiency, health, and safety of the workers, for heat control (in many industrial settings the heat released cannot be absorbed or offset by means of normal ventilation methods), and for control of gases, vapors, and fumes. Furthermore, government regulations must be observed for ventilation rates and other requirements for hazardous substances.

In a typical *pharmaceutical plant,* different medicines are manufactured in different rooms of the same building. Efficient ventilation is required to avoid cross-contamination of the products and removal of undesirable airborne particles.

In general, during the *manufacture of many industrial products,* volatile particulate and compounds are released. In many cases it is desirable to remove those particles to prevent health hazards. Examples are: ferroalloy manufacture (it has many dust- or fume-producing steps: raw material handling, mix delivery, crushing, grinding, sizing, and furnace operations), glass industry, production of general organic and inorganic chemicals (e.g., fertilizer, petroleum derivatives), and many other common industrial processes.

Semiconductor Industry

Probably the most innovative and expanding area in which the indoor contamination control is applied is the manufacture of semiconductor products. A *semiconductor* is a material, such as silicon, which conducts electricity better than an insulator such as glass, but not as well as a conductor such as copper. Because semiconductors act as conductors under certain conditions but act as insulators under different conditions, they can be used to control the flow of electricity. The tiny electronic circuits that are made on the thin disks of silicon called *wafers* are called *chips* and are separated and packaged individually. The term *semiconductor* also refers to the complete electronic component that is formed during manufacturing. Such components help many everyday products work. Examples include auto antilock brakes, computers, satellites, microwaves, pacemakers, cell phones, fax machines, telephones, stereos, traffic lights, and hearing aids.

Semiconductor manufacturing companies use heat and chemicals to transform wafers into different electronic components. The manufacturing area where the wafers are processed are called *wafer fabs*. The main integrated-circuit (IC) chip manufacturing steps are as follows:

1. *Material preparation.* The goal for this first step is to grow the wafer. The wafer is a round silicon disk that will have all of the processing performed on it.
2. *Diffusion and ion implant.* Diffusion and ion implant are used to place desired impurities into the wafer. These impurities (or dopants) are used to change the electrical characteristics of the wafer.
3. *Oxidation.* Oxidation refers to the process of growing a layer of silicon dioxide onto the wafer surface. This layer of silicon dioxide serves two purposes: (a) the silicon dioxide acts as an insulator to electricity (this insulator is important for a semiconductor to achieve the correct electrical characteristics); (b) the silicon dioxide layer also serves to protect the inner part of the wafer.
4. *Photolithography.* Photolithography is the selective process that actually allows patterning of the desired circuit onto the wafer.
5. *Etch.* Immediately after photolithography etching is used to remove unwanted material from the wafer by tracing the pattern onto the wafer using photoresistences.

6. *Wafer testing, assembly, and packaging.* The wafers need to go through different forms of testing. Finally, the wafer needs to be separated and assembled into different individual chips.

Many of these wafer processes are performed in clean rooms in order to protect the delicate wafer surface. People who work in the clean rooms also need to wear special outfits, called "bunny suits," to protect the wafers from being contaminated by bits of hair, skin, or dust. However, the most attention is focused on the ventilation systems of such environments. Specific guidelines have been developed to improve occupational safety and product quality in the semiconductor industry. An example is the *Safety Guideline for Semiconductor Manufacturing Equipment* developed by the SEMATECH's member companies. It gives specific attention to proper design of the manufacturing systems. It states: "These systems should optimize the use of air flow, as far as practical directing escaping chemicals so they do not impinge on the equipment or expose personnel." Such guidelines state the goals that must be achieved in design of the ventilation system and recommends tests to verify its effectiveness. Unfortunately, as happens in all the current ventilation guidelines, it gives only general recommendations about how to achieve such goals. In most cases these recommendations are far from being sufficient to guarantee the air cleanliness required.

Since improving an already existing contamination control system requires long investigation times and high costs, an efficient way to tackle this problem is during the design phase. In this chapter, a CFD/VR combined method is described to overcome the limitations of current design assistant tools for contaminant-free environment design.

10.3 CFD AND VR FOR CONTAMINATION CONTROL

As stated before, ventilation is the most widely used method for removing indoor airborne particles. Increasing the ventilation rate above certain levels inevitably causes an increase in energy use and may negatively affect the comfort of the occupants. It can also interfere with the manufacturing operations carried on within the room, especially in fine-tolerance manufacturing. Consequently it is not always possible to obtain the required air cleanliness by focusing only on the ventilation rate. A better solution is to adjust the configuration of the ventilation system (position and number of inlets and outlets, ventilation rate) and of the room (position of objects) — we call these *room parameters* — so as to obtain an effective airflow pattern within the room. The room is to be designed in such a way that the contaminants emitted are clustered efficiently into the incoming airflow and thrown out from the outlets before they can contaminate the manufacturing process.

During the design phase the room configuration that guarantees the most efficient contaminant removal has to be found. Experimental investigations of all the design alternatives cannot always be performed because of their high costs and long times. CFD techniques have recently provided a fast and reliable alternative to experiments: They allow us to simulate the indoor environment numerically on a PC or workstation and therefore to evaluate the proposed solutions faster and more cheaply than by using experimental methods. CFD is the process of representing a fluid flow problem by mathematical equations based on fundamental laws of physics, and solving those partial differential equations numerically to predict the variation of relevant parameters within the flow field. CFD techniques also present several disadvantages that have limited their application in contaminant-free indoor design. They require a high level of skill and knowledge for interpreting the CFD output and, based on it, setting opportunistically all the variables of the design phase. In fact, industrial planners have to be able to create a computer model of the manufacturing room, simulate its indoor environmental behavior with CFD computer models, read the CFD results critically, and, based on them, adjust the geometry and the boundary conditions. This cycle would be repeated until a close-to-optimal room configuration is found. This process is currently iterative and often requires weeks or months of labor-hours to find the best mathematical model and computational solution. In fact, the setup of a CFD simulation requires several long, tedious steps (building the CAD model, generation of mesh, setup of physical and computation parameters). Data postprocessing is tedious and difficult as well. Intuitive and efficient design-assistant tools are needed to overcome these limitations.

Virtual reality has only recently been proposed as a method to compensate for CFD limitations in contaminant-free environment design, and only a few limited applications are available. Unfortunately, the complicated setup, the unfriendly interface, and long computation time of current CFD software packages hinder application of the CFD and VR to the room layout design (we call it *CFD/VR-aided design*). It has not been possible so far to completely integrate a CFD simulator and a VR environment. This is the reason why the three CFD/VR applications we found in the literature did not represent an efficient design tool. In all of them CFD simulations of the indoor environment are performed separately, and in the traditional way, on a desktop computer. Then the CFD results (basically, one or more text files containing data depicting the vectorial fields of interest) are loaded and visualized in the VR environment. The long setup and computational times, which is the main problem that prevents the evaluation of all the design alternatives, still persist.

In this chapter, a novel method for overcoming this problem is developed. It is based on what we call the *VR preprocessing step:* It allows us to reduce drastically the number of design alternatives to consider. Furthermore, the final *CFD/VR assistant tool* for fast and easy design of contaminant-free rooms in the VR environment has been conceptualized.

The intuitive visualization/interaction features offered by VR techniques have already been coupled with CFD airflow computation to assist design of contaminant-free environments. Two pioneering works in CFD/VR-aided design are reviewed. The first work is a project by NCSA (National Computational Science Alliance, formerly National Center for Supercomputing Applications at the University of Illinois at Urbana–Champaign) in collaboration with United Technologies Research Center. Data depicting the airflow in a conference room have been obtained using a CFD code developed by D. Tafti at NCSA. The data set has then been used for a VR simulation of the motion of massless particles in a conference room. The final aim is to visualize the particle motion superimposed on the turbulent flow field in a VR environment. This has been the first step of this project, which is focused on developing a tool for intuitive VR/CFD-aided design of conference rooms. The second work in this direction is at Argonne National Laboratory, which has led to the development of a VR system for the interactive simulation and visualization of particle motion in a flow. This system has then been applied to the industrial problem of designing pollution and slag control systems for commercial boilers and incinerators, and the prototype is called Boilermaker. This software package allows displaying on the VR device the particle motion in real time. A real-time interface cannot be combined with CFD simulations, which require relatively long setup and computation time. Therefore in this package the only design parameter that can be considered variable is the injection location. It can reasonably be assumed that changing the particle injection location does not change the airflow pattern. In this way the CFD simulation can be run only once for testing the airflow description used for all boiler configurations. A real-time interface is limited by time concerns. Therefore, it cannot take into account important design parameters (such as position of objects, of air fans, air intake rate) which change the airflow pattern and then require a new CFD computation for each configuration.

In this chapter, a new approach of establishing the CFD/VR design interface is described. It is shown that the dependence of effective contaminant removal from an enclosed space on the typical design parameters has a rational trend and can be described mathematically. The aim of this work is to do this and use it as a preprocessing step before the actual VR/CFD-aided room design. The number of possible solutions is enormous. This number can be reduced drastically by knowing the way the room parameter settings affect the contaminant removal. Subsequently, a CFD/VR tool can be employed for choosing the best solution among a restricted set of possibilities. The approach described is compared to Boilermaker in Table 10.1.

TABLE 10.1 COMPARISON OF THE DESCRIBED APPROACH TO BOILERMAKER

Features	Boilermaker (Diachin et al., 1998)	Parametric VR Study
Environment considered	Industrial boilers	General contaminant-free rooms
Contaminant source position	Variable	Variable (partially)
Inlet/outlet position	Fixed	Variable
Objects position	Fixed	Variable
Air intake rate	Fixed	Variable
Evaluation criteria	CFD simulation + VR displaying of results	Mathematical "screening" + CFD simulation + VR display
Number of candidate solutions that can be evaluated	Small (due to time concerns)	Large (due to a preprocessing /screening step)

10.4 DESIGN OF EXPERIMENT: PARAMETERIZATION OF ROOM CONFIGURATION

The VR enabling parameterization for CFD predictions is conceived in two steps:

1. *Computation.* In the first step several different room configurations are set up and a CFD prediction of airflow and contaminant behavior is conducted for each of them. Different configurations are obtained by varying the parameters of a room design phase: namely, position of objects, particle source, inlets and outlets, and value of the air intake rate.

2. *Data elaboration.* The results obtained are organized and elaborated to rank all configurations. The goal is to find a way in which the ventilation effectiveness (deducible from the CFD results) is related to room parameters.

To select an efficient room layout, the fundamental guidelines in room ventilation (ASHRAE, CDC) have been followed in choosing the position of all room parameters. A common recommendation of ventilation guidelines is to place the outlet close to the floor in case the inlets are located on the ceiling. The air velocity is usually greater near the air supply system openings. A common situation is when insufficient air velocity in regions far from the inlet limits the ability to drag airborne particles. This can result in contaminant accumulation. Also, for more efficient contaminant removal, the air exhaust might be located close to the sources so as either to enforce particle dispersion into the room or to enforce particle clustering along a desired path. Locating inlets on the ceiling is quite common in modern buildings, and it is recommended by many guidelines. The guidelines also recommend placing inlets and outlets in such a way that makeup air from the inlets passes through the contaminant source before being exhausted from the outlet. In choosing and setting up the room layout for parametric study, these recommendations are generally followed.

Most existing models ignore the impact of objects inside the room, such as furniture. They assume objectless rooms, which is a very simplistic and unacceptable assumption. The presence of an object increases the deflection of flow direction and flow turbulence level around the object. Also, the final result with more than one object is not direct superimposition of the impact of individual objects; in other words, the result is not directly additive.

On this basis, a typical room layout has been chosen (Figs. 10.1, 10.2) and a parametric study has been performed on it. The room has dimensions of 6 m × 9 m × 4 m. A pair of inlets is located on the ceiling, and an outlet is located on the front wall; their dimensions are 0.5 m × 0.5 m. Two objects are placed in the room: object 1, which is the main contaminant source and has dimensions 2.5 m × 1.5 m × 1.5 m, and object 2, of dimensions 1.5 m × 1.5 m × 2.2 m. This is a common configuration that can be found in many real situations in which optimal contaminant removal is desired.

Two sets of particles are injected in the room. The first set consists of 650 particles injected around and above object 1. The second set consists of 850 particles injected uniformly throughout the room. This assumption is valid in many cases. For example, in hospital operating or isolation rooms, the particle source is usually confined to or around beds (patient or operating staff), but contaminant particles can also be generated elsewhere within the room. In pharmaceutical or chemical manufacturing, particles can be generated near equipment, but they can also be generated elsewhere in the room.

The inlets supply a total of 6 ACH (air changes per hour). This rate is recommended by CDC for general rooms and is usually the minimum rate that might guarantee a good contaminant removal. It is known from past works and also later in this chapter that insignificant changes in the air supply rate do not change the airflow velocity pattern significantly. This means that the results found for ACH = 6 are still valid for other inlet rates.

The two-object configuration analyzed here reflects many real situations. For example, it is very common in a hospital isolation or operating room to have a bed (object 1), where the patient (particle source) mostly stays, and a big furniture or medical machine (object 2). Many chemical-processing rooms can also reasonably be described by this configuration. Also, hospital waiting rooms usually have this layout; the main concern is removal of hazardous particle emitted close to health care workers, in the interviewing area, which is usually the main desk (object 1).

The starting point for the parametric study is the *standard configuration,* depicted in Figures 10.1 and 10.2. The two inlets are placed on the left side of the ceiling and the outlet is placed in the middle of the front wall, 0.5 m from

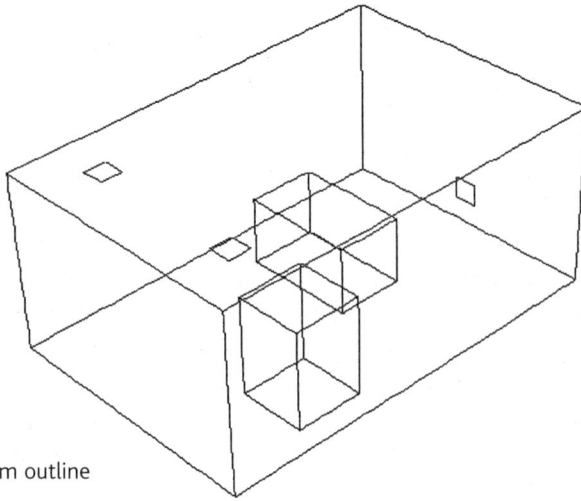

Figure 10.1 Room outline (isometric view).

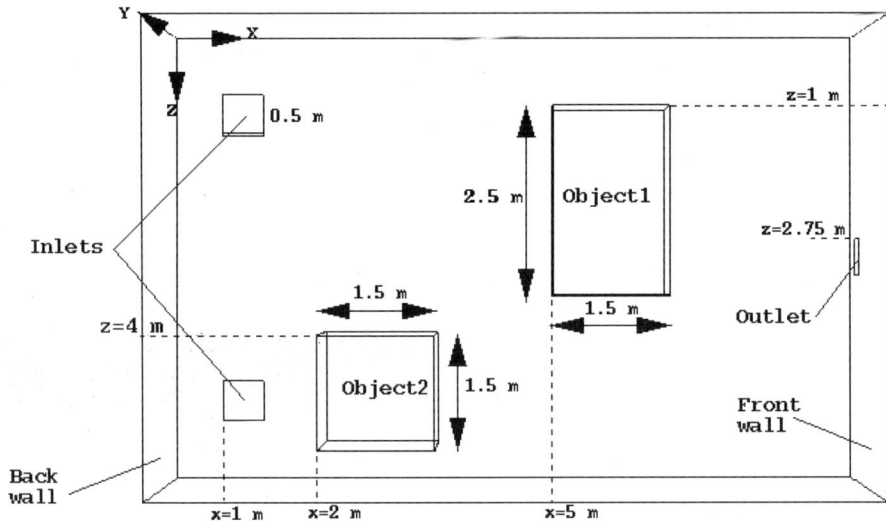

Figure 10.2 Room outline: standard configuration (plane view).

the floor. Objects 1 and 2 are located 2 and 5 m from the back wall, respectively. The following are the room parameters that are considered variables:

- x coordinates of the inlet couple. The two inlets are supposed to lay always on the ceiling at the same z location, but with variable x location.
- z and y coordinates of the outlet, which is always on the front wall (see Figure 10.2).

- Location of both objects.
- Air supply rate (i.e., number of air changes per hour (ACH) supplied through the inlets).

As noted before, two particle injection sets were considered: one located around object 1 (set 1) and one dispersed all over the room (set 2). It is noted that the contaminant removal effectiveness might be different when particles are injected uniformly throughout the room and when they are injected at a particular object's location. For this reason the objective function is developed in such a way that it is possible for the user to select if the particle injections occur primarily around the objects or if they have a random location. In most cases the designer knows that the injections occur around one object but is also concerned with contaminant removal from the entire room. A weighted score is adopted to take this factor into account. The objective of a contaminant removal system is to remove particles from the room in the shortest possible time; this means that particles should have the lowest possible residence time in the room. Another objective is to prevent some particles from being trapped in the room.

Given the following notation:

avg_1 = average residence time of particle injection set 1
avg_2 = average residence time of particle injection set 2
$trap_1$ = number of set 1 particles trapped
$trap_2$ = number of set 2 particles trapped
$max\,t$ = maximum residence time above which the particle is considered trapped
n_1 = number of particles of injection set 1 = 650 particles
n_2 = number of particles of injection set 2 = 850 particles
w_{avg1} = weight of avg_1
w_{trap1} = weight of avg_2
w_{avg2} = weight of $trap_1$
w_{trap2} = weight of $trap_2$

The score for a general room configuration is

$$S = 100 \cdot \left(w_{avg1} \frac{max\,t - avg_1}{max\,t} + w_{trap\,1} \frac{n_1 - trap_1}{n_1} + w_{avg2} \frac{max\,t - avg_1}{max\,t} + w_{trap2} \frac{n_2 - trap_2}{n_2} \right)$$

With the following constraints:

(10.1)

$$0 \le w_i \le 1$$

$$\sum_i w_i = 1$$

(10.2)

$$avg_{1,2} \le max\,t$$

As a result of equations (10.1) and (10.2), S has a value between 0 and 100; the larger the value, the better the ventilation effectiveness.

It has been assumed that for this case, as it happens in most real cases, the goal is to achieve an effective removal of particles generated around object 1 (set 1), but at the same time good removal is required from all locations within the room (set 2). So weights of 50% and 20%, respectively, have been assigned to the average time and number of trapped particles of particle set 1 ($w_{avg}1$ and $w_{trap}1$ respectively) and 20% and 10% to the average time and the number of trapped particles of set 2 (w_{avg2} and w_{trap2}). In this way, the results shown here are valid for a wide range of cases, even for those in which exact locations for contaminant source are not well defined. The number of different configurations required to ascertain ventilation efficiency versus room parameters is enormous. An efficient way to study the effect of several factors simultaneously is to plan matrix experiments using orthogonal arrays, which guarantee statistically that the conclusions from such experiments are valid over the entire experimental region spanned by the control factors and their settings. To perform a matrix experiment, an appropriate orthogonal array has to be determined. Here the number of rows of the orthogonal array represent the number of experiments to be performed. For an array to be a workable choice, the number of rows must be at least equal to the degrees of freedom required for the study. The number of degrees of freedom is equal to the number of factors involved plus one (representing the degree of freedom associated with the overall mean). The number of columns represents the maximum number of factors that can be studied using that array.

In our case there are six factors, so the matrix has to have at least six columns. The total numbers of degrees of freedom is equal to the sum of the number degrees of freedom of each factor, which is equal to the number of levels chosen for that factor minus one. A factor level corresponds to one particular value that the factor can assume during the experiments. Five levels for each factor are used, therefore, the total number of degrees of freedom is equal to $6 \cdot (5-1) = 24$. Then the orthogonal matrix has to have at least 24 rows. A standard orthogonal array $L_{25}(5^6)$ was chosen to obtain a minimum configuration set to simulate (Table 10.2).

A property of the orthogonal matrix assures that this set of simulations is enough for characterizing the effects of all the six parameters. The goal of this work is to obtain a graphical and mathematical expression of contaminant removal dependence on room parameters. To achieve this goal, many more configurations had to be simulated. More than 100 different configurations have been set up and simulated using a CFD software, Fluent. Data regarding residential time (interval time between the particle emission and its exhaust through the outlet) and the fate of the injected particles have been collected and elaborated. An appropriate score, given by equation (10.1), has been assigned to each configuration. The data obtained have been analyzed to establish correlation between room parameters and ventilation effectiveness.

TABLE 10.2 ACTUAL ORTHOGONAL ARRAY $L_{25}(5^6)$ USED FOR DEFINITION OF
THE ROOM CONFIGURATIONS

		Columns				
Simulation	*ACH*	*x-inlets* [m]	*x-object 1* [m]	*x-object 2* [m]	*y-outlet* [m]	*z-outlet* [m]
1	2	1	0.2	0.2	0.1	0.1
2	2	2	2	1	1	2.75
3	2	3	3.5	2	2	3.3
4	2	4.5	5	4	3	4
5	2	5.5	6	6	3.5	5
6	5	1	2	2	3	5
7	5	2	3.5	4	3.5	0.1
8	5	3	5	6	0.1	2.75
9	5	4.5	6	0.2	1	3.3
10	5	5.5	0.2	1	2	4
11	10	1	3.5	6	1	4
12	10	2	5	0.2	2	5
13	10	3	6	1	3	0.1
14	10	4.5	0.2	2	3.5	2.75
15	10	5.5	2	4	0.1	3.3
16	15	1	5	1	3.5	3.3
17	15	2	6	2	0.1	4
18	15	3	0.2	4	1	5
19	15	4.5	2	6	2	0.1
20	15	5.5	3.5	0.2	3	2.75
21	20	1	6	4	2	2.75
22	20	2	0.2	4	2	2.75
23	20	3	2	0.2	3.5	4
24	20	4.5	3.5	1	0.1	5
25	20	5.5	5	2	1	0.1

10.5 RESULTS

The first investigation considers the effect of moving objects within the room.
It is assumed that both objects have a fixed z and that only their x coordinate
can be varied. The two object positions have been varied and simulations of
the configurations obtained have been performed. The results are shown in
Figures 10.3 and 10.6. Figure 10.3 shows that the closer object 1, the main
particle source, is moved to the outlet, the more effective is the particle
removal (Figures 10.4 and 10.5). In general, it is observed that the closer the
particle source is to its outlet, the straighter the trajectories, the smaller the residence times, and the fewer the trapped particles. From the data obtained it is
reasonable to assume that there is an exponential dependence between room
score and x-object 1.

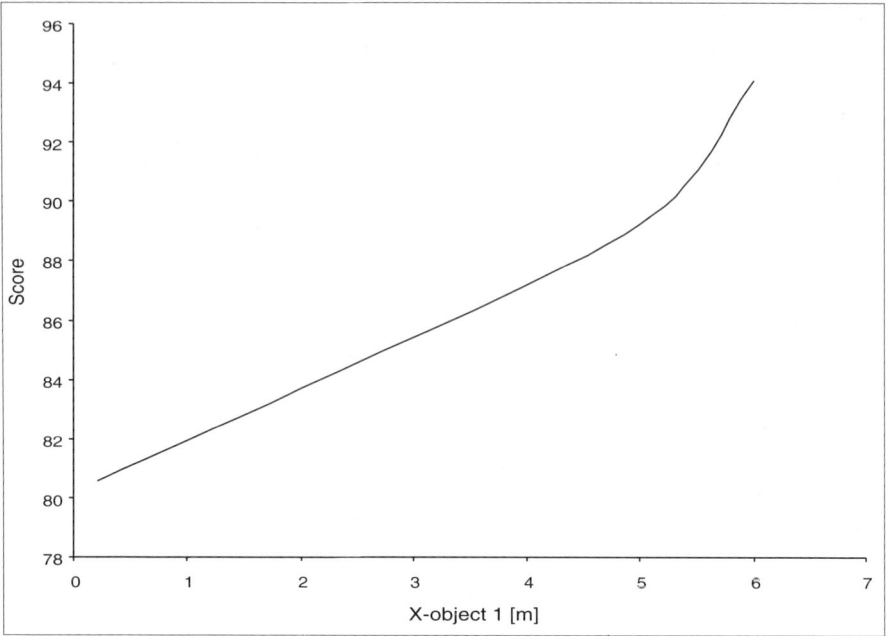

Figure 10.3 Effects of moving object 1 (main particle source).

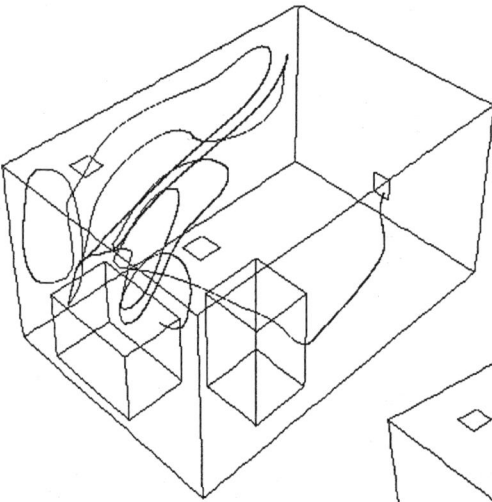

Figure 10.4 Trajectory of a particle injected above object 1 in case of bad object 1 positioning.

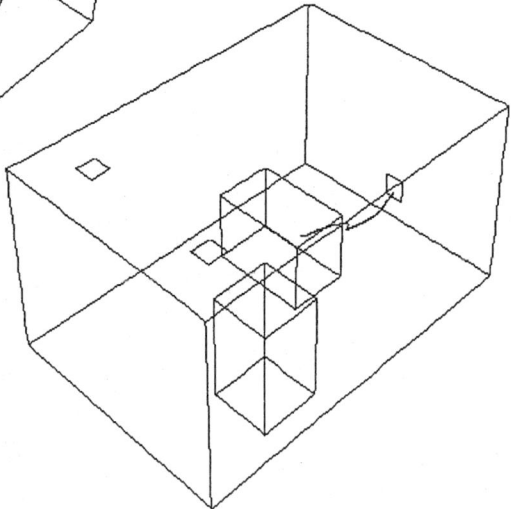

Figure 10.5 Trajectory of a particle injected above object 1 in case of good object 1 positioning.

The room layout score shows a constant-trend behavior when x-object 2 is moved, as depicted in Figure 10.6. The function obtained has the same general shape in any case, but has shown a strong dependence of some of its features (range, inclination) on the value of x-object 1. This can be seen in Figure 10.6, where dependence of the score on x-object 2 for two different values of x-object 1 has been shown. Furthermore, it is evident from the two previous graphs that moving object 1 (particle source) is more efficient (the score variation is larger in Figure 10.3 than in Figure 10.6) and more straightforward (Figure 10.3 is a monotonic function, Figure 10.6 is a sinusoidal one) than moving object 2. In many applications moving the particle source object is not possible, nor is it very convenient. There are several reasons why the source object should sometimes be located in a given region, and its location cannot be considered a variable in the design phase. This can happen in some chemical processing rooms, when the reactor (source object) must have a fixed position: for example, because it is supplied with processing or cooling material from the outside and the receiving position cannot be changed. Another example is the operating room, in which the location of the operating bed is constrained by the position of all other operating machines. So, knowing the optimal positioning of the second object could be the only way of improving contaminant removal. Because of the symmetry of the room geometry, the z-positions of objects 1 and 2 can be swapped, keeping the results still valid.

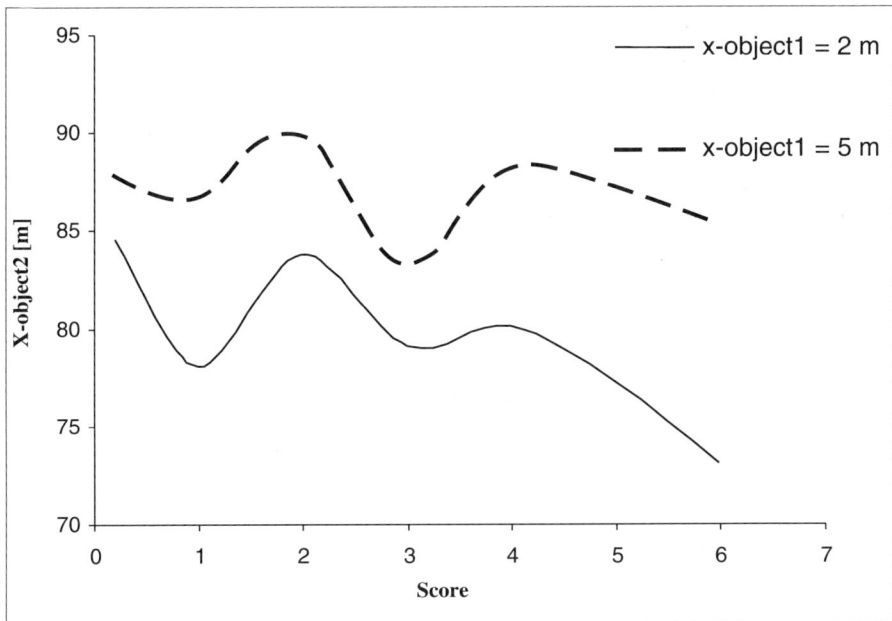

Figure 10.6 Effects of moving object 2.

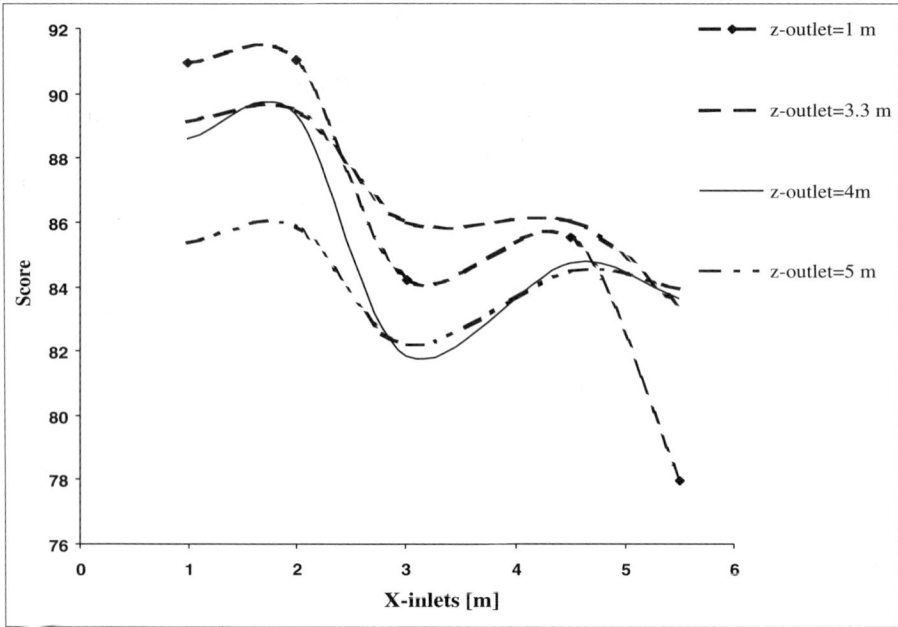

Figure 10.7 Room score versus x-inlets (both inlets have same x-location).

Another parameter that is found to affect the room score significantly is the x location of the two inlets. Figure 10.7 shows the room score dependence on location of x-inlets. All the curves obtained using different z-outlets present a common trend: the score decreases as the inlet approaches the outlet. A sinusoidal trend exists which is also superimposed with a monotonically decreasing trend. Two local maxima are present. The first one (x-inlets $\simeq$ 1.5 m) is due to good positioning of the inlets: close enough to the bottom wall for clustering all the particles, but not so close as to weaken the airflow in the region close to the outlet. The second maximum (x-inlets $\simeq$ 4.5 m) is due to the fact that the inlets are right before the main particle source; therefore, there is an optimal clustering of the first set of particles. In this case particles are clustered immediately into the airflow and directed toward the outlet; furthermore, their residence time has a major weight in the final room score. The score starts to decrease for high x-inlets because in these configurations the main particle injections (set 1) are never between the inlets and the outlet; an airflow shortcut is created. The air coming out from the inlet flows through the outlet without passing by the main particle source (object 1). From Figure 10.7 it can also be seen that the score dependence on x-inlet varies for different values of z-outlet. However, the general trend of the equation remains constant.

The effectiveness of an opportunistic placement of the outlet has also been investigated. As mentioned before, the outlet is always located on the front wall, so the variables in this case are its x and y coordinates (x-outlet and

y-outlet). The effects of varying the height of the outlet have been investigated as well. Simulations have been carried out on various *x*-inlets/*z*-outlet configurations in which *y*-outlet has been increased progressively. Figure 10.8 contains two cases showing the main features of the score dependence on outlet

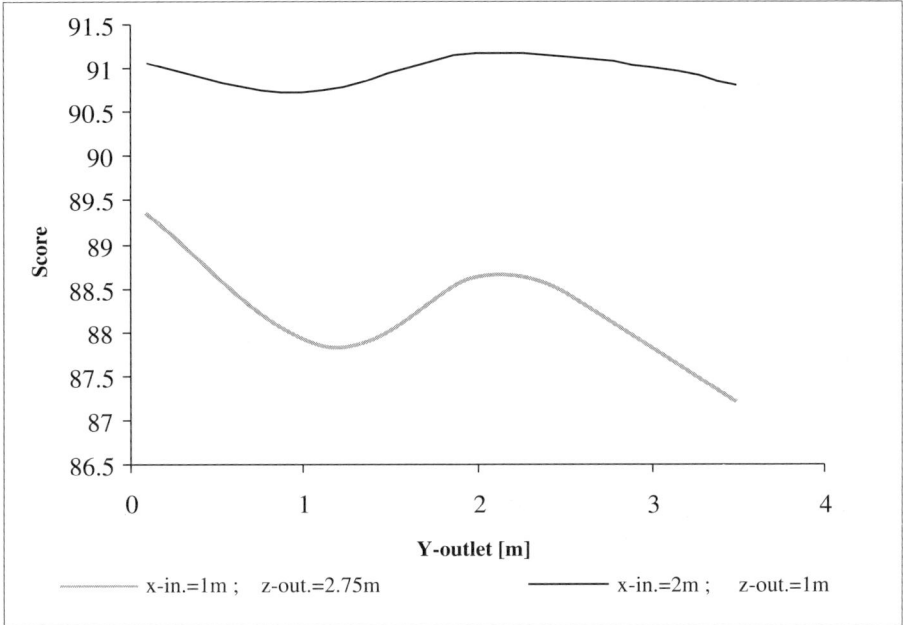

Figure 10.8 Effects of outlet height on room score (i.e., ventilation efficiency).

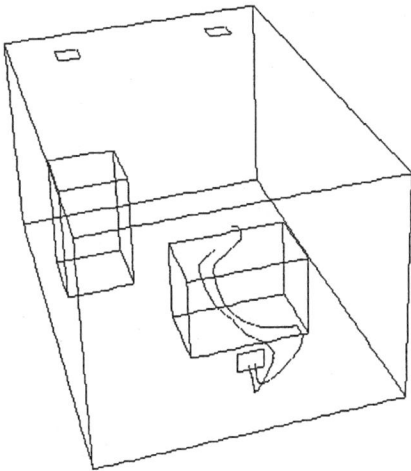

Figure 10.9 Bad outlet height: distorted trajectories of particles injected above object 1.

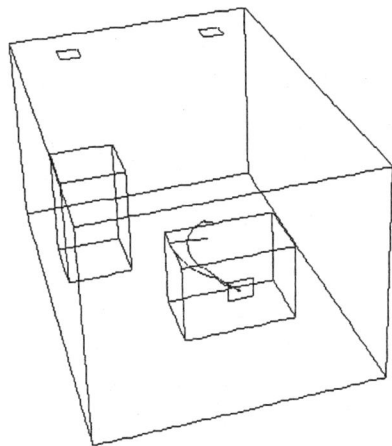

Figure 10.10 Good outlet height: straight trajectories of particles injected above object 1.

height. The main factor determining the shape of the curve is the location of the outlet (z-outlets). When the outlet is in the middle of the wall (z-outlet = 2.75 m) the curve has a rounded shape with a minimum at y-outlet ≃ 1.5 m and a maximum y-outlet ≃ 2 m, due to the fact that the outlet is right above the height of the particle injections. As soon as the outlet is moved toward the side-walls (z-outlet starts increasing or decreasing) the shape of the curves start changing, keeping the same sinusoidal trend but getting straighter.

Figures 10.9 and 10.10 show the various trajectories that two particles, injected at the same point, have when the inlet height has both an undesirable and a desirable value. These two situations correspond to the relative minimum and maximum shown in Figure 10.8.

The score dependence on z-outlet has been investigated as well. Unfortunately, this dependence does not present a simple behavior, and it cannot be described by a simple one- or two-variable function. However, the influence of this parameter has not been neglected; its effect on room score is included in the previous three graphs. As shown, the amount of influence of the other parameters, such as inlet and outlet positions or x-object, depends on the z-outlet value in the ways described previously.

The air supply rate is one of the most important parameters for good contaminant removal. Figure 10.11 gives the room score versus the number of ACH (air changes per hour) supplied. It is observed that the score = $f(ACH)$ has a

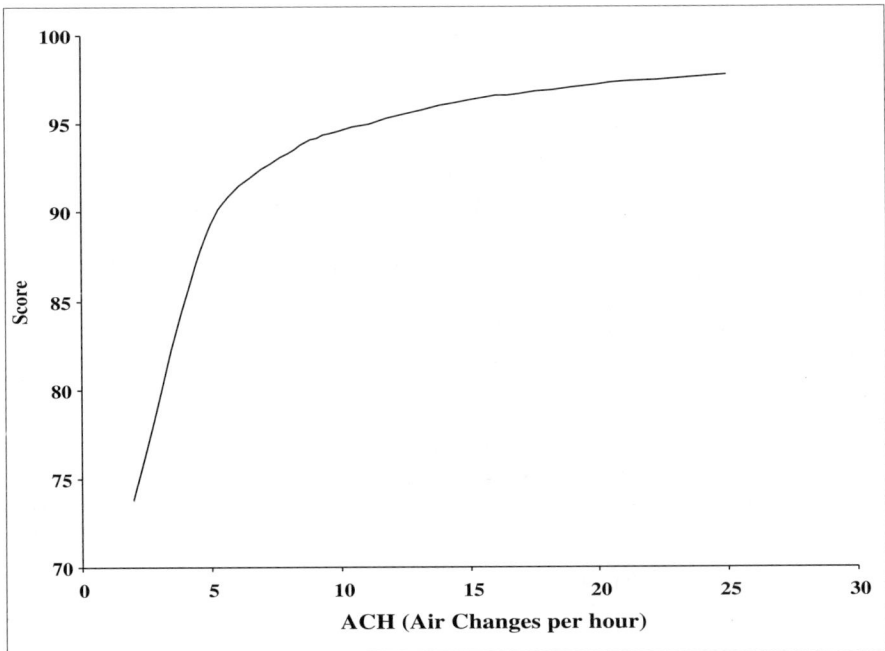

Figure 10.11 Effects of air supply rate on contaminant removal.

trend that tends toward the best score (of 100) as ACH increases. It is a function of the form $f(x) = \dfrac{A \cdot x}{B \cdot x + C}$, where x corresponds to the number of ACH.

Figure 10.12 shows airflow patterns in the plane $x = 3$ m (also shown in the color section). The same picture has been obtained for every value of the supply rate, with all the other parameters kept fixed: The airflow pattern is always the same, even though its magnitude changes. The apparently equal magnitude of the vectors in the various cases is due only to the automatic scaling factor included in the CFD graphic postprocessor. This shows a known property of the airflow pattern within the room: It does not depend on the supply rate (i.e., the incoming air velocity). The particle trajectories are then the same for every reasonable ACH, but their residence times are different, so are the room scores. This is because the greater the value of ACH, the faster is the particle motion within the room. This implies that the score dependence on the other room parameters is independent of the supply rate.

An analytical approach to the above is described next.

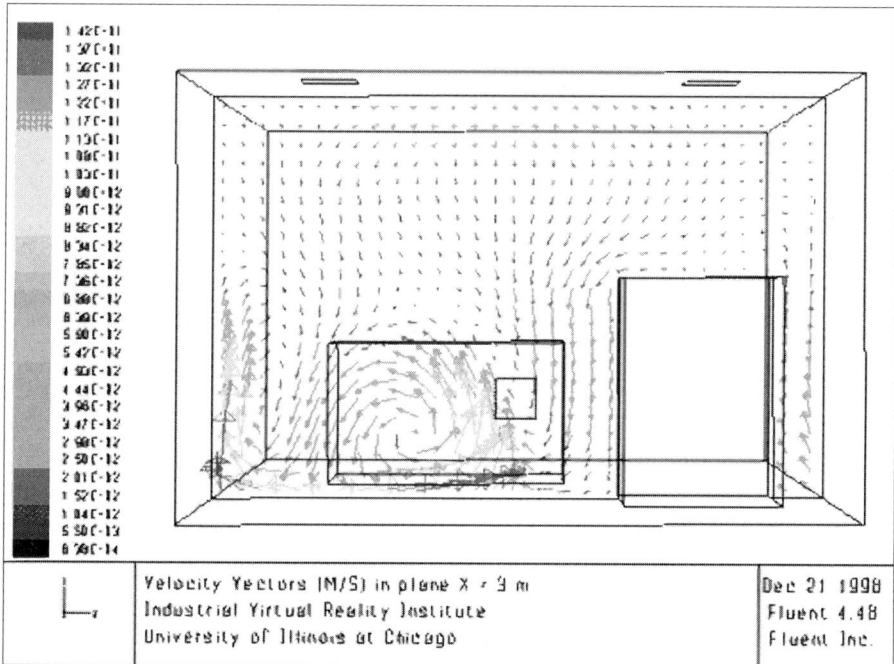

Figure 10.12 Air velocity fields in planes $x = 3$ m; the airflow pattern is equal for every air supply rate (also shown in the color section).

10.6 ANALYTICAL APPROACH

In Section 10.5, it was shown that the dependencies of the ventilation efficiency on each room parameter (such as position of inlets, outlets, objects, ventilation rate) have consistent trends. Such trends have been observed to be affected only marginally by the settings of the other parameters and to have the shape of fundamental mathematical functions or simply combinations of them. For evaluating the ventilation efficiency of a typical contaminant-free room, a scoring system has been established, CFD simulations have been performed, and data have been gathered and organized scientifically to show graphically the effects of changing room parameters. The work of Diachin et al. (1998) can be consulted for details on the choice of room layout, the room parameters studied and the trends observed. The results gathered and the trend observed earlier have formed the starting point for the development of the mathematical expression of the room score dependence on design variables. In fact, such an equation would allow a real-time screening among all possible solutions, for example displaying in the VR environment an index of the contamination removal effectiveness as the user interactively changes the room parameters.

An analytical expression is a general and useful means of representing the room layout score dependence on the various parameters. It allows one to merge the score dependencies on different parameters and come up with a mathematical expression of the ventilation efficiency with respect to the typical room design parameters. Curve-fitting techniques are used to obtain equations from a set of data points obtained from CFD simulations. The method is an empirical one in which the user has to try to find a mathematical relationship, with unknown constants, that fits the data well. Because of errors in the CFD simulations due to inaccurate modeling of the turbulent flow and approximations used for numerically solving the partial differential equations involved, such relationships have two components: a mathematical part, describing the relationship, and an error part, describing any random variation about the relationship. Then for each simulation we have: $simultation_{i,j} = modelfunction, + error_{i,j}$; or, in mathematical terms,

$$y_{ij} = g\left(x_{ij};\theta\right)+\epsilon_i \tag{10.3}$$

The indices i and j indicate the simulated configuration in which the variable x_i assumed the jth among the m possible values. y is the $n \times m$ vector containing the room layout scores of all the simulated configurations; x is the $n \times m$ vector containing the values of the n room variables in question (x-inlets, y-outlet, z-outlet, x-object 1, x-object 2, ACH). In this case, the number of variables is $n = 6$ and m is equal to the number of values assumed by each variable. θ is a p-dimensional vector of parameters that are generally unknown and need to be

estimated from the data. The known function $g(\mathbf{x},\theta)$ describes the way the n room parameters affect the score (i.e., the efficiency of room ventilation). The error ϵ accounts for the difference between the CFD simulation and the mathematical model.

The following assumptions are made:

1. The mean value, or expectation, of the error component ϵ is zero, $E(\epsilon) = 0$. This implies that there is no systematic departure of the observation y from the model $g(\mathbf{x},\theta)$, y is just as likely to lie above $g(\mathbf{x},\theta)$ as below it, so that the average departure is zero.
2. The ϵ_{ij} are statistically independent. This means that the accuracy of a simulation is not affected by the errors observed in the other simulations.
3. The distribution of ϵ does not vary with x (i.e., the error process is "stable"). There is no tendency of the departure to vary with x.
4. The ϵ_{ij} are assumed to be normally distributed: the error has a normal distribution. This assumption is reasonable: in fact, it has been observed that for many engineering data the random error appears nearly normal (i.e., usually they are roughly bell-shaped).

The least squares method allows us to estimate θ from the data pairs $(\mathbf{x},\mathbf{y})$. The general method consists of finding the value of θ that minimizes the sum of the squares of errors, given by:

$$s(\theta) = \sum_{j=1}^{n}\sum_{i=1}^{n}\left[y_{ij} - g\left(x_{ij};\theta\right)\right]^2 \tag{10.4}$$

The problem of minimizing equation (10.4) is not mathematically complex, but requires knowledge of values of y_{ij} for each pair (i, j). This means that CFD simulations should be carried on each possible configuration, which would make this entire study infeasible. In fact, even if we consider only five possible values for each variable x ($m = 5$), the total number of possible configurations would be enormous. It is unreasonable to carry on such a number of CFD simulations, and each of them requires relatively long setup and computing times.

To simplify the problem and especially to reduce the number of CFD simulations required, one crucial assumption has been used in this work: there is no interaction between the variables x_i. This means that the way one variable affects the room layout score does not depend on the particular values assumed by the other five variables. Mathematically speaking, this means that for each variable set $\left[\overline{x}_1,...,\overline{x}_{m-1},\overline{x}_{m+1},...,\overline{x}_n\right]$, the function $S(\overline{x}_1,...,\overline{x}_m,...,\overline{x}_n)$

has the same mathematical form. In other words, S is the sum of $n = 6$ functions, each of them is a function of only one variable. This is quite a strong assumption, and of course, it is a source of error. However, as shown in the first

part of the chapter, it is still true that the score dependence on one single parameter has a constant shape apart from the values assumed by the other five parameters and can therefore be depicted by the same type of mathematical functions.

This last assumption allows us to assume that the score function has the form

$$S = \sum_{i=1}^{n} f_i(x_i) \qquad (10.5)$$

The following exceptions to the previous assumption are considered:

- For $i = 4$ (Z-outlet) the corresponding function is 0, since no constant behavior has been observed varying Z_0. This variable has been included, instead, in the core of the other functions, as shown later.
- For $i = 5$ (object 1) and $i = 6$ (object 2) the two corresponding functions are merged together in a single function of the form $f(x_5, x_6) = f(x_{obj1}, x_{obj2})$ since the effectiveness of positioning object 2 is strongly related to the corresponding position of object 1 (as described earlier). Then the two functions depicting the dependence on X_{obj1} and X_{obj2} have been combined together in one single equation.

These assumptions allow one to obtain an expression for the function $S(\mathbf{x})$ from a reasonable number of CFD simulations of different room configurations. The only requirement needed is that all the functions $f(x_i)$ must have at least one common point. In other words, they must all include at least one common configuration, which has been chosen to be the standard one. The meaning of this last assumption will be clear later, when the general equation is presented.

So, to find the overall equation $S = S(x_1, x_2, \ldots, x_n)$, mathematical expressions of the dependence of S on each single variable have been computed, using sets of configurations that contain the standard one. In these cases the problem is reduced to minimizing the function

$$s(\theta) = \sum_{i=1}^{n} [y_i - g(x_i; \theta)]^2 \qquad (10.6)$$

where x is the vector containing all the values of the room variable in question used for the CFD simulations (all the other variables are assumed to have their standard values) and y contains the room layout scores corresponding to such configurations; θ is the p-dimensional vector containing the parameters to be determined by minimizing the error function (10.5). It is worth stressing the fact that the equations found to describe score dependence to each single variables cannot provide, in many cases, an exact value for the predicted room score, but they are useful for investigating the effects of changing one variable while keeping the others fixed. However, they accurately describe the score dependence on one parameter and, combined in one equation, they can yield a complete mathematical description of the score function.

Air Changes per Hour (ACH)

The air supply rate is the most effective variable. In fact, looking at the graphs shown in the preceding section, it is evident that the largest variations of the room score are achieved varying the number of ACH. The function $S = S(ACH)$ (when all other variables assume their standard value, i.e., the value they have in the standard configuration) is shown in the preceding section. The plot of the simulated data clearly has a hyperbolic trend which tends asymptotically to the ideal score (100). The same trend has also been observed for different configurations in which the inlet rate has been gradually varied. Therefore, the analytical function to fit the set of data has been chosen of the form

$$S_{ACH}(ACH) = \frac{ACH}{c_1 \cdot ACH + c_2} \tag{10.7}$$

The constants c_1 and c_2 have been determined using the least squares method. The equation found is

$$S_{ACH}(ACH) = \frac{1000 \cdot ACH}{9.9 \cdot ACH + 7.1} \tag{10.8}$$

Equation (10.7) gives a mathematical expression of the variation of room score due to the air supply rate. The values of c_1 and c_2 used comply with the score values found simulating the case in which the room is in its standard configuration and varying only the air supply rate. The values of the two constants in equation (10.8) have been determined for this configuration in the light of the way the $S(ACH)$ term is used to generate the final equation (10.19), describing the room score versus all the parameters. This is shown later in the chapter. The same method has been followed in determining the constants for dependencies on other variables. Figure 10.13 shows the close match of experimental and analytical results.

Inlet Location

Another way to largely affect the room score has been found to be varying the x-location of the pair of inlets. The score curves present a sinusoidal shape and a constant decrease as x-inlets (X_i) increases, as mentioned in the data analysis section. The best type of equation to fit to this kind of data is a sum of a negative-slope line and a sinusoidal function:

$$S_{X_i}(X_i) = c_3 + c_4 X_i + k_1 \cos(a_1 X_i + \phi_1) + k_2 \sin(a_2 X_i + \phi_2) \tag{10.9}$$

This is shown in Figure 10.14, which also shows the method used for fitting this type of data. First, a straight line that passes through the extreme points of the data curve has been determined, which means that c_3 and c_4 have been calculated. Subtracting the line

$$y_1 = c_3 + c_4 x \tag{10.10}$$

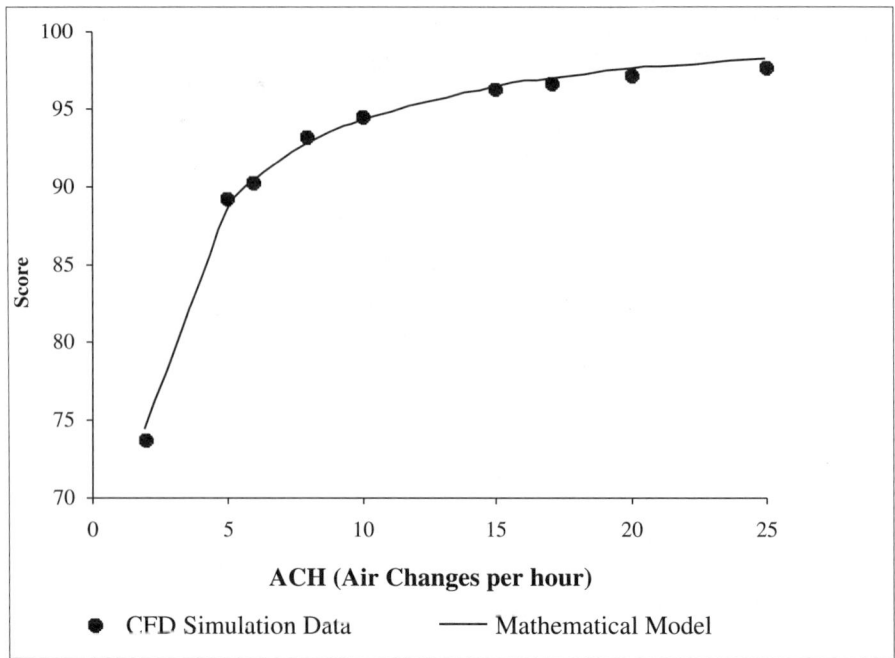

Figure 10.13 Data-fitting result for dependence of the score on the number of ACH.

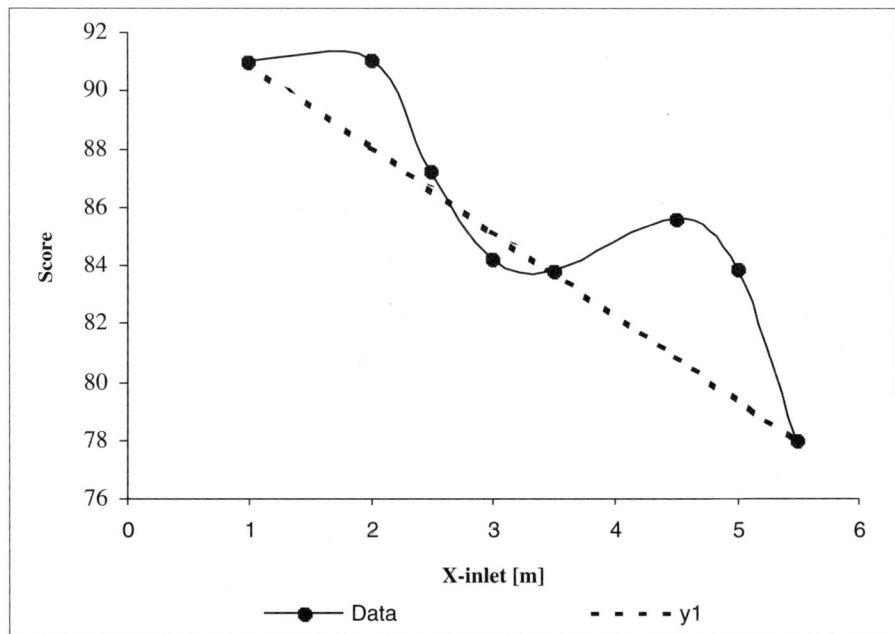

Figure 10.14 Plot of the CFD data set versus x-inlets for a configuration having standard values for all other parameters except x-inlets, i.e., (6, X_i, 0.1 m, 2.75 m, 5 m, 2 m). The dashed line is the straight line, first term of the function $S = S(X_i)$.

from the room score data set, a new data set, which has the shape of a sinusoidal function centered at $y = 0$, has been obtained. These data have been fitted to the function:

$$y_2 = k_1 \cos(a_1 X_i + \phi_1) + k_2 \sin(a_2 X_i + \phi_2) \qquad (10.11)$$

In this way, the remaining terms have been determined, taking into account their dependence on Z_o. In fact, it has been observed that the range of this function and the amplitude of its sinusoidal part depend primarily on the value of Z_o (Z-outlet); therefore, some of the constants in equation (10.9) have been determined as functions of Z_o. Figure 10.15 shows that the analytical function fits the data set quite accurately. This picture refers to a configuration in which all the other parameters have their standard value [i.e., it is the plot of $S = S(6$ ACH, X_i, 0. 1 m, 2.75 m, 5 m, 2 m)].

The range studied has been $X_i \in [0, 5.5 \text{ m}]$. The effects of using values of X_i greater than 5.5 m have not been studied because they have been observed to affect the contaminant removal negatively. In fact, placing the inlets too close to the outlet creates an airflow shortcut between inlets and outlet, so the airflow does not pass through the zones where the particles are emitted. This is

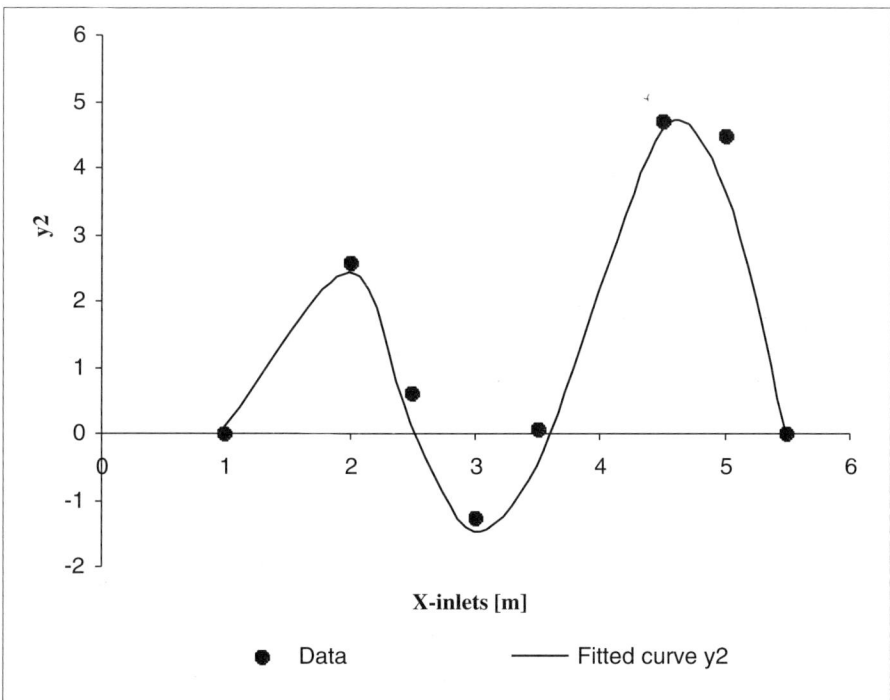

Figure 10.15 Curve fitting for the sinusoidal part of $S = S(X_i)$.

shown clearly by the decrease of the score as X_i increases after 5 m in all the cases studied in the preceding section.

The final expression of this function is the sum of y_1 and y_2:

$$S_{X_i}(X_i) = c_1 + c_2 X_i - k_1 \cos\left[2X_i - (X_{obj1} - 5)\right] + 2\sin(a_1 X_i) \qquad (10.12)$$

This function has a relative minimum right before the location of object 1. This complies with the CFD prediction results in the preceding section.

The parameters in previous equations take Z_o dependence into account:

$$c_1 = -1.668 Z_o + 95.7$$
$$c_2 = 0.56 Z_o - 3.254$$
$$k_1 = 2.73 - 0.29 Z_o + 2.675 \cdot \cos(3Z_o - 5) + 1.95 \cdot \sin(1.54 Z_o - 0.046) \qquad (10.13)$$
$$a_1 = -0.19 Z_o^{3} + 1.623 Z_o^{2} - 3.962 Z_o + 3.539$$

Figure 10.16 shows the analytical model versus CFD data set for x-inlet dependence.

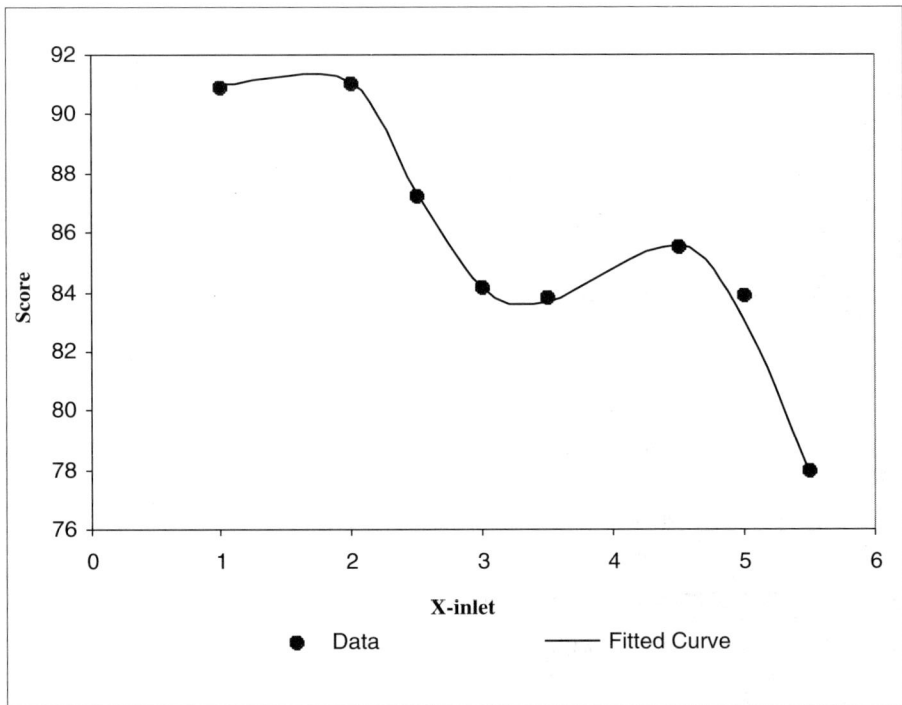

Figure 10.16 Analytical model versus CFD data set for x-inlet dependence.

Outlet Height (Y-Outlet)

The same fitting technique used for the Xi dependence has been used to determine an analytical expression of the score dependence on y-outlet (Y_o). In fact, both present the same trend. So, applying the same procedure, the following expression has been found:

$$S_{Y_o}(Y_o) = c_3 + c_4 Y_o + k_2 \sin(1.8\, Y_o) \tag{10.14}$$

where, again, the parameters depend on Z_o:

$$c_3 = 92.074 - 0.9448 Z_o$$
$$c_4 = -0.3386 Z_o + 0.24 \tag{10.15}$$
$$k_2 = -0.275 Z_o - 0.04$$

Figure 10.17 shows the comparison between the analytical form of $S(Yo)$ and the CFD prediction for two sets of data.

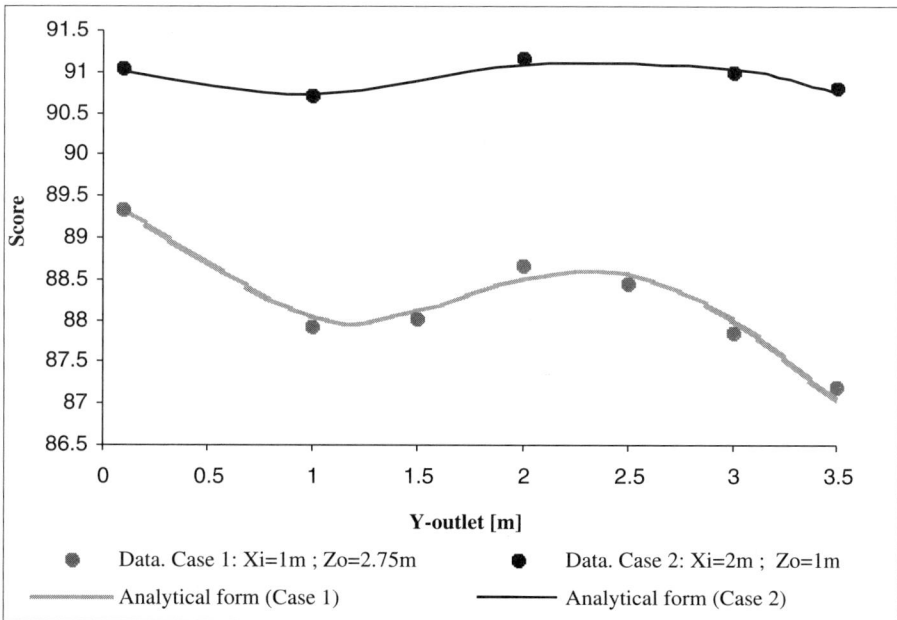

Figure 10.17 Analytical model versus CFD data set for y-outlet dependence.

Object Location (X-object$_{1,2}$)

The position of object 1, the main particle source, strongly affects the room score. The trend is clearly exponential and, with curve-fitting techniques, its mathematical expression has been determined as

$$S_{X_{obj1}}(X_{obj1}) = c_4 + 2.8035\, e^{0.288 \cdot X_{obj1}} \tag{10.16}$$

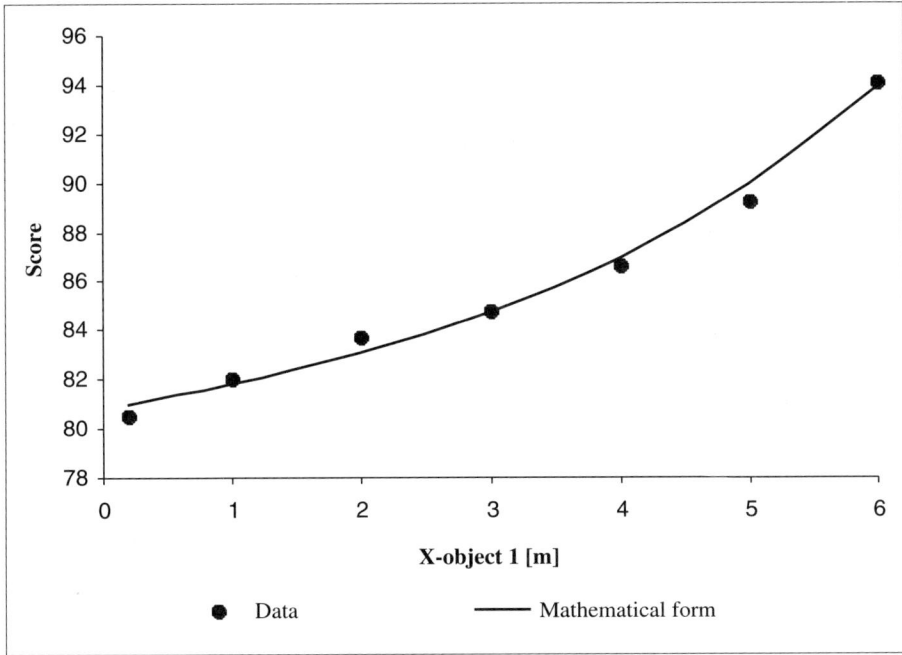

Figure 10.18 Analytical model versus CFD data set for x-object 1 dependence.

Figure 10.18 shows the results from the analytical model and compares it to the CFD data for x coordinate of object 1.

The value of c_4 is the score when X_{obj1} tends to zero plus 2.8035; therefore, it depends on the other parameters and should be computed from previous equations. A value of 78 gives an acceptable approximation, but to obtain more accurate results this equation should be coupled with the others, containing dependence on other parameters that does not have their standard values, as will be done later in this section. The score dependence on X_{obj2} shows a sinusoidal trend that can be described coupling a polynomial equation and an equation of the form (10.9).

These fundamental forms of the X_{obj1} and X_{obj2} dependencies have been the starting points for obtaining the following two-variable equation, by means of a two-variable curve fitting of the data set:

$$S\left(X_{obj1}, X_{obj2}\right) = 80.547 + 0.2258 X_{obj1}^2 - 0.3706 X_{obj2}^2 + 0.3315 X_{obj1} X_{obj2} + 0.0513 X_{obj1}$$

$$+ 0.266 X_{obj2} + 0.5455 e^{-0.0542 X_{obj1}} + 0.5833 \cdot \cos\left(-2.759 X_{obj2} - 0.5532\right)$$

$$+ 0.4033 \, \sin\left(-0.0277 X_{obj2} - 0.0474\right) \tag{10.17}$$

General Equation

The final goal of this work has been the development of a general mathematical formulation of the room score as a function of the six parameters. To do that, the assumptions made before are needed: The equation has to be of the form (10.3) and it has a constant trend along lines parallel to the x_i axes.

One of the assumptions made is that all the equations developed in previous sections contain one common point, which is the one corresponding to the standard configuration $P_{STD} \equiv (\mathbf{x_{STD}}) \equiv (6\text{ACH}, 1\text{m}, 0.1\text{m}, 2.75\text{m}, 5\text{m}, 2\text{m})$.

The general equation is of the form:

$$S(x_1,...,x_n) = S_{X_1}(x_1) + \sum_{i=2}^{n} \left[S_{x_i}(x_i) - S_{x_i}(x_{iSTD}) \right] \tag{10.18}$$

where $x_1 \equiv$ ACH and x_{iSTD} corresponds to the value of the variable x_i at the standard point P_{STD}.

Equation (10.17) allows us to obtain an approximation of the room score function. Basically, the way it works can be explained by an example: calculate the score at the point $\overline{P} = (\mathbf{x}) = \left(\overline{\text{ACH}, X_i, Y_o, Z_o, X_{obj1}, X_{obj2}} \right)$. Starting from $S = 0$, we move along the hyperbolic curve $S_{ACH}(\text{ACH})$ until we reach the point $\left(\overline{\text{ACH}}, X_{iSTD}, Y_{oSTD}, Z_{oSTD}, X_{obj1STD}, X_{obj2\,STD} \right)$. From this point, we move toward the final point P along the curves $S_{xi}(x_i)$ in the planes $x_2, \ldots, x_{i-1}, x_{i+1}, \ldots, x_n$, moving on each of them one per time, since they are assumed to have constant shape, and therefore constant mathematical expression, in all the possible planes $x_2, \ldots, x_{i-1}, x_{i+1}, \ldots, x_n$.

Substituting the expressions of $S(X_i)$ into equation (10.16) and using trigonometric properties, the final equation is obtained:

$$S\left(\text{ACH}, X_i, Y_o, Z_o, X_{obj1}, X_{obj2}\right) =$$

$$= 2.9 + \frac{1000 \cdot \text{ACH}}{9.9 \cdot \text{ACH} + 7.1} - 2.613Z_o + c_4 Y_o + k_2 \sin(1.8 Y_o)$$

$$+ {}_2X_i - k_1 \cos\left[2X_i - \left(X_{obj1} - 5\right)\right] + 2\sin(a_1 X_i)$$

$$+ 0.2258 X_{obj1}^2 - 0.3706 X_{obj2}^2 + 0.3315 X_{obj1} X_{obj2}$$

$$+ 0.0513 X_{obj1} + 0.266 X_{obj2} + 0.5455 e^{-0.0542 X_{obj1}}$$

$$+ 0.5833 \cos\left(-2.759 X_{obj2} - 0.5532\right) + 0.4033 \sin\left(-0.028 X_{obj2} - 0.0474\right)$$

$$\tag{10.19}$$

where the expressions of the constants a_i, c_i, and k_i are those given earlier in this chapter. Because of the strong underlying assumptions, equation (10.19) yields an approximate value of the room score. It is subject to errors due to the

CFD turbulent airflow model and the scoring system chosen. However, this mathematical model is still very effective for quickly comparing and evaluating the enormous number of alternatives.

10.7 APPLICATION OF VIRTUAL REALITY

Equation (10.19) gives the room layout score for every value assumed by the six design parameters. It is used as a criterion for screening the enormous number of room layout alternatives. This step is described next. In a VR environment, the user is able to change the room parameters and, in real time, see the global effects of the change through the room layout score, which is updated automatically as a parameter is moved. The results obtained previously are used as a basis for rationally changing the parameters, in other words, they provide the user with knowledge of how a change could affect the room score. An illustrative example is provided next.

The results given before in this chapter can be summarized briefly as follows. An increase in the air intake rate (ACH) improves the air cleanliness tremendously, but at the same time negatively affects the comfort level of an occupant. Moving object 1 (main particle source) causes a similar improvement in removal and does not affect comfort. So this should be the first preferred parameter to move unless other constraints hinder it. Inlet position is the third most influencing parameter. As shown, it has a decreasing sinusoidal shape with a maximum before the position of object 1. So optimal positions can be found close to the back wall (low values of X_i) or close to object 1. The dependencies on outlet and object 2 positions have a sinusoidal trend but are less predictable and have a smaller range than previous ones. The optimal position of these parameters can tentatively be found in the VR preprocessing phase and, eventually, verified in the subsequent CFD/VR step.

This operation is very fast and intuitive due to the efficiency of the VR interface and display. Figure 10.19 (also shown in the color section) shows the user immersed in the CAVE during this phase. The Computer Assisted Virtual Environment (CAVE™, formerly known as Cave Automatic Virtual Environment) is a multiperson, room-sized, high-resolution, three-dimensional video and audio virtual reality environment, originally developed at University of Illinois at Chicago and currently comarketed by Pyramid Systems and Silicon Graphics. It provides insight and understanding of three-dimensional immersive VR solution spaces not available when using a two-dimensional screen display mechanism. The user interacts with the VR environment using a three-dimensional mouse called a *wand*. A tracking system is used to get the user's head and hand orientation, to update the image displayed accordingly. This system makes extremely easy and intuitive the process of changing the room

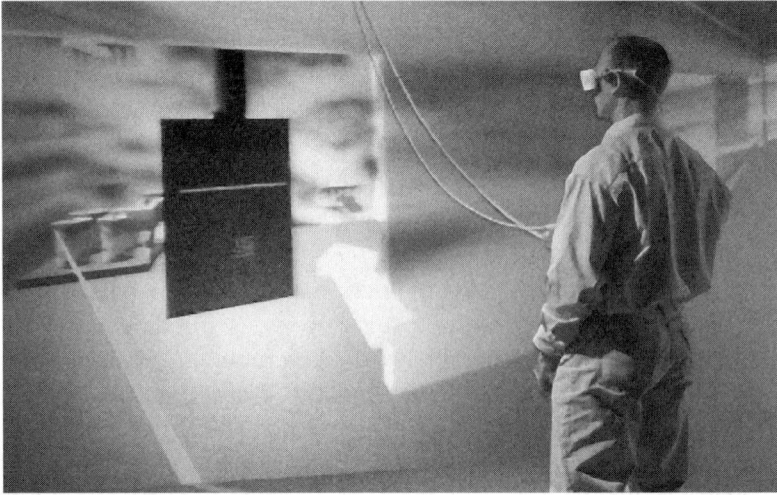

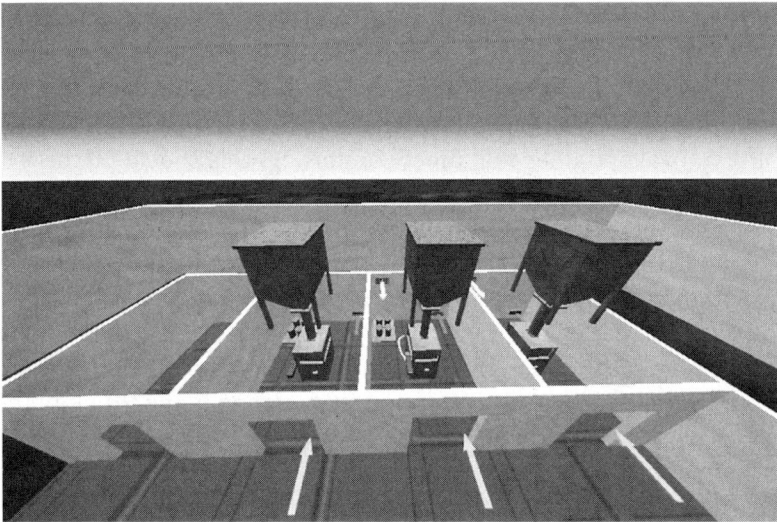

Figure 10.19 Preprocessing step (top image also shown in the color section). The user is immersed in the VR representation of the room (in this specific case, a pharmaceutical tablet press room) and can easily change parameters simply by "dragging and dropping" them. The various arrows schematically depict different airflow parameters to the user.

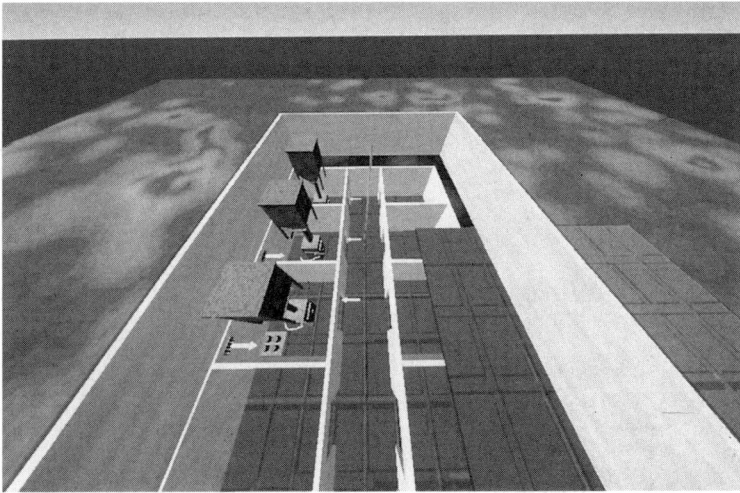

parameters and allows quick screening of all possible solutions, to focus on a small set of preferred solutions to be analyzed further. This is done by means of real-time evaluation of the room layout, using equation (10.19).

The mathematical model of the contaminant removal effectiveness cannot always guarantee high accuracy. This is because of the simplifying assumptions underlying equation (10.19), which have been observed to be reasonable for a general characterization of the phenomenon but cannot provide high accuracy. Furthermore, equation (10.19) takes into account the fate of both particles emitted around object 1 (with a high weight in the final score) and those emitted uniformly throughout the room (with a lower weight). In this

Figure 10.20 Representation of the airflow in a recirculation region using fixed tetrahedral darts (also shown in the color section). *Source:* Lori Freitag, Argonne National Laboratory, and Bill Michels, Nalco Fuel Tech.

case the designer is interested only in removing just particles emitted; for example, throughout the room, the actual score could be slightly different, even though equation (10.19) can still be used for a general screening of solutions.

For these reasons, the best solution or solutions can be found by means of an immersive VR analysis of the actual CFD prediction of the restricted set of alternatives found in the preprocessing step. The VR display allows us to overcome the gap affecting the traditional CFD preprocessors: the incapability of giving an insight into the phenomenon. In fact, in most cases direct analysis of the CFD output cannot provide a clear overview of particle motion. It is usually very hard to understand how the presence of obstacles, position of air exhausts, and location of sources affect particle paths if based only on the analysis of numerical data or using traditional postprocessors. It is even more difficult to find out from those data hints how parameters might be changed to obtain the desired contaminant particle trajectories to guide them easily out of the room.

In the VR environment, the user can physically walk in the room and "see" the airflow pattern. An example of airflow representation in the CAVE is given in Figure 10.20 (also shown in the color section): the airflow velocity vectors are depicted using tetrahedral darts; the direction and magnitude of each dart depicts direction and magnitude of the airflow velocity field at that point. Furthermore, the user is able to inject a set of particles at some points of the room and see their trajectories instantaneously. In fact, a real-time integration of particle motion equation is possible with current computer graphics processors, as has been done by the Trackpack software developed by Diachin et al. (1998) for their Boilermaker system.

Ideally, the designer should be able to "follow" the particles within the domain and thus "see" what their trajectory is and where it ends up, then try to modify the room parameters so as to drive the particles in the right direction. If the user is physically immersed in a three-dimensional model of the room, is able to navigate along the particle trajectories and at the same time is able to visualize the airflow velocity vector field at point of interest, he or she would be able to intuitively understand which objects cause undesirable particle trajectories and where they can be placed to prevent obstruction of the contaminant flow out of the room; where the supply air exhaust should be placed to capture a large number of emitted particles or to eliminate stagnant zones; where the inlet should be placed for effective clustering of particles; and how the parameters can be changed to avoid stagnant zones and contaminant accumulation. After setting up the required changes to the room configuration in the VR environment (Figure 10.19) quickly and easily, the user can run the CFD simulation again and analyze the effectiveness of the changes brought to the domain. Proceeding in this way, the solution can be improved iteratively until a nearly optimal configuration of the contamination control system can be found.

A similar result can take many person-weeks of traditional CFD analysis. Furthermore, the solution obtained with the VR tool is expected to be more efficient, due to the ease with which the critical points of the room can be zoomed in and the rapidity with which the needed changes can be done in VR. Another reason for which a traditional CFD analysis cannot usually provide a solution as good as the one obtained with this VR tool is the limited number of "test sessions" (i.e., number of iterations) that this "relatively long" method is allowed to perform. In a VR environment, the complete setup–trajectory computation–result displaying–analysis of result cycle takes only several minutes, even for subjects not familiar with VR tools. This means that the laborious step of setting up the new configuration and computing the new boundary conditions simply comes down to "moving" or "placing" objects in a VR environment. Furthermore, the hard phases of critically interpreting the CFD output data become a simple visit to the VR simulated room.

10.8 CONCLUSIONS

The efficiency of a contaminant removal system is highly dependent on some parameters that are usually variables during the designing phase. The complicated setup and long computation time of current CFD software packages hinder application of this VR/CFD-aided design to the room layout design. This makes it difficult to conduct the CFD/VR analysis of all possible solutions that designers have at their disposal.

In this chapter, a method for overcoming this problem by using a VR preprocessing step has been described. A mathematical expression of the dependence of contaminant removal effectiveness on room parameters has been developed. It provides enough accuracy for efficient screening of the large set of alternatives in real time in a VR environment. Subsequently, the selected set of preferred solutions is investigated further by VR analysis of the CFD simulation. This provides insight for optimal positioning of all parameters for room air cleanliness. This two-step CFD/VR-aided design technique allows one to design room layout taking into account air cleanliness issues in a fast and effective way, overcoming the problems that affect the integration of CFD and VR technique in contaminant-free environments.

Future research in this area focuses mainly on the investigation of the effects of other design parameters on contaminant removal. Such parameters are (1) inlet/outlet size and number; (2) size and number of objects, (3) presence of heat sources (e.g., people or chemical reactors), and (4) placement of inlets/outlets on different walls. Various types of rooms should be investigated to generalize the results further.

REFERENCES

Diachin, D., L. Freitag, D. Heath, J. Herzog, and W. Michels, "Interactive Simulation and Visualization of Massless, Massed, and Evaporating Particles," *IIE Transactions,* Vol. 30, No. 7, pp. 621–628, 1998.

Giallorenzo, V., "Improving Particle Flow in Facility Design Using Computational Fluid Dynamics and Virtual Reality," M.S. thesis, University of Illinois at Chicago, 1999.

APPENDICES

A1

B-SPLINE CURVE FITTING

PARAMETRIC CUBIC CURVE REPRESENTATION

Cubic polynomials provide a compact and flexible representation of spatial piecewise smooth curves. The easiest manageable way of representing curves modeled by cubic polynomials is the parametric form, through which the curve equations that define a curve segment $Q(t) = [x(t), y(t), z(t)]$ can be written as follows:

$$x = x(t) = a_1 t^3 + b_1 t^2 + c_1 t + d_1$$
$$y = y(t) = a_2 t^3 + b_2 t^2 + c_2 t + d_2 \qquad \text{(A1.1)}$$
$$z = z(t) = a_3 t^3 + b_3 t^2 + c_3 t + d_3$$

where t is the parameter and usually varies between 0 and 1. (Although t can vary within any range, for computational simplicity and without loss of generality it is convenient to set the range as the [0,1] interval; the only constraint imposed is that intervals of variation of parameter t have to have the same size over all curve segments, in case of uniform curves). The coefficients of the

289

polynomials, a_i, b_i, c_i, $i = 1, 2, 3$, can be determined by writing the boundary and continuity conditions between curve segments. The coefficients have different values for each segment.

Let us define the following matrices:

$$\mathbf{T} = [t^3 \ t^2 \ t \ 1] \quad \text{and} \quad \mathbf{C} = \begin{bmatrix} a_1 & a_2 & a_3 \\ b_1 & b_2 & b_3 \\ c_1 & c_2 & c_3 \\ d_1 & d_2 & d_3 \end{bmatrix} \tag{A1.2}$$

So we can rewrite the curve equation as

$$Q(t) = \mathbf{TC} \tag{A1.3}$$

The derivative of $Q(t)$ is the *parametric tangent* vector of the curve, which has the following form:

$$\frac{d}{dt}Q(t) = Q'(t) = \left[\frac{d}{dt}x(t) \ \ \frac{d}{dt}y(t) \ \ \frac{d}{dt}z(t)\right] = \frac{d}{dt}\mathbf{TC} \tag{A1.4}$$

The tangent vector $Q'(t)$ is the velocity of a point on the curve with respect to the parameter t. Similarly, the second derivative of $Q(t)$ is the acceleration. If a camera is moving along a parametric cubic curve in equal time steps and records a picture after each step, the tangent vector gives the velocity of the camera along the curve. The camera velocity and acceleration at join points should be continuous, to avoid jerky movements in the resulting animation sequence. It is this continuity of acceleration across the join point in Figure A1.1 that makes the C_1 curve continue farther to the right than the C_2 curve before bending around to the endpoint.

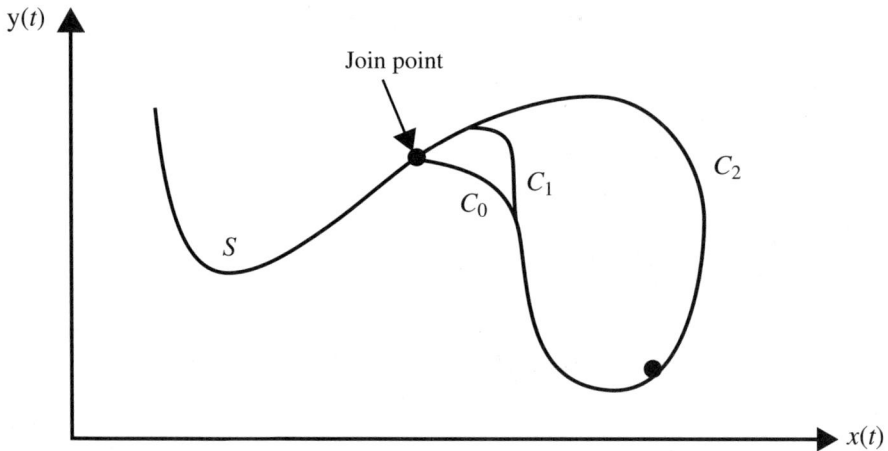

Figure A1.1 Curve segment S joined to segments C_0, C_1, and C_2, with 0, 1, and 2 degrees of parametric continuity, respectively. (Note that the visual distinction between C_1 and C_2 is slight at the join, but obvious away from the join.)

If two curve segments join, the curve has C_0 geometric continuity, and if the directions and magnitudes of the tangent vectors of two adjacent curve segments are equal, the curve displays C_1 continuity. In general, when for two joining curve segments the direction and magnitude of $\dfrac{d^n}{dt^n}[Q(t)]$ through the nth derivative are equal at the join point, the curve is called C_n *continuous*. In most cases we define curves with up to C_2 degree of continuity (when the curvatures at the join point are equal), which provide us with a complete representation for most applications. Furthermore, we can split the matrix **C,** defined in equation (A1.3), as the product between two matrices: **M,** a 4×4 matrix called the *basis matrix,* and **G,** which is a four-element column vector of geometric constraints, called the *geometry vector.* The matrix **M** is constant for every curve segment but has different forms for the different curve types, such as Bezier, Hermite, or splines. The values of the elements of **G** depend on the coordinates of the points through which the curve is constrained to pass (called *control points*). So the general form of the equation of a cubic curve segment is

$$Q(t) = \mathbf{TMG} \tag{A1.5}$$

UNIFORM B-SPLINES

The regular (natural) splines are curves that interpolate (pass through) a sequence of control points. These curves are similar to the curves described in the preceding section. The disadvantage of this kind of representation is that the polynomial coefficients of the curve segments are dependent on all the control points; hence moving a control point affects the shape of the entire curve, leading to longer computation time and a lower degree of flexibility. The way to overcome this disadvantage is to design a curve whose polynomial coefficients depend on just a few control points (the property of *local control*). These curves are named *B-splines*. They do not interpolate (do not pass through) the control points. Their property is C_1 and C_2 continuity at the join points. The peculiarity of these curves is that control points are shared between segments, and a curve segment depends on *four* control points.

Let us consider that a cubic B-spline approximates a set of $m + 1$ control points $P_0, P_1, \ldots, P_m$, $m \ge 3$, with a curve consisting of $m - 2$ cubic polynomial curve segments $Q_3, Q_4, \ldots, Q_m$. Again, for the purpose of simplicity, we consider parameter t as varying between 0 and 1. Each curve segment Q_i is defined by four control points. For example, segment Q_3 (see Figure A1.2) is defined by control points P_0 to P_3. Apart from the control points, we define *knots,* which are the join points between adjacent curve segments. So for each spline segment, t has 0 value at the first knot and 1 value at the second knot. In Figure A1.2, the notions of control points and knots for a two-dimensional

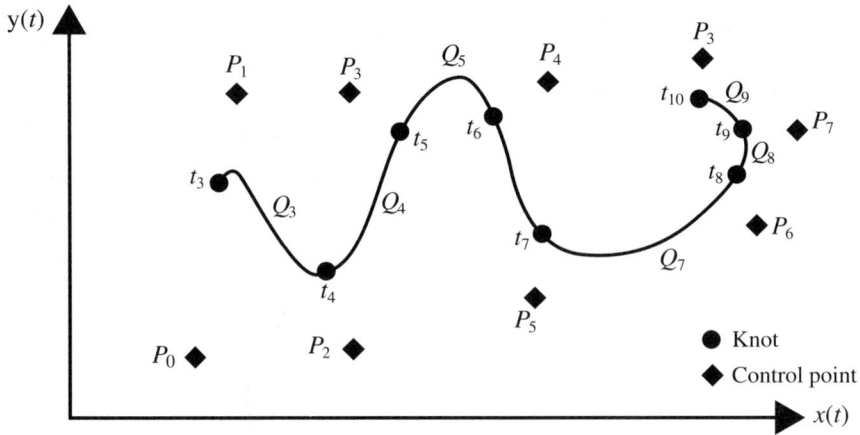

Figure A1.2 Example of a B-spline curve.

curve are illustrated. The curve shown in Figure A1.2 consists of the segments Q_3 through Q_9. For example, segment Q_3 has the control points P_0 through P_3, segment Q_4 has the control points P_1 through P_4, and so on. So we can see that some of the control points are shared between the segments, hence the property of local control. We can deduce that one control point affects at most four successive curve segments, since two adjacent segments share three control points. Having introduced the notion of knots, we can specify that a uniform curve means that the knots are spaced at equal intervals of parameter t (in our case $[0,1]$).

The matrix **M** (the basis matrix) is constant and has the following form:

$$\mathbf{M} = \frac{1}{6} \begin{bmatrix} -1 & 3 & -3 & 1 \\ 3 & -6 & 3 & 0 \\ -3 & 0 & 3 & 0 \\ 1 & 4 & 1 & 0 \end{bmatrix} \tag{A1.6}$$

Considering a segment Q_i defined by the control points P_{i-3}, P_{i-2}, P_{i-1}, and P_i, the geometry vector $\mathbf{G}_i$ for segment Q_i is

$$\mathbf{G_i} = \begin{bmatrix} P_{i-3} \\ P_{i-2} \\ P_{i-1} \\ P_i \end{bmatrix}, \quad 3 \leq i \leq m \tag{A1.7}$$

where P's represent the coordinates of the control points. Obviously, there is a geometry vector **G** for each x, y, and z coordinates of the control points.

The expressions of matrices **T, M,** and **G** are plugged in equation (A1.5) and the equations of the polynomials in t representing the curve segment are obtained. The coefficients of the polynomials are determined from the equation expressing the continuity of the B-spline. Let us consider two adjacent curve segments (i and $i + 1$) joining at the point P, a knot of the curve. In P, parameter t has 1 value for the first segment and 0 value for the second segment. Let us consider the x components of the adjacent segments. The continuity conditions are written as follows:

- C^0 *continuity:* $x_i (t = 1) = x_{i+1} (t = 0)$

- C^1 *continuity:* $\dfrac{d}{dt} x_i (t = 1) = \dfrac{d}{dt} x_{i+1}(t = 0)$

- C^2 *continuity:* $\dfrac{d^2}{dt^2} x_i (t = 1) = \dfrac{d^2}{dt^2} x_{i+1}(t = 0)$

From the equations above, the coefficients of the cubic polynomials are derived in a straightforward manner. The same equations hold if we consider the y and z components of the adjacent segments.

The slope of a planar curve [$x(t),y(t)$] at some point can be defined as follows:

$$\left[\frac{dy}{dx} \right]_{t=t^1} = \left[\frac{\frac{dy}{dt}}{\frac{dx}{dt}} \right]_{t = t^1} \tag{A1.8}$$

In practice, when the geometric features are sought at some points of interest, the slope is computed usually at the knots, so it is enough to plug $t = 0$ in equation (A1.1). Another feature of maximal interest in practical applications is the curvature at some point. Again, it is enough to compute the curvature at the knots (where $t = 0$). The formula for the curvature can be derived as follows, using the well-known formula for computing the Gaussian curvature for planar curves:

$$C_V(0) = \frac{d^2 y}{dx^2} \Bigg/ \left[1 + \left(\frac{dy}{dx} \right)^2 \right]^{3/2} = \left(\frac{dx}{dt} \frac{d^2 y}{dt^2} \right) - \frac{d^2 x}{dt^2} \frac{dy}{dt} \Bigg/ \left\{ \left[\left(\frac{dx}{dt} \right)^2 + \left(\frac{dy}{dt} \right)^2 \right]^{3/2} \right\} \tag{A1.9}$$

If we consider the first two equations of (A1.1) (which in fact represent a planar curve), the curvature at $t = 0$ knot can be computed using the following equation:

$$C_v(0) = 2 \frac{c_1 b_2 - c_2 b_1}{(c_1^2 + c_2^2)^{3/2}} \tag{A1.10}$$

A2

PSEUDOINVERSE METHOD FOR OVERDETERMINED SYSTEMS OF LINEAR EQUATIONS

In general, a system of linear equations can be written in matrix form as

$$\mathbf{A}\mathbf{x} = \mathbf{b} \tag{A2.1}$$

where $\mathbf{A}$ is an $m \times n$ matrix of coefficients, $\mathbf{x}$ is the $n \times 1$ vector of unknowns, and $\mathbf{b}$ is an $m \times 1$ vector of right-hand side elements. When the number of equations is greater than the number of unknowns ($m > n$), the system is said to be *overdetermined* and therefore the matrix $\mathbf{A}$ is noninvertible. Such a system does not have an exact solution and the best we can do is to minimize the errors in the least-squares sense.

The error in solving the system is defined as follows:

$$\mathbf{e} = \mathbf{b} - \mathbf{A}\mathbf{x} \tag{A2.2}$$

The squared error is written as follows:

$$\mathbf{e}^{\mathrm{T}}\mathbf{e} = (\mathbf{b} - \mathbf{A}\mathbf{x})^{\mathrm{T}}(\mathbf{b} - \mathbf{A}\mathbf{x}) \tag{A2.3}$$

The squared error is typically minimized by zeroing the derivative of the squared error with respect to the unknowns. This derivative is computed below:

$$\frac{d(\mathbf{e}^T\mathbf{e})}{d(\mathbf{x})} = (0 - \mathbf{AI})^T(\mathbf{b} - \mathbf{Ax}) + (\mathbf{b} - \mathbf{Ax})^T(0 - \mathbf{AI})$$

$$= (-\mathbf{AI})^T(\mathbf{b} - \mathbf{Ax}) + (\mathbf{b}^T - (\mathbf{Ax})^T)(-\mathbf{AI}) = -\mathbf{IA}^T(\mathbf{b} - \mathbf{Ax}) + (\mathbf{b}^T - \mathbf{x}^T\mathbf{A}^T)(-\mathbf{AI})$$

$$= -\mathbf{IA}^T\mathbf{b} + \mathbf{IA}^T\mathbf{Ax} - \mathbf{b}^T\mathbf{AI} + \mathbf{x}^T\mathbf{A}^T\mathbf{AI}$$

Above, $\mathbf{I}$ is the identity matrix. Note that $\mathbf{b}^T\mathbf{AI}$ and $\mathbf{x}^T\mathbf{A}^T\mathbf{AI}$ are 1×1 matrices. The transpose of a 1×1 matrix is equal to itself, so we can transpose these expressions and we get:

$$\frac{d(\mathbf{e}^T\mathbf{e})}{d(\mathbf{x})} = -\mathbf{IA}^T\mathbf{b} + \mathbf{IA}^T\mathbf{Ax} - \mathbf{IA}^T\mathbf{b} + \mathbf{IA}^T\mathbf{Ax} = 2\mathbf{I}(\mathbf{A}^T\mathbf{Ax} - \mathbf{A}^T\mathbf{b}) = 0$$

from where

$$\mathbf{A}^T\mathbf{Ax} = \mathbf{A}^T\mathbf{b} \Rightarrow \mathbf{x} = (\mathbf{A}^T\mathbf{A})^{-1}\mathbf{A}^T\mathbf{b} = \mathbf{A}^*\mathbf{b} \qquad (A2.4)$$

In equation (A2.4), $\mathbf{A}^* = (\mathbf{A}^T\mathbf{A})^{-1}\mathbf{A}^T$ is called the pseudo-inverse of matrix $\mathbf{A}$ — hence the term *pseudoinverse* method.

A3

INTRODUCTION TO KALMAN FILTERING

For the discussion that follows, we assume that the reader has some basic knowledge of probabilities and statistics. The purpose of Kalman filtering is to optimally recover the desired information given a set of noisy observations. It is appropriate to introduce the technique of Kalman filtering in the context of dynamic systems, where it originated. The behavior of a dynamic system can be characterized by the evolution of a set of variables, called *state variables*. In general, the state variables are not directly observable, but rather, measurements are performed that are functions of the state variables. The measurements are corrupted by noise. Also the system's behavior may be corrupted by random disturbances. The objective is to reduce as much as possible the effect of noise and to estimate the state variables optimally from the noisy measurements. We describe next the Kalman filtering technique as applicable to discrete systems, but the content of this description can easily be adapted for continuous systems. We present first the standard Kalman filter, designed for linear systems, followed by an extension of the filter for nonlinear systems.

A discrete system is a system whose state is modified at discrete time steps. In general, a discrete dynamic system (whose behavior is characterized by n state variables) can be modeled in the following form:

$$\mathbf{x}_{k+1} = \mathbf{\Phi}_k \mathbf{x}_k + \mathbf{w}_k \tag{A3.1}$$

where $\mathbf{x}_k$ is an $(n \times 1)$ process state vector at time $\mathbf{t}_k$, $\mathbf{\Phi}_k$ an $(n \times n)$ matrix relating $\mathbf{x}_k$ to $\mathbf{x}_{k+1}$, and $\mathbf{w}_k$ an $(n \times 1)$ process noise vector, assumed to be white noise (in control theory, noise is termed *white* when it has a constant spectral density function).

The relationship between the observable parameters (measurements) and the state variables is written as follows (assume that the number of measurements is m):

$$\mathbf{z}_k = \mathbf{H}_k \mathbf{x}_k + \mathbf{v}_k \tag{A3.2}$$

where $\mathbf{z}_k$ is an $(m \times 1)$ vector of measurements at time t_k; $\mathbf{H}_k$ an $(m \times n)$ matrix expressing the ideal (noiseless) relationship between $\mathbf{z}_k$ and $\mathbf{x}_k$, at time t_k; and $\mathbf{v}_k = (m \times 1)$ measurement noise vector, assumed also to be white noise and uncorrelated with the process noise. The elements of noise vectors $\mathbf{w}_k$ and $\mathbf{v}_k$ are uncorrelated over time, but there may be a mutual correlation at specific points in time among their elements. This mutual correlation is expressed by means of covariance matrices, denoted $\mathbf{Q}_k$ for $\mathbf{w}_k$ and $\mathbf{R}_k$ for $\mathbf{v}_k$.

Kalman filtering starts with an initial estimate of the state vector at some point in time t_k. This initial estimate is denoted as $\hat{\mathbf{x}}_k^-$, where the "hat" symbol denotes estimate and superscript sign denotes that the estimate is being made before assimilating the current measurement. The estimation error is defined as

$$\mathbf{e}_k^- = \mathbf{x}_k - \hat{\mathbf{x}}_k^- \tag{A3.3}$$

The estimated error has associated with it a covariance defined by

$$\mathbf{P}_k^- = E[\mathbf{e}_k^- \mathbf{e}_k^{-T}] = E[(\mathbf{x}_k - \hat{\mathbf{x}}_k^-)(\mathbf{x}_k - \hat{\mathbf{x}}_k^-)^T] \tag{A3.4}$$

where $E[\cdots]$ denotes expected value. A possible way to integrate the current measurement (by *current* we mean at time t_k) is through the following equation:

$$\hat{\mathbf{x}}_k = \hat{\mathbf{x}}_k^- + \mathbf{K}_k(\mathbf{z}_k - \mathbf{H}_k \hat{\mathbf{x}}_k^-) \tag{A3.5}$$

where $\mathbf{K}_k$ is a blending factor, called *Kalman gain,* and $\hat{\mathbf{x}}_k$ is the updated estimate, after the new measurement has been assimilated. We will not develop here the expression of the Kalman gain (for more details, see "Further Reading"). We mention though that the development of the expression of the Kalman gain is based on the minimization of the mean-squared error in the estimated state vector. A popular formula for the Kalman gain is given by

$$\mathbf{K}_k = \mathbf{P}_k^- \mathbf{H}_k^T (\mathbf{H}_k \mathbf{P}_k^- \mathbf{H}_k^T + \mathbf{R}_k)^{-1} \tag{A3.6}$$

It is customary to call $\hat{x}_k$ the *a posteriori estimate* of the state vector, after integrating the new measurement. The a posteriori error covariance matrix is given by

$$\mathbf{P}_k = E[\mathbf{e}_k \mathbf{e}_k^T] = E[(\mathbf{x}_k - \hat{\mathbf{x}}_k)(\mathbf{x}_k - \hat{\mathbf{x}}_k)^T] = (\mathbf{I} - \mathbf{K}_k \mathbf{H}_k)\mathbf{P}_k^- \quad\quad (A3.7)$$

where $\mathbf{I}$ is the $n \times n$ identity matrix.

To proceed to the next step ($k + 1$), the filter needs the initial estimate for this step. Since the noise $\mathbf{w}_k$ has zero mean and is uncorrelated timewise with previous process noise vectors, its contribution to the new estimate can be ignored. The new initial estimate and its error covariance matrix are given by:

$$\hat{\mathbf{x}}_{k+1}^- = \Phi_k \hat{\mathbf{x}}_k$$
$$\mathbf{P}_{k+1}^- = \Phi_k \mathbf{P}_k \Phi_k^T \quad\quad (A3.8)$$

Now we can review how Kalman filtering works:

1. Initialize the state vector estimate and its error covariance.
2. Compute the Kalman gain [equation A3.6].
3. Update the estimate by integrating the measurement [equation A3.].
4. Compute the error covariance for the updated estimate [equation A3.7].
5. Predict the new state vector and covariance matrix [equations A3.8].
6. Repeat steps 1 to 5 until no more measurements are available, or as needed.

From the discussion above, we can infer the recursive nature of the Kalman filter. This makes it very efficient for computer implementation, especially for systems with a large number of state variables, due to the fact that it incorporates all past estimations into a single equation.

So far we have treated the Kalman filter only for systems that can be modeled by linear equations. A natural question would arise if Kalman filtering can be extended to nonlinear systems and still be optimal. A nonlinear discrete system can be characterized by the following equation:

$$\mathbf{x}_{k+1} = f(\mathbf{x}_k) + \mathbf{w}_k \quad\quad (A3.9)$$

where $f = (f_1, f_2, \ldots, f_n)$ is a multidimensional function expressing the nonlinear relationship between $\mathbf{x}_{k+1}$ and $\mathbf{x}_k$. Similarly, the measurements can be nonlinear functions of the state variables:

$$\mathbf{z}_k = h(\mathbf{x}_k) + \mathbf{v}_k \quad\quad (A3.10)$$

where h is a nonlinear function that describes the measurement-state variables relationship. To adapt the Kalman filtering to nonlinear systems, we have to find a way to linearize functions f and h. A way to do this is to employ the Taylor series expansion of the functions around the current estimate $\hat{\mathbf{x}}_k$ and ignoring the terms with the order higher than or equal to two:

$$f(\mathbf{x}_k) = f(\hat{\mathbf{x}}_k) + \mathbf{J}_k^f(\hat{\mathbf{x}}_k)(\mathbf{x}_k - \hat{\mathbf{x}}_k)$$
$$h(\mathbf{x}_k) = h(\hat{\mathbf{x}}_k) + \mathbf{J}_k^h(\hat{\mathbf{x}}_k)(\mathbf{x}_k - \hat{\mathbf{x}}_k) \quad\quad (A3.11)$$

where

$$\mathbf{J}_k^f = \begin{bmatrix} \dfrac{\partial f_1}{\partial x_{1k}} & \dfrac{\partial f_1}{\partial x_{2k}} & \cdots \\[2mm] \dfrac{\partial f_2}{\partial x_{1k}} & \dfrac{\partial f_2}{\partial x_{2k}} & \cdots \\[2mm] \cdot & \cdot & \cdots \\ \cdot & \cdot & \cdots \end{bmatrix}$$

is the Jacobian of f at step k (x_{ik}, $i = 1, 2, \ldots, n$ are the elements of $\mathbf{x}_k$) and

$$\mathbf{J}_k^h = \begin{bmatrix} \dfrac{\partial h_1}{\partial x_{1k}} & \dfrac{\partial h_1}{\partial x_{2k}} & \cdots \\[2mm] \dfrac{\partial h_2}{\partial x_{1k}} & \dfrac{\partial h_2}{\partial x_{2k}} & \cdots \\[2mm] \cdot & \cdot & \cdots \\ \cdot & \cdot & \cdots \end{bmatrix}$$

is the Jacobian of h at step k. In equations (A3.11), $f(\hat{\mathbf{x}}_k)$ and $h(\mathbf{x}_k)$ are known quantities. Equations (A3.9) and (A3.10) then become

$$\mathbf{x}_{k+1} - f(\mathbf{x}_k) = \mathbf{J}_k^f (\mathbf{x}_k - \hat{\mathbf{x}}_k) + \mathbf{w}_k$$
$$\mathbf{z}_k - h(\mathbf{x}_k) = \mathbf{J}_k^h (\mathbf{x}_k - \hat{\mathbf{x}}_k) + \mathbf{v}_k$$

(A3.12)

We can apply the same procedure as in the case of linear systems for the system described by equations (A3.12). This version of Kalman filtering is called *extended Kalman filtering* (EKF). There are many successful applications where EKF has been applied, even though it is more likely than the standard Kalman filter to diverge in an unusual situation.

FURTHER READING

Kalman filtering technique is presented in an article by Kalman (1960). A good Kalman filtering reference is Brown (1993).

REFERENCES

Kalman, R. E., "A New Approach to Linear Filtering and Prediction Problems," *Transactions of the ASME — Journal of Basic Engineering,* vol. 82, pp. 35–45, March 1960.

Brown, R. G., *Introduction to Random Signal Analysis and Kalman Filtering,* Wiley, New York, 1983.

A4

KALMAN FILTER FOR HAND AND HEAD TRACKING

We use the constant-acceleration model for tracking the hand position. This model considers slight changes in acceleration as white noise. If we denote the vector of state variables as **x,** the discrete-time state equation is

$$\mathbf{x}(k+1) = \mathbf{F}\mathbf{x}(k) + \mathbf{w}(k) \tag{A4.1}$$

where $\mathbf{x}(k + 1)$, $\mathbf{x}(k)$ is the 9×1 state vector at time steps $k + 1$ and k, respectively, $\mathbf{F}$ is the 9×9 state transition matrix and $\mathbf{w}(k)$ is the 9×1 process white noise vector, zero mean and with known covariance matrix $\mathbf{Q}$. If we consider only the state variables corresponding to one coordinate of motion (say, x), the state vector looks as follows:

$$\mathbf{x} = (x, \dot{x}, \ddot{x})^{\mathrm{T}} \tag{A4.2}$$

Since the velocity and acceleration obey Newton laws, the state transition matrix (corresponding to x) looks as follows:

$$F = \begin{bmatrix} 1 & \Delta t & \frac{1}{2}\Delta t^2 \\ 0 & 1 & \Delta t \\ 0 & 0 & 1 \end{bmatrix}$$

(A4.3)

In equation (A4.3), Δt is the time increment between two consecutive observations. In reality, F is expanded to accommodate all three coordinates of motion. Process noise covariance matrix Q looks as follows (again, only for the coordinate x):

$$Q = E[w(k)w'(k)] = \begin{bmatrix} \Delta t^5/20 & \Delta t^4/8 & \Delta t^3/6 \\ \Delta t^4/8 & \Delta t^3/3 & \Delta t^2/2 \\ \Delta t^3/6 & \Delta t^2/2 & \Delta t \end{bmatrix} q$$

(A4.4)

In equation (A4.4), q is the constant process noise variance (determined a priori), $E[\cdots]$ denotes mathematical expectation, and the prime denotes transposition. As with F, matrix Q is expanded to accommodate all three coordinates of motion.

The measurement equation of the Kalman filter is written as follows:

$$z(k+1) = Hx(k+1) + v(k+1)$$

(A4.5)

where $z(k + 1)$ is the 3×1 measurement vector (hand/head positions along x, y, and z axes, as reported by the tracker), H is the measurement matrix that relates measurements to the corresponding state variables (in this case, since measurements are also state variables corresponding to position, $H = I_3$, where I_3 is the 3×3 identity matrix) and $v(k + 1)$ is the 3×3 measurement noise vector, zero mean and with covariance matrix R, determined a priori. Based on process and measurement equations, the Kalman filter runs recursively.

A5

VIRTUAL REALITY MODELING LANGUAGE

Although there are a number of programming and modeling languages using the scenegraph concept (such as Performer™, Inventor™, Java3D™), Virtual Reality Modeling Language (VRML) has been chosen here because it is easy to learn some of the basic concepts, because some of the browsers supporting it, such as, Cosmo can be freely downloaded, and because it is widely used, especially for smaller applications.

OVERVIEW

Although there are a number of good references on VRML (www.vrml.org), we provide a brief overview of some of the basic concepts here. The VRML Web site can provide further details. VRML is an object-oriented language. The language is structured through a number of files, each file containing a collection of VRML objects. The terms node and object are often used interchangeably in VRML. A node is a fundamental building block of a VRML file. Some nodes are objects (e.g., Cylinder, Box, Elevation Grid). Other nodes (e.g., Shape) are used as containers to hold related nodes. Each node contains fields, which hold the data for the node by specifying the various attributes of the node. Every VRML file starts with the header

```
#VRML V2.0 utf8
```

The utf8 specification refers to an ISO standard for text strings known as UTF-8 encoding.

VRML makes use of a number of defaults, which is very common in object-oriented programming. For example, the program file

```
#VRML V2.0 utf8
Shape {
  appearance Appearance {
   material Material {}
   }
  geometry      Cylinder {}
}
```

has the fields appearance, material, and geometry; while the nodes Material and Cylinder take on default values since nothing is specified. In VRML, node names are capitalized, field names are in lowercase. Including an empty Material node applies the default color (light gray) to the object. (If no Material node is included, the object will be black and will not be visible.) The Cylinder has a default radius of 1 and a height of 2. All of its parts are visible. It is in the default position with an upright axis centered on the screen.

On the other hand, here is an example of a Material node that specifies an object that is purple and shiny:

```
material {
  diffusecolor .5 0 .5
  shininess     .5
}
```

For simplicity, some fields are omitted from the example nodes. The diffusecolor field contains three values, for the *red, green,* and *blue* components of the color. In this system, often referred to as *RGB,* a value of (0,0,0) specifies black (i.e., no color), and a value of (1,1,1) specifies white. A value of (0.5,0,0.5) thus specifies 50 percent red, no green, and 50 percent blue, which results in a medium purple color. The shininess field takes a value between 0.0 and 1.0. Lower values produce soft, diffuse reflections, and higher values produce more focused, sharper highlights. The default coordinate system used in VRML is shown in Figure A5.1. The default viewpoint position is (0,0,10).

In almost all cases, one has to group objects, combining them into scenes. Some grouping nodes in VRML are Group, Transform, LOD, Anchor,

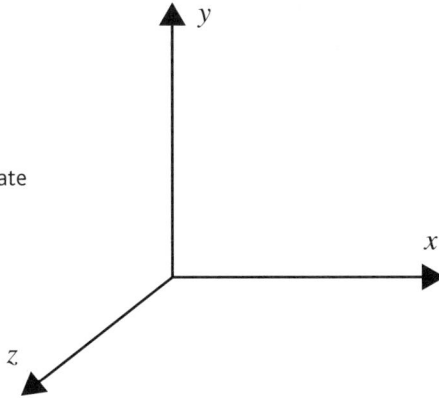

Figure A5.1 Default coordinate system in VRML.

and Inline nodes. Inline node allows one to access an external file to the scene. It allows one to organize the program into different files. This helps in better organization of large programs. Other Web sites can also be accessed in your program file by using inlines. Anchor nodes can be used to create hyperlinks to other scenes, to html pages, to sounds, and to movie files located anywhere on the Web. Anchors are more versatile than inlines, most of the Inline information normally has to be available locally.

DEF...USE allows one to reuse copies of the same object at multiple places. To use slightly varying copies of an object, each with some customized parameters, custom node types need to be created by prototyping using the command PROTO. A prototype creates a new node type, not a node. Instances of this node type will provide the nodes. Field declarations and default values of fields can be set. Fields of prototyped nodes can be used as values for fields of existing nodes (e.g., in Example 3 below, diffuseColor IS topColor sets the diffuseColor field of Material node to topColor of TwoColorTable node). Instead of defining the prototypes in the main scene file, one can indicate that a prototype is defined fully in an external file using EXTERNPROTO. One still declares the interface, but one does not give default values for fields in the externalprototype declaration. EXTERNPROTO is just a reference to a PROTO defined in another file. This method works a little like an Inline node. If the file contains more than one prototype, the URL should indicate which one (e.g., "filename.wrl#PrototypeNodeName").

The prototype definition creates a barrier between the inside and outside of the node. One needs to have a public interface between the inside and outside of a node. This is achieved through field and event declarations. Field has already been explained. Besides declaring fields, the public interface can declare the events that a prototype can send or receive, using eventIn and eventout declarations. The distinction between the field declarations and the eventIn and eventout declarations is an important one. A field can be given an initial value when you create an instance of the node but cannot be changed subsequently. An event is, in most cases, simply a request to change the value of a field — an incoming event asks the prototyped node to change one of its own field values, while an outgoing event is a request from the proto-typed node asking some other node to change one of its field values.

One can also specify a special kind of field called an `exposedfield`, which combines an ordinary field with a pair of events, one incoming event and one outgoing event. Note that the term `exposedfield` describes a par-ticular kind of field; the word is not a field name, and it never occurs in a VRML file except in prototype definitions. The node descriptions often include fields labeled "exposed field." This term is a convenient shorthand to describe a field for which there are two implicitly defined associated events: a `set_incom-ing` event that lets you set the field's value and a `_changed` outgoing event that sends out the new value when the field's value changes. For instance, the translation field of the `Transform` node is an exposed field, so the events `set_translation` and `translation_changed` are implicitly defined; if one wants to change the value of the `translation` field, one sets up a route to `set_translation,` and if one wants another node to be noti-fied whenever the value of translation changes, one sets up a route from `translation_changed.` For simplicity and ease of typing, the `set_` and `_changed` parts of the event names for exposed fields can be left out; if one routes to or from the name of an exposed field, the browser understands to use the appropriate `set_` or `_changed` event name. For instance, if XFORM1 and XFORM2 are `Transform` nodes, the following ROUTE state-ments are equivalent, the latter being a shorthand notation:

```
ROUTE XFORM1.translation_changed TO XFORM2.set_transla-
tion
ROUTE XFORM1.translation TO XFORM2.translation
```

Note that the latter shorthand notation works only with exposed fields. When in doubt, use the full event name.

One of the important uses of VRML is animation. A simple animation involves a trigger and a target. The trigger in VRML is usually a `Sensor` node. When a sensor is triggered — whether by a user's action (e.g., `TouchSensor` node) or by the passage of time (e.g., `TimeSensor` node) — it generates an event, which can then be redirected into another node's incoming event by using routes. If one routes the event into a standard node, the event changes a field value. If one routes it into a `Script` node or a complex prototyped node, one can perform all sorts of processing on the data before passing them on to their final destination. Like other nodes, a `Script` receives incoming events and generates outgoing events; unlike other nodes, a `Script` can perform some processing before producing those outgoing events. The part of the node that does that processing is called a *script* (or sometimes a *program*) and can be written in any programming language that a browser supports (e.g.,

JavaScript, Java). Note that one cannot define an exposed field in a `Script`; an exposed field is simply shorthand for a field plus an associated `set_incoming` event and `_changed` outgoing event, so if one wants the same effect in `Script,` one has explicitly to list the field, the set_ event, and the _changed event, and then handle those events in the script like any other events.

Some of the other important nodes and features are outlined next.

1. An `IndexedFaceSet` node consists of a list of vertices (in the form of a `Coordinate` node in the `coord` field) and a list of polygons (in the `coordIndex` field). While drawing an indexed face set the browser draws each polygon by connecting the points in a given order, connecting the last vertex back to the first one and then filling the resultant polygon. A vertex number of -1 is used to separate one polygon from the other in `coordIndex`. An `IndexedLineSet` node operates similarly for lines, instead of polygons. A `PointSet` node operates similarly for a collection of points. The `colorPerVertex` field of `IndexedFaceSet` (or `IndexedLineSet`) allows one to apply color to each face or to each vertex. Specifying colors per vertex allows one to achieve sophisticated effects economically because the browser interpolates colors between the vertices and across the faces to achieve smooth color gradation throughout (Gouraud Shading). The `colorPerFace` field of `IndexedFaceSet,` on the other hand, applies uniform color to each face and is useful for sharp contrast from one face to the next. Similarly, by using the `normalPerVertex` field, the renderer computes a different normal for each point on the surface to create a smooth variation between the normals at the vertices. This interpolation causes the shading to appear smooth. Specifying normals per vertex allows you to achieve curved effects without using lots of tiny polygons (which are expensive in terms of performance). The crease angle is a tolerance angle (in radians) that is compared to the normals for two adjacent faces of a shape. If the angle between two normals is less than or equal to the specified value for creaseangle, the edge between the two adjacent polygonal faces is smoothly shaded. (The faces will share the same normal.) If the angle between two normals is greater than the value specified for creaseangle, the edge is faceted, and a separate normal is calculated for each face. A texture map is specified as a two-dimensional image that extends between 0 and 1 coordinates in the horizontal (*s*-axis) and vertical (*t*-axis) directions.

2. To represent terrain features — from mountains to tiny irregularities in the ground surface — one can use the `ElevationGrid` node. This node provides a compact way to represent ground that varies in height over an area. The node specifies a rectangular grid and the height of the

ground at each intersection in that grid. The `xdimension` and `zdi-mension` fields specify the number of grid points in the x and z directions, respectively, defining a grid of `zdimension` by `xdimension` lines in the xz plane. Similarly, `Extrusion` node is another important feature with `crossSection` and `spine` (path definition) being important fields.

3. The `Sound` node can be used to insert audio clips. The common format is .wav. `startTime`, `stopTime`, and `loop` are some of the important fields, their meanings are self-explanatory.

Finally, the recommended organization of a VRML file is outlined:

- A VRML file starts with the following header (required): #VRML 2.0 utf8.
- Global nodes such as `Worldlnfo` can be placed anywhere in the file; usually, they appear at the beginning of the file.
- Prototypes are usually placed at the beginning of the file. A prototype must be defined before you can create an instance of the prototyped type.
- The scene hierarchy usually comes next in the file. Most scene hierarchies have one node as a root node, but that is not required; a scene hierarchy is allowed to have any number of nodes as top-level nodes.
- Scripts and interpolators can be placed anywhere in the file. Often, they are placed after the scene hierarchy.
- Routes must appear after the nodes and fields to which they refer. Often, these are placed after the scene hierarchy, for easy reference.
- Time sensors can appear either before or after the scene hierarchy.

SIMPLE EXAMPLES

The following examples are adapted from the < www.vrml.org > Web site. To use the examples in VRML, create a file for the code given in each example, add a filename suffix .wrl, and read them in a VRML browser.

Example 1. Box and Sphere

This example contains a simple scene defining a view of a red sphere and a blue box, lit by a directional light:

```
#VRML V2.0 utf8
Transform {
   children [
     NavigationInfo { headlight FALSE } # We'll add our
own light
     DirectionalLight {          # First child
         direction 0 0 -1        # Light illuminating
the scene
     }
```

```
Transform {    # Second child - a red sphere
      translation 3 0 1
      children [
        Shape {
          geometry Sphere { radius 2.3 }
          appearance Appearance {
            material Material { diffuseColor 1 0 0 }
# Red
          }
        }
      ]
    }

    Transform { # Third child - a blue box
      translation -2.4 .2 1
      rotation       0 1 1   .9
      children [
        Shape {
          geometry Box {}
          appearance Appearance {
            material Material { diffuseColor 0 0 1 }
# Blue
          }
        }
      ]
    }

  ] # end of children for world
}
```

Example 2: Instancing (Sharing)

Reading the following file results in three spheres being drawn. The first sphere defines a unit sphere at the origin named "Joe," the second sphere defines a smaller sphere translated along the $+x$ axis, and the third sphere is a reference to the second sphere and is translated along the $-x$ axis. If any changes occur to the second sphere (e.g., radius changes), the third sphere will change, too. (Note that the spheres will be unlit because no appearance is specified.)

```
#VRML V2.0 utf8
Transform {
  children [
    DEF Joe Shape { geometry Sphere {} }
    Transform {
      translation 2 0 0
      children   DEF Joe Shape { geometry Sphere {
radius .2 } }
    }
```

```
      Transform {
         translation -2 0 0
         children      USE Joe
      }
   ]
}
```

Example 3: Use of Prototype

This example involves a simple table with variable colors for the legs and top, using a prototype:

```
#VRML V2.0 utf8
PROTO TwoColorTable [ field SFColor legColor   .8 .4 .7
                      field SFColor topColor .6 .6 .1
]
{
   Transform {
      children [
         Transform { # table top
           translation 0 0.6 0
             children
                Shape {
                   appearance Appearance {
                      material Material { diffuseColor IS
topColor }
                   }
                   geometry Box { size 1.2 0.2 1.2 }
                }
         }

         Transform { # first table leg
           translation -.5 0 -.5
             children
                DEF Leg Shape {
                   appearance Appearance {
                      material Material { diffuseColor IS
legColor }
                   }
                   geometry Cylinder { height 1 radius .1 }
                }
         }
         Transform { # another table leg
           translation .5 0 -.5
             children USE Leg
         }
         Transform { # another table leg
           translation -.5 0 .5
             children USE Leg
         }
```

```
        Transform { # another table leg
          translation .5 0 .5
            children USE Leg
        }
      ] # End of root Transform's children
    } # End of root Transform
  } # End of prototype
```

```
# The prototype is now defined. Although it contains a
# number of nodes, only the legColor and topColor
# fields are public. Instead of using the default legColor
# and topColor, this instance of the table has red legs
# and a green top:
```

```
TwoColorTable {
  legColor 1 0 0 topColor 0 1 0
}
NavigationInfo { type "EXAMINE" }    # Use the Examiner
viewer
```

Example 4: Use of Script Node

This Script node decides whether or not to open a bank vault given openVault and combinationEntered messages. To do this, it remembers whether or not the correct combination has been entered. The Script node is combined with a Sphere, a TouchSensor, and a Sound node to show how it works. When the pointing device is over the sphere, the combinationEntered eventIn of the Script is sent. Then, when the Sphere is touched (typically, when the mouse button is pressed), the Script is sent the openVault eventIn. This generates vaultUnlocked eventOut, which starts a click sound. Note that openVault eventIn and vaultUnlocked eventOut are of type SFTime, which allows them to be wired directly to TouchSensor or TimeSensor. Here is the example:

```
#VRML V2.0 utf8
DEF OpenVault Script {
    # Declarations of what's in this Script node:
    eventIn SFTime openVault
    eventIn SFBool combinationEntered
    eventOut SFTime vaultUnlocked
    field SFBool unlocked FALSE

    # Implementation of the logic:
    url "javascript:
        function combinationEntered(value) { unlocked =
value; }
        function openVault(value) {
        if (unlocked) vaultUnlocked = value;
    }"
}
```

```
Shape {
    appearance Appearance {
        material Material { diffuseColor 1 0 0 }
    }
    geometry Sphere { }
}

Sound {
    source       DEF Click AudioClip {
                 url "click.wav"
                 stopTime 1
    }

    minFront     1000
    maxFront     1000
    minBack      1000
    maxBack      1000
}

DEF TS TouchSensor { }

ROUTE TS.isOver TO OpenVault.combinationEntered
ROUTE TS.touchTime TO OpenVault.openVault
ROUTE OpenVault.vaultUnlocked TO Click.startTime
```

Example 5: Use of Geometric Properties

The following `IndexedFaceSet` (contained in a Shape node) uses all four of the geometric property nodes to specify vertex coordinates, colors per vertex, normals per vertex, and texture coordinates per vertex (note that the material sets the overall transparency):

```
#VRML V2.0 utf8
Shape {
    geometry IndexedFaceSet {
        coordIndex [ 0, 1, 3, -1, 0, 2, 3, -1 ]
        coord Coordinate {
            point [ 0 0 0, 1 0 0, 1 0 -1, 0.5 1 0 ]
        }
        color Color {
            color [ 0.2 0.7 0.8, 0.5 0 0, 0.1 0.8
0.1, 0 0 0.7 ]
        }
        normal Normal {
            vector [ 0 0 1, 0 0 1, 0 0 1, 0 0 1 ]
        }
        texCoord TextureCoordinate {
            point [ 0 0, 1 0, 1 0.4, 1 1 ]
        }
    }
}
```

```
    appearance Appearance {
        material Material { transparency 0.5 }
        texture  PixelTexture {
            image 2 2 1 0xFF 0x80 0x80 0xFF
        }
    }
  }
}
```

Example 6: Directional Light

A directional light source illuminates only the objects in its enclosing grouping node. The light illuminates everything within this coordinate system, including the objects that precede it in the scenegraph, as shown below:

```
#VRML V2.0 utf8
Group {
    children [
        DEF UnlitShapeOne Transform {
            translation -3 0 0

            children Shape {
                appearance DEF App Appearance {
                    material Material {
                        diffuseColor 0.8 0.4 0.2
                    }
                }
                geometry Box { }
            }
        }

        DEF LitParent Group {
            children [
                DEF LitShapeOne Transform {
                    translation 0 2 0

                    children Shape {
                        appearance USE App
                        geometry Sphere { }
                    }
                }

                # lights the shapes under LitParent
                DirectionalLight { }
                DEF LitShapeTwo Transform {
                    translation 0 -2 0

                    children Shape {
                        appearance USE App
                        geometry Cylinder { }
                    }
                }
```

```
            ]
        }

        DEF UnlitShapeTwo Transform {
            translation 3 0 0
            children Shape {
                appearance USE App
                geometry Cone { }
            }
        }
    ]
}
```

Example 7: PointSet

This simple example defines a PointSet composed of three points. The first point is red (1,0,0), the second point is green (0,1,0), and the third point is blue (0,0,1). The second PointSet instances the Coordinate node defined in the first PointSet, but defines different colors:

```
#VRML V2.0 utf8
Shape {
    geometry PointSet {
        coord DEF mypts Coordinate {
            point [ 0 0 0, 2 2 2, 3 3 3 ]
        }
        color Color { color [ 1 0 0, 0 1 0, 0 0 1 ] }
    }
}
Transform {
    translation 2 0 0
    children Shape {
        geometry PointSet {
            coord USE mypts
            color Color { color [ .5 .5 0, 0 .5 .5, 1
1 1 ] }
        }
    }
}
```

Example 8. Level of Detail

The LOD node is typically used for switching between different versions of geometry at specified distances from the viewer. However, if the range field is left at its default value, the browser selects the most appropriate child from the list given. It can make this selection based on the performance or perceived importance of the object. Children should be listed with the most detailed version first, just as for the normal case. This "performance LOD" feature can be combined with the normal LOD function to give the browser a selection of children from which to choose at each distance.

In this example, the browser is free to choose either a detailed or a less detailed version of the object when the viewer is closer than 10 m (as measured in the coordinate space of the LOD). The browser should display the less detailed version of the object if the viewer is between 10 and 50 m and should display nothing at all if the viewer is farther than 50 m. Browsers should try to honor the hints given by authors, and authors should try to give browsers as much freedom as they can to choose levels of detail based on performance. For best results, ranges should be specified only where necessary and LOD nodes should be nested with and without ranges.

```
#VRML V2.0 utf8
LOD {
    range [ 10, 50 ]
    level [
        LOD {
            level [
                Shape { geometry Sphere { } }
                DEF LoRes Shape { geometry Box { } }
            ]
        }
        USE LoRes,
        Shape { } # Display nothing
    ]
}
```

Example 9: Color Interpolator

This example interpolates from red to green to blue in a 10-s cycle:

```
#VRML V2.0 utf8
DEF myColor ColorInterpolator {
    key         [   0.0,     0.5,      1.0 ]
    keyValue    [ 1 0 0,   0 1 0,    0 0 1 ] # red,
green, blue
}
DEF myClock TimeSensor {
    cycleInterval 10.0        # 10 second animation
    loop TRUE    # infinitely cycling animation
}
Shape {
    appearance Appearance {
        material DEF myMaterial Material { }
    }
geometry Sphere { }
}
ROUTE myClock.fraction_changed TO myColor.set_fraction
ROUTE myColor.value_changed TO
myMaterial.set_diffuseColor
```

INDEX

317